Technology Development for Adsorptive Heat Energy Converters:

Emerging Research and Opportunities

Kostyantyn M. Sukhyy
Ukrainian State University of Chemical Engineering, Ukraine

Elena A. Belyanovskaya
Ukrainian State University of Chemical Engineering, Ukraine

Mikhailo P. Sukhyy
Ukrainian State University of Chemical Engineering, Ukraine

A volume in the Advances in Chemical and Materials Engineering (ACME) Book Series

IGI Global
PUBLISHER of TIMELY KNOWLEDGE

Published in the United States of America by
IGI Global
Engineering Science Reference (an imprint of IGI Global)
701 E. Chocolate Avenue
Hershey PA, USA 17033
Tel: 717-533-8845
Fax: 717-533-8661
E-mail: cust@igi-global.com
Web site: http://www.igi-global.com

Library of Congress Cataloging-in-Publication Data

Names: Sukhyy, Kostyantyn M., 1976- author. | Belyanovskaya, Elena A., 1972- author. | Sukhyy, Mikhailo P., 1943- author.
Title: Technology development for adsorptive heat energy converters : emerging research and opportunities / by Kostyantyn M. Sukhyy, Elena A. Belyanovskaya, and Mikhailo P. Sukhyy.
Description: Hershey, PA : Engineering Science Reference, an imprint of IGI Global, [2020] | Includes bibliographical references and index. | Summary: "This book focuses on technology development for adsorptive heat energy converters"-- Provided by publisher.
Identifiers: LCCN 2019059913 (print) | LCCN 2019059914 (ebook) | ISBN 9781799844327 (hardcover) | ISBN 9781799851684 (paperback) | ISBN 9781799844334 (ebook)
Subjects: LCSH: Heat exchangers. | Heat storage devices. | Energy conversion.
Classification: LCC TJ263 .S88 2020 (print) | LCC TJ263 (ebook) | DDC 621.402/5--dc23
LC record available at https://lccn.loc.gov/2019059913
LC ebook record available at https://lccn.loc.gov/2019059914

This book is published in the IGI Global book series Advances in Chemical and Materials Engineering (ACME) (ISSN: 2327-5448; eISSN: 2327-5456)

British Cataloguing in Publication Data
A Cataloguing in Publication record for this book is available from the British Library.

Advances in Chemical and Materials Engineering (ACME) Book Series

ISSN:2327-5448
EISSN:2327-5456

Editor-in-Chief: J. Paulo Davim, University of Aveiro, Portugal

MISSION

The cross disciplinary approach of chemical and materials engineering is rapidly growing as it applies to the study of educational, scientific and industrial research activities by solving complex chemical problems using computational techniques and statistical methods.

The **Advances in Chemical and Materials Engineering (ACME) Book Series** provides research on the recent advances throughout computational and statistical methods of analysis and modeling. This series brings together collaboration between chemists, engineers, statisticians, and computer scientists and offers a wealth of knowledge and useful tools to academics, practitioners, and professionals through high quality publications.

COVERAGE

- Toxicity of Materials
- Materials to Renewable Energies
- Multifuncional and Smart Materials
- Fatigue and Creep
- Sustainable Materials
- Industrial Chemistry
- Nanomaterials
- Thermo-Chemical Treatments
- Electrochemical and Corrosion
- Chemical Engineering

IGI Global is currently accepting manuscripts for publication within this series. To submit a proposal for a volume in this series, please contact our Acquisition Editors at Acquisitions@igi-global.com or visit: http://www.igi-global.com/publish/.

The Advances in Chemical and Materials Engineering (ACME) Book Series (ISSN 2327-5448) is published by IGI Global, 701 E. Chocolate Avenue, Hershey, PA 17033-1240, USA, www.igi-global.com. This series is composed of titles available for purchase individually; each title is edited to be contextually exclusive from any other title within the series. For pricing and ordering information please visit http://www.igi-global.com/book-series/advances-chemical-materials-engineering/73687. Postmaster: Send all address changes to above address.

Titles in this Series

For a list of additional titles in this series, please visit:
http://www.igi-global.com/book-series/advances-chemical-materials-engineering/73687

Handbook of Research on Advancements in Supercritical Fluids Applications for Sustainable Energy Systems
Lin Chen (Institute of Engineering Thermophysics, Chinese Academy of Sciences, China & University of Chinese Academy of Science, China)
Engineering Science Reference • © 2021 • 821pp • H/C (ISBN: 9781799857969) • US $415.00

Inspiration and Design for Bio-Inspired Surfaces in Tribology Emerging Research and Opportunities
Hisham Abdel-Aal (Drexel University, USA)
Engineering Science Reference • © 2021 • 367pp • H/C (ISBN: 9781799816478) • US $195.00

Advanced Applications of 2D Nanostructures
Subhash Singh (National Institute of Technology Jamshedpur, India) and Kartikey Verma (Indian Institute of Technology, Kanpur, India)
Engineering Science Reference • © 2020 • 300pp • H/C (ISBN: 9781799829911) • US $225.00

Innovative Applications of Layered Materials and Structures
Sarmila Sahoo (Heritage Institute of Technology, Kolkata, India)
Engineering Science Reference • © 2020 • 300pp • H/C (ISBN: 9781799835875) • US $225.00

Using Lasers as Safe Alternatives for Adhesive Bonding Emerging Research and Opportunities
Barbara Ewa Ciecińska (Rzeszow University of Technology, Poland)
Engineering Science Reference • © 2020 • 248pp • H/C (ISBN: 9781799846345) • US $175.00

For an entire list of titles in this series, please visit:
http://www.igi-global.com/book-series/advances-chemical-materials-engineering/73687

IGI Global
PUBLISHER of TIMELY KNOWLEDGE

701 East Chocolate Avenue, Hershey, PA 17033, USA
Tel: 717-533-8845 x100 • Fax: 717-533-8661
E-Mail: cust@igi-global.com • www.igi-global.com

Table of Contents

Preface

Energy companies around the world are faced with two of the main challenges such as energy saving and reducing emissions of harmful substances into the environment. World community has developed long-termed initiatives aimed to mitigate the situation arising from emissions greenhouse gases and global warming (the Montreal and Kyoto Protocols, the Paris Agreement within the United Nations Framework Convention on Climate Change (UNFCCC)). These initiatives promote the replacement of traditional fuels (oil, gas, coal) with renewable energy sources (solar energy, wind, geothermal, tidal inflow) and development of measures to enhance efficiency of heat from industry, transport and households (re-use of heat from exhaust gases, energy storage, etc.).

The most promising way to save fuel and use renewable sources is heat energy storage. The heat energy storage is decisive for the implementation of such energy saving measures as: harmonization of heat generation and consumption regimes, expansion of using secondary and non-traditional renewable energy sources. There are four main mechanisms of heat energy accumulation, such as: sensible heat energy storage, phase-change heat energy storage, thermochemical heat energy storage and adsorptive heat energy storage.

The sensible heat storage and latent or phase-change storage has been studied in-depth because of their simplicity. However, heat storage density based on these mechanisms is strongly limited with properties of the applied materials. Thermochemical mechanism based on reversible chemical reaction can be considered as one of the most promising methods of heat energy storage. Nevertheless, despite the high heat energy storage, thermochemical materials require extremely high driving temperature and don't exhibit stability of their properties during long term cycling.

The urgency of the system of energy storage based on adsorption phenomena is caused by their high energy density and insignificant loss of heat and the stability during long time and manifold looping. Moreover, adsorptive conversion of heat generally needs of environmentally safe working fluids, it being of great potential to save fossil fuel. Nevertheless, technology development of adsorptive heat conversion meets a wide range of obstacles.

Despite the multiplicity of works focused on adsorptive storage and conversion of low-potential heat, still there is a gap between laboratory prototype and commercial device. The current latest advances in the field of adsorptive heat energy conversion cannot be considered as quite satisfactory because of poor properties of adsorptive materials used. For example, low maximal adsorption or adsorbate uptake leads to high volume of adsorbent to supply required thermal load. Potential of heat source is strongly limited by regeneration temperatures which are greater than 100°C.

Recent researches have been focused on selection among existing sorbents, or synthesis and research of new materials. However, maximal adsorption (adsorbate uptake) of conventional adsorbent does not exceed 0.5 kg sorbate per 1 kg of sorbent which results in high volumes of adsorment required for the performance of adsorptive heat converter.

Further progress in adsorptive heat energy conversion will become possible only when use of innovative adsorbents with additional properties used. Composites 'Salt inside Porous Matrix' are recognized as promising materials for adsorptive heat conversion due to their high adsorption ability to common working fluids (water, methanol and ammonia). Despite the large number of publications describing adsorbents 'Salt inside Porous Matrix' and their applications, the data are disembodied and sometimes contradictory.

Another obstacle for commercialization of adsorptive heat energy conversion is due to the absence of procedures to calculate the performance of adsorptive unit. In spite of two-mode performance of adsorptive heat conversion unit, most of mathematic models are devoted to only one phase of operating the adsorptive heat storage device, i.e. adsorption. Most of suggested mathematical models are devoted to adsorptive units based on conventional adsorbents. At that, their results are rather contradictory. Furthermore, efficiency of such devices is estimated only by one step of operating in spite of the fact that regeneration of adsorbent requires considerable amount of heat energy. It strongly inhibits integral estimation of efficiency in conditions of systems of heat supply, ventilation and air-conditioning and impedes transition from laboratory prototype to full-scale devices and units.

So, all listed lacunae stipulate the further study of processes of adsorptive heat conversion directed to overpass the gap between laboratory prototype and commercial samples of heat converters which will promote implementation of technology of adsorptive heat transformation in systems of district and distributed heating, ventilation and air-conditioning. In the furtherance of this aim the next tasks ought to be set:

- To develop the techniques of obtaining the adsorbents with enhanced adsorptive properties;
- To determine the correlation of adsorbents and design parameters of adsorptive heat conversion unit;
- To develop mathematical model and algorithm for integral estimation of the efficiency of adsorptive heat energy converter;
- To evaluate affection on efficiency of adsorptive heat conversion unit by its operation parameters.

Accordingly, promising ways of solutions of the task of industrial introduction of adsorptive heat energy conversion and original researches of the authors are analyzed in present work.

The first chapter, 'State of the Art of Materials for Adsorptive Heat Energy Conversion', written by Kostynatyn M. Sukhyy and Elena A. Belyanovskaya, is focused on state of the art of materials for adsorptive heat energy conversion basic principles for substantiation of working pair choice. Types of heat storage materials based on heat storage mechanism were compared. Sensible heat mechanism of thermal energy concerns with increasing the temperature of the material. Phase-change mechanism of heat energy storage concerns with alternating reversible processes of phase-changing. Generally, they are mainly melting-crystallization. Thermo-chemical heat energy storage mechanism is based on reversible chemical reactions. Limitations of conventional sensible heat storage are shown to be lowest density of heat energy storage determined by sensible heat of materials, which lead to large mass storage units and additional needs of large areas and building volumes, calculated according to heat storage density, constant changing the temperature when discharged, the need for a large overheating of heat heat storage media. The main drawbacks of phase-change materials are instability of properties of heat-accumulating substances in multiple cycles of crystallization - melting, degradation in time, corrosion activity, the need for developed surfaces of heat exchange and environmental danger. Application of thermal chemical storage materials is strongly limited by high operating temperatures of thermal chemical storage materials which are unacceptable for systems of district heating and decentralized heat supply due to Sanitary Regulations, impropriety for multifold cycling because of irreversibility of a wide range of chemical reactions. Perspectiveness of adsorptive heat energy storage and conversion is shown. Operating temperatures and heat storage density of conventional adsorptive materials are shown to be intermediate between phase-change and thermal chemical heat storage materials. Properties of probable adsorptive heat storage materials were analysed according with literary data. Low adsorptive capacities of conventional adsorbents are shown to result in low heat of adsorption and heat energy storage density. Salts forming crystalline hydrate occur to exhibit rather high energy storage density of crystalline hydrate, but their application is strongly inhibited not only by physical and chemical instability along with the corrosive activity of these salts at high temperatures, instability in multifold cycling, degradation in time, and an underdeveloped heat exchange surface. Modification of conventional adsorbents with salt can be considered as promising solution. Composites 'salt inside porous matrix' are shown to be promising alternative to conventional adsorbents. Main advantages of these mateials are low regeneration temperature and high adsorptive capacity. Industrial introduction of composite adsorbents 'salt inside porous matrix' is strongly inhibited by complex technology of their production based on rather expensive dry and wet impregnation of porous media by crystalline hydrate solutions. As alternative, sol gel method for obtaining composite adsorbents 'silica gel – crystalline hydrate' developed by authors can be used. The adsorptive properties of the obtained composite adsorbents 'silica gel – sodium sulphate' and 'silica gel – sodium acetate' are shown to be a non-linear combination of characteristics of silica gel and massive salt. The key distinction of kinetics of adsorption of water vapor with massive salts and composites obtained with sol gel method is shown to be difference limitative stage of process. The adsorption of water with massive crystalline hydrates is complicated by kinetic limitations. For composite adsorbents rate-determining stage is water transport through the pore system. Composites 'slilica gel – crystalline hydrate' are shown to be promising materials for adsorptive heat energy storage and conversion. Coefficient of performance was shown to be strongly affected by proper adsorbent choice.

In second chapter, 'Basic Principles for Substantiation of Working Pair Choice', written by Kostyntyn M. Sukhyy and Elena A. Belyanovskaya, are analysed basic principles for substantiation of working pair choice. The main requirements to adsorbate are the next: low cost, easy of obtaining, small molecular size to facilitate the adsorption effect; high latent heat of evaporation and small volume in liquid state; high thermal conductivity; low viscosity, thermal stability with adsorbent in the operating temperature range; chemical stability in the working temperature range; non-toxicity for animals and human, non-aggressiveness and incombustibility; low pressure saturation (slightly above atmospheric pressure) at normal operating temperature; the absence of environmental problems. Water conforms to these requirements. The crucial requirements to adsorbent are formulated as follows: the ability to adsorb large amounts of adsorbate when cooled to environment temperature, and give a high cooling effect; high values of maximal adsorption; desorption of the major portion of adsorbate (ideally all) when heated by an accessible source of heat; low heat capacity; good heat conductivity, short cycle time; no deterioration and loss of adsorption capacity over time or use; non-toxicity, non-aggressiveness; chemical physical compatibility with the selected adsorbate; low cost and wide availability. Properties of various types of adsorbents such as crystalline hydrates, zeolites, silica gels were compared. Promising media for heat storage and conversion are shown to be composite adsorbents 'salt inside porous matrix'. Characteristics of thermodynamic cycles of heat conversion were analysed. The ways of improving the coefficient of performance were shown to be affected by a proper choice of an adsorbent.

The third chapter, 'Structure and Properties of Composite Adsorbents Salt Inside Porous Matrix', written by Mikhailo P. Sukhyy and Elena A. Belyanovskaya is focused on structure and properties of composite adsorbents 'salt inside porous matrix'. Properties of composite adsorbents 'salt inside porous matrix', such as 'zeolite – crystalline hydrate', 'vermiculite – crystalline hydrate', 'silica gel – crystalline hydrate' were analysed. Main advantages of composite adsorbents concern with higher adsorptive capacity and lower regeneration temperature as compared with host matrix. Adsorptive properties of composite materials are shown to be significantly enhanced by introduction of salts in host matrix such as zeolite, vermiculite or silica gel, water uptake by composite adsorbent being increased when the salt content in it rised. The drawback of most of existing impregnation technologies is impossibility of obtaining composite with salt content more than 40 – 60% along with complexity. Sol gel method is presented as an alternative for conventional impregnation methods. Characteristics of adsorbents 'silica gel – sodium sulphate' synthesized according to sol gel method developed by authors were considered. The composite 'silica gel – sodium sulphate' composition and structure were studied with IR-spectroscopy and wide-angle X-ray scattering. Adsorptive properties of crystalline Na_2SO_4 when allocated in silicon oxygen matrix are shown to result from dispersion up to nanoscale. Adsorptive capacities and heat of adsorption of composites 'silica gel – sodium sulphate' and 'silica gel – sodium acetate' are shown to exceed the value calculated from the linear superposition of the sorption capacities of the sorbent and massive salt almost by 30%. Their adsorption properties are non-linear combination of properties of silica gel and salt. The formation of an unique structure which promoted an increase in the rate of reaction between crystalline hydrates and water vapor in the developed pores of the silicon-oxygen matrix is confirmed. It results in increasing the heat of adsorption and the heat energy storage density. Strong difference of water sorption kinetic curves of composite 'silica gel – sodium sulphate' and massive sodium sulphate is revealed. The correlation of their composition, structure, water adsorption kinetic and operating characteristic as heat storage material is stated.

The fourth chapter. 'Design of Adsorptive Heat Storage Devices' (authors are Kostynatyn M. Sukhyy and Elena A. Belyanovskaya), concerns with prospects of application of adsorptive heat storage devices, principles of operating the adsorptive heat storage systems, design of adsorptive storage devices and main factors determining the design of adsorptive heat storage unit. Ways to use adsorptive heat storage devices in heat supply systems were analyzed. Basic principles of performance of heat storage devices were considered. Adsorptive heat storage units working in close and open modes were compared. Constructions of adsorptive heat storage units of open and closed types were described. An efficient algorithm for calculating the volume of the adsorptive thermal energy storage device for a decentralised heat supply system of a private house is suggested by authors. The following procedure of computation involves: calculation of thermal load for heating including the determination of thermal losses through external fences, thermal losses due to infiltration and internal heat dissipation, the evaluation of maximal adsorption, adsorption heat, and determination of adsorbent mass and adsorbent volume. The maximal adsorption value is suggested to be calculated by the characteristics of the adsorbent, i.e. its maximal adsorption or in the absence of data for a composite adsorbent, as a linear superposition for a mechanical mixture. The adsorbent mass is suggested to calculate as a ratio of a thermal load for heating and heat of adsorption. The adsorbent volume is calculated as ratio of mass and density of adsorbent. An evaluative calculation of the heat load for a private house was carried out with the proposed algorithm. Mass and volumes of conventional silica gels were compared with composite adsorbents 'silica gel – sodium sulphate' and 'silica gel – sodium acetate' obtained by sol gel method developed by authors. Mass and volume of silica gels occur to surpass the suggested composite at least by 1.5 – 5 times. This is shown to result from higher maximal adsorption and heat of adsorption of suggested composite adsorbents. The optimal composition of the composite adsorbents 'silica gel – sodium sulphate' and 'silica gel – sodium acetate' was determined according to the minimal volume of the layer of heat storage material. Both the lowest volume values and the highest efficiency of composites with a mass ratio of silica gel and Na_2SO_4 or CH_3COONa 20: 80 are explained by the maximum value of adsorption heat. Suggested composite adsorbent 'silica gel – sodium sulphate' and 'silica gel – sodium acetate' are shown to be promising for heat supply systems.

The fifth chapter, 'Performance of Adsorptive Heat Storage Devices for Heat Supply', written by Elena A. Belyanovskaya and Mikhailo P. Sukhyy, is focused on modelling of performance of adsorptive heat storage devices and estimation of performance of heat storage devices. Two groups of models of adsorptive heat storage units suggested by previous researchers are analyzed. The first one is focused on predicting the heat energy storage density which based on Dubinin-Polanyi theory. The second one is devoted to analyzing the kinetic of adsorption processes and performance of the adsorber or adsorptive-desorptive reactor filled with traditional adsorbent or salt which forms crystalline hydrates. The key drawback of both groups of models results from considering only one stage of exploitation of adsorptive heat storage device in spite of its operating in two-stage mode, i.e. alternating discharge (adsorption) and charge (regeneration). It inhibits estimation of efficiency of adsorptive heat storage device basing on full operating cycle and its involving in heat supply system. Two algorithms for estimation of operating parameters are suggested by authors for closed-type and open-type heat storage devices. The algorithm for calculation of operating parameters of closed type adsorptive heat storage device is proposed to involve the next stages: calculation of the mass transfer coefficient, adsorption, useful heat, i.e. heat of adsorption, determination of the heat input, it being calculated as heat inputs for heating the adsorbent, device housing, water in the tank, evaporation of water in the tank, heating of the adsorbed water and desorption. Then efficiency factor is calculated. The operating characteristics of a closed-type heat energy storage device were studied when the composite adsorbent 'silica gel – sodium sulphate' used. The effect of the humid airflow velocity on the efficiency factor is taken into account by introducing a coefficient equal to the value of the adsorption. An increase in the efficiency coefficient was stated when the velocity and relative humidity of the airflow. It is shown that the humid air flow temperature practically does not affect its value. Having been used the suggested algorithm, the optimal operational parameters of an adsorptive heat storage device of a closed type based on a composite adsorbent 'silica gel – sodium sulphate' for a private house heating system are revealed to be humid air velocities of 0.6 – 0.8 m/s and relative humidity 40 – 60%. When these operational data applied, the efficiency coefficient is shown to reach the maximum values (about 55%). Algorithm of calculation of operating parameter of open-type heat storage device involves computation of mass transfer cofficient, adsorption, useful heat (heat of adsorption), heat input for heating the adsorbent, device casing, water in the humidifier, evaporation of water, heating the adsorbed water, desorption, and calculating efficiency coefficient. Performance of open-type heat storage device based on the composite adsorbent 'silica gel – sodium sulphate' is estimated. The optimal operating conditions of the heat accumulating device which allow operating with maximal magnitudes of efficiency coefficients 53 – 57% are stated to be humid airflow speed of 0.6 – 0.8 m/s and relative humidity of 40 – 60%. Correlation between efficiency factors obtained by experiments and calculated with suggested algorithm is confirmed. The possibility of reducing the power consumption when heat storage devices applied in 2,4 – 90 times versus decentralized heating systems on basis of solid fuel boiler, gas boiler and electric boiler is stated when open-type sorptive heat storage device used.

The sixth chapter, 'Technology of Heat and Moisture Regeneration for Ventilation Systems', by Elena A. Belyanovskaya and Kostyantyn M. Sukhyy, is devoted to technology of heat and moisture regeneration for ventilation systems. Recent progress in adsorptive technologies for air dehumidification, heating and conditioning is analyzed. Results of original researches of authors on adsorptive heat and moisture regeneration are presented. The design of adsorptive heat-moisture regenerator for ventilation systems is given. Its operation and the results of field tests are described. The technology of regeneration of low-potential heat and moisture by composite sorbent 'silica gel – sodium sulphate' is proposed. Experimental plots of temperature, absolute and relative humidity at the inlet and the outlet of the apparatus and between cassettes with the composite are given. Correlation of flows switch-over time, airflow rate and temperature drop is stated. The relationships temperature efficiency factor vs. dimensionless temperature drop and moisture efficiency factor vs. absolute humidity dimensionless drop are derived with fair accuracy for engineering calculation. Ability of purposeful modification of the above-mentioned characteristics within broad ranges by changing the half-cycle time, the size of the granules of the adsorbent and its amount is revealed. The mathematical model and algorithm for determining the basic parameters of adsorptive regenerator operating processes are developed. The proposed algorithm involves calculation of the volume of air passed through the layer of adsorptive heat-storage material, the concentration of water in the airflow at the outlet of the regenerator, the adsorption, the heat of adsorption, the final temperature of the cold air, the air temperature after mixing the cold air from the street and the warm air in the room at the warm end of the regenerator during inflow, calculation of the final concentration of water in the flow at the cold end of the regenerator, the volume of air passing through the layer of heat-accumulating material, adsorption and heat of adsorption, the final temperature of the air at the cold end of the regenerator, the air temperature after mixing of the cold air from the street and the warm air from the room at the cold end of regenerator during outflow, determining the temperature efficiency coefficient, summarized adsorption and maximal adsorption time. The correlation of air temperatures near the warm and cold end of the regenerator, as well as the temperature efficiency factors calculated according to the proposed algorithm and obtained by experimental way is confirmed. The mathematical modeling of the processes of operation of adsorption regenerators based on composites 'silica gel – sodium sulphate' and 'sodium acetate' in the conditions of the typical ventilation system of residential premises is carried out. The dependences of the temperature efficiency factor vs. the time of switching air flows and the velocity of air flow, as well as the temperatures of external and internal air under stationary conditions are shown. An optimal composition of composite adsorbents is stated to be 20% of silica gel and 80% of salt, i.e. sodium sulphate or sodium acetate. Due to higher value of maximal adsorption composite 'silica gel – Na_2SO_4' is shown to be required in half as much as compared with 'silica gel – CH_3COONa'.

The seventh chapter, 'Design and Performance of Adsorptive Heat Pumps', written by Elena A. Belyanovskaya and Mikhailo P. Sukhyy, is devoted to design and performance of adsorptive heat pumps. State-of-the-art of the adsorptive heat pumping is analyzed. It involves analyzing the operating principles of adsorptive heat pumps, comparing the properties of adsorbents used, bed specifications and operating conditions. Original construction of the adsorptive heat pump designed by authors for independent heat supply systems or hot water supply of buildings is given. The composites 'silica gel – sodium sulphate' or 'silica gel – sodium acetate' were proposed to be used as adsorbents. Discharging was performed in a day time, when heat pump supplied heating system with water warmed to 45–35°C. The regeneration stage proceeded at night from 0.00 to 8.00 a.m. Efficiency of suggested adsorptive heat pump is estimated by two methods: as ratio of adsorption heat to sum of desorption heat and external heat supplied to adsorbent during its heating up to regeneration temperature (coefficient of performance of cycle) and as ratio of heat of adsorption to heat supplied by solar collector (net coefficient of performance). Coefficients of energy performance of cycle of suggested heat pump were stated to be 2.084 when composite 'silica gel – sodium sulphate' used and 2.021 when 'silica gel – sodium acetate' used. Seasonal dependence of net coefficient energy performance for suggested adsorptive heat pump based on composites 'silica gel – sodium sulphate' and 'silica gel – sodium acetate' is revealed. Correlation of coefficients of energy performance of adsorptive heat pump and composite sorbents properties (sorption capacity and regeneration temperature) is stated. Insignificant decreasing of coefficients of energy performance when 'silica gel – sodium acetate' used is explained by lower sorptive capacity as compared to 'silica gel – sodium sulphate'. Suggested heat pump application perspectives are shown for heat supply systems to result from traditional energy sources independence and environmental advantages. Adsorptive heat pumps development challenges, major limitations for commercialization of adsorptive heat pumping and requirements to ongoing innovations are analysed.

The eighth chapter, 'Construction and Operating Parameters of Adsorptive Chillers', by Mikhailo P. Sukhyy and Elena A. Belyanovskaya, is focused on the design and performance of adsorptive chillers. Basic types of design and operating principle of adsorptive chillers were analyzed. Performance of one-, two-, three- and four-bed solar power adsorptive chillers are compared. Operational processes of adsorptive chillers based on composite adsorbents were studied. The correlation between the adsorbent composition and the coefficient of energy performance of the adsorptive chiller was revealed. An optimal composition of adsorbent 'silica gel – sodium sulphate' is stated to be of 20% silica gel and 80% sodium sulphate. The maximal values of the coefficient of performance of cycle of studied solar adsorptive chiller about of 1.14 are stated for composites containing about 20 wt. % silica gel and 80 wt% sodium sulphate. As a result of decreasing of adsorbent mass, the coefficient of performance is shown to increase as sodium sulphate content in the composite increased. The coefficient of performance of the adsorptive chiller was shown to strongly depend on regeneration process parameters of the composite used. The growth of the coefficient of performance is stated to result from decreasing the difference between adsorbent temperature and regeneration temperature from 85 to 55°C. The basic factors affecting the net coefficient of energy performance of the adsorptive solar refrigerator were stated daily solar radiant flux alongside with composition of the adsorbent and difference between adsorbent temperature and temperature regeneration. Net coefficients of performance of solar adsorptive refrigerator based on composite 'silica gel – sodium sulphate' were stated to change from 0.25 to

0.34 during operating period. Utilization of the adsorption heat is suggested to warm the heat carrier which applied to heat adsorbent during regeneration. The ways of improving the design and performance of adsorptive solar chillers are suggested. The first one includes the introduction of solar collectors made of cellular polycarbonate plastics in the design of adsorptive solar chiller. Instantaneous efficiency coefficient were calculated as special thermal performance-solar radiant flux surface density ratio, optical efficiency factor is determined as special thermal performance-solar radiant flux surface density ratio at the equal temperatures of heat transfer medium and environment, reduced heat loss factor being calculated as the product of solar collector efficiency factor and net heat loss coefficient. The environmental test of developed collectors PSK-AV2-3, PSK-AV1-2, PSK-AV2-1, PSK-VS1-2, PSK-VS2-2, PSK-VS2-3, PSK-ST10-PW were conducted. The correlation of their results with laboratory tests when the thermohydraulic stand applied is shown. Relative accuracy of laboratory and environment tests was shown to be not exceeding 5 – 7%. The optical efficiency factor and the coefficient of thermal losses of polymeric solar collectors were determined. On the basis of the dependencies of the efficiency of the solar collectors vs. the reduced temperature, optimal designs of the polymeric solar collectors for the adsorption chilling solar systems are determined to be depended on the temperature of the regeneration temperature of the sorbents. As the temperatures of the regeneration of composite adsorbent ranged from 50°C to 60°C, appliance of the collectors PSK-AV2-1, PSK-CT10-PW occur to be expedient, and PSK-AB2-3, PSK-VS2-3, PSK-AB1-2, PSK-VS2-2 and PSK-VS1-2 are revealed to be more efficient when regeneration temperatures increased over 80 °C. Thermotechnical characteristics of designed polymeric solar collectors are shown to surpass conventional metal and vacuum collectors. The perspectives of polymeric solar collectors in the design of adsorptive chilling solar plants were shown. Another way to improve the performance of adsorptive solar chillers concerns with equipping it with a photosensitive element and an electric drive, which will allow changing the angle of slope of the adsorber to the horizon depending on the intensity of the solar radiation.

The present book can be useful for design the efficient heat supply, ventilation and air-conditioning systems based on adsorptive heat conversion devices or involved adsorptive heat conversion units and adsorptive chilling plants. It can help to fill the gap between laboratory prototype and commercial samples of heat conversion devices and enhance implementation of technology of adsorptive heat transformation in systems of district and decentralized heating, ventilation, and air-conditioning and also in developing energy efficient chillers.

The target audience can be experts in heat power engineering, heat supply, ventilation, air-conditioning, and refrigeration engineering. It can be interesting for experts in processes and apparatus of chemical engineering. It can be useful for students studying in heat power engineering, heat supply, ventilation, air-conditioning, and refrigeration engineering.

ACKNOWLEDGMENT

This work was supported by Ministry of Education and Science of Ukraine (grant number 0119U002243).

Section 1

The Main Challenges to Materials for Adsorptive Heat Storage and Conversion

Chapter 1
State-of-the-Art Materials for Adsorptive Heat Energy Conversion

ABSTRACT

The chapter is focused on state of the art of materials for adsorptive heat energy conversion basic principles for substantiation of working pair choice. Types of heat storage materials based on heat storage mechanism were compared. Sensible heat mechanism of thermal energy is based on increasing the temperature of the material. Phase-change mechanism of heat energy storage concerns with alternating reversible processes of phase-changing. As a rule, they are mainly melting-crystallization. Thermo-chemical heat energy storage mechanism is based on reversible chemical reactions. Limitations of conventional sensible heat storage are shown to lowest density of heat energy storage determined by sensible heat of materials, which led to large mass storage units and additional needs of large areas and building volumes, calculated according to heat storage density, constant changing the temperature when discharged, the need for a large overheating of heat storage media. The main defects of phase-change materials are instability of properties of heat-accumulating substances in multiple cycles of crystallization – melting, degradation in time, corrosion activity, the need for developed surfaces of heat exchange and environmental danger. Commercilisation of thermal chemical storage materials is strongly limited by high operating temperatures of thermal chemical storage materials, which are unacceptable for systems of district heating and decentralized heat supply due to sanitary

DOI: 10.4018/978-1-7998-4432-7.ch001

regulations, impropriety for multifold cycling because of irreversibility of a wide range of chemical reactions. Perspective of adsorptive heat energy storage and conversion is shown. Interval of operating temperatures and heat storage density of conventional adsorptive materials are shown to be intermediate between phase-change and thermal chemical heat storage materials. Properties of probable adsorptive heat storage materials were analysed according with literary data. Low adsorptive capacity of conventional adsorbents results in low heat of adsorption and heat energy storage density. Salts forming crystalline hydrate occur to exhibit rather high energy storage density of 1.9–2.7 GJ/m^3 of crystalline hydrate, but their application is strongly inhibited not only by physical and chemical instability along with the corrosive activity of these salts at high temperatures, but instability in multifold cycling, degradation in time, and an underdeveloped heat exchange surface. As engineering solution, modification of conventional adsorbents with salt can be considered. Composites 'salt inside porous matrix' is shown to be promising alternative to conventional adsorbents. Main advantages of these materials are low regeneration temperature and high adsorptive capacity. Crucial impediments of industrial introduction of composite adsorbents 'salt inside porous matrix' is shown to be complex technology of their production based on rather expensive dry and wet impregnation of porous media by crystalline hydrate solutions. As an alternative, sol gel method for obtaining composite adsorbents 'silica gel – crystalline hydrate' developed by authors is suggested. The adsorption properties of the obtained composite adsorbents 'silica gel – sodium sulphate' and 'silica gel – sodium acetate' are shown to be non-linear combinations of characteristics of silica gel and massive salt. The key distinction of kinetics of adsorption of water vapor with massive salts and composites obtained with sol gel method is shown to be difference limitative stage of process. The adsorption of water with massive crystalline hydrates is shown to be complicated by kinetic limitations. For composite adsorbents limiting stage is water transport through the pore system. Composites 'slilica gel – crystalline hydrate' are shown to be a promising material for adsorptive heat energy storage and conversion.

NOMENCLATURE

A – adsorption, g/g;
D – coefficient of diffusion;

T – temperature, K;
τ – time, sec.

PRINCIPLES OF HEAT STORAGE

The ability to use renewable energy sources and efficiency upgrading of fossil fuels combustion is strongly affected by the operational parameters of heat energy transformation. A heat energy storage and conversion is a promising technical solution to bridge the gap between the supply and demand of thermal energy from renewable energy sources. A heat energy storage system would store the intermittent energy when it is available to be used when it is needed.

Nevertheless, the high ñuctuations in the energy flux of renewable sources occur to be a major challenge for systems to store and use this resource effectively (Hadj Ammar, Benhaoua, & Bouras, 2016). Solar energy is an example of an intermittent energy source where energy flux varies depending on the time of day and season. The main requirements to heat energy storage and conversion processes are a little thermal energy loss and a high extraction efficiency of the stored energy (Kato, 2007).

The two most extensively used and studied types of heat energy storage are sensible heat and latent heat. Sensible heat storage based on increasing temperature of heat storage media. As examples of such heat energy storage, hot water storage (hydro-accumulation), underground thermal energy storage (aquifer, borehole, cavern, ducts in soil, pit), and rock ðlled storage (rock, pebble, gravel) can be used (Koohi-Fayegh & Rosen, 2020). As solid sensible heat energy storage, brick, reinforced concrete and sand are suggested (Sarbu & Sebarchievici, 2018). Latent heat storage, i.e. phase-change heat storage, based on reversible processes of emission or absorption heat which accompanying phase-change. As example of phase-change heat storage materials, paraffin waxes, fatty acids, esters, alcohols, glycols, salts or salts hydrates can be used. Interval of operating temperatures of organic phase-change materials is of 4 – 150°C (Koohi-Fayegh & Rosen, 2020). Operating temperatures of inorganic phase-change materials vary in more wide range from 8 – 900°C. Both types are simple and perspective for energy storage. However, both the latent heat and sensible heat systems lose thermal energy over time to the surroundings and therefore, appeared to be currently unsuitable for storing thermal energy on a long-term basis (Scapino, Zondag, Van Bael, Diriken, Rindt, 2017).

Thermo-chemical storage systems, materials and processes are at the initial stages of development. They are based on chemical reactions which require or release thermal energy. Performance of thermo-chemical storage device involves three steps: endothermic dissociation, storage of reaction products, and exothermic reaction of the dissociated products. The various types of thermal energy storage materials will be discussed individually, with an emphasis placed on adsorptive and thermo-chemical energy storage. Typical parameters of heat energy storage systems are given in the Table 1.

Table 1. Typical parameters of heat energy storage systems

Heat Energy Storage System	Capacity, KWh/t	Power, MW	Efficiency, %	Storage Period	Cost (€/ kWh)
Sensible (hot water)	10 – 50	0.001 – 10	50 – 90	Days/months	0.1 – 10
Phase-change material	50 – 150	0.001 – 1.0	75 – 90	Hours/months	10 – 50
Thermochemical (Chemical reactions)	120 – 250	0.01 – 1.0	75 – 100	Hours/days	8 – 100

(Sarbu & Sebarchievici, 2018)

Despite the lowest cost of kWh, sensible heat energy storage system exhibits the lowest capacity and efficiency. Obviously, these parameters are strongly affected by properties of heat storage materials.

The energy storage in the form of sensible and latent heat is the most studied technology at a high stage of development (Hadj Ammar et al., 2016; Kato, 2007; Sharma, Tyagi, Chen, & Buddhi, 2009). Sensible heat mechanism of thermal energy concerns with increasing the temperature of the material (Sharma et al., 2009; Nielsen, 2003). The storage energy density is calculated as the product of the specific heat of these materials and the temperature change. The main applications of sensible heat storage are accumulation of hot water (hydro-accumulation), underground heat energy (aquifer, wells, and caves), sensible accumulation (rock, pebble, gravel) (Nielsen, 2003; Lanahan & Tabares-Velasco, 2017). The main disadvantages of this method are low density of energy accumulation limited by sensible heat of materials, which resulted in large mass and additional needs of large areas and building volumes, calculated on the accumulation of 1 GJ of heat, constant temperature change when the battery discharged, the need for a large overheating of heat accumulation material. So, working temperatures interval and heat energy storage density of sensible heat storage materials are the lowest between all the types of heat storage materials (Figure 1).

Figure 1. Energy density (MJ/m³) from several materials used as thermal energy storage. Highlighted: Phase change materials, sorption heat storage materials, and chemical reaction heat storage materials
(N'Tsoukpoe, Liu, & Le Pierres., 2009).

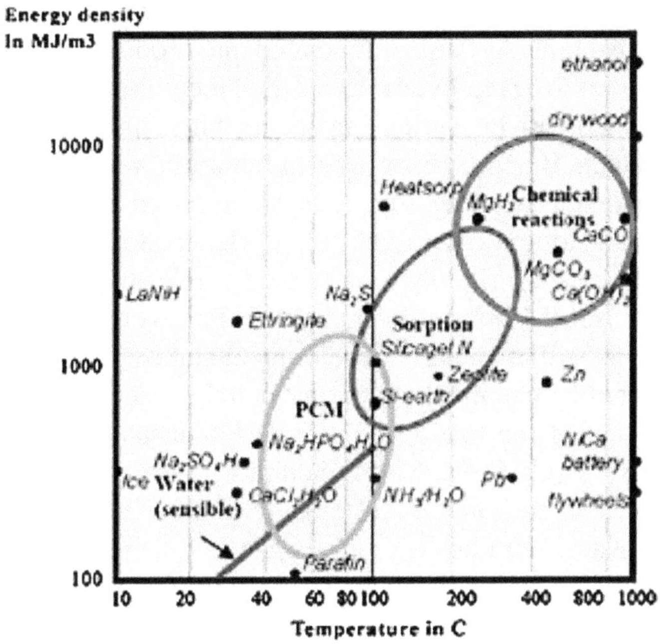

Phase-change mechanism of heat energy storage is based on periodical reversible processes of phase-changing, it being generally melting-crystallization. Energy storage is equal to the amount of heat absorbed or emitted by the material when its aggregate state changed (Nielsen, 2003).

Studies of various types of phase-change materials reveals that heat energy storage density can in 5-14 times more as compared with sensible materials (Zalba, Marrın, Cabeza, & Mehling, 2003). As phase-change materials can be considered paraffin and a lot of crystalline hydrates. Despite the cheapness of phase-change materials, the instability of properties of heat-accumulating substances in multiple cycles of crystallization - melting (Kato, 2007), degradation in time (N'Tsoukpoe et al., 2009), corrosion activity, the need for developed surfaces of heat exchange and environmental danger stimulate the research and development another mechanism of heat energy storage.

According to energy storage density, thermochemical materials seem to be the most promising for accumulating of heat energy. It recovers heat through a chemically reverse reaction:

$$A = B + C \tag{1}$$

Charge step the endothermic dissociation reaction which is the heat storage step, chemical reactant (A) being dissociated into products (B and C). One of the products (C) has to be removed to avoid their recombination. The discharge stage corresponded with heat releasing occurs when, after the reintroduction of (C), the products B and C react to form the initial reactant (A).

Thermochemical heat energy storage systems are characterized by highest storage density and long-term storage period. Heat is stored until the reactants remain separate for an infinitely long time. So, releasing heat occurs when required by the user (Mastronardo et al, 2017; Pardo et al., 2014). However, this technology is still at an early development status.

A few reversible chemical reactions can be used depending on the temperature range of the heat to be stored. For instance, hydroxides, like calcium hydroxide (Ca(OH)2/CaO) (Schmidt, Szczukowski, Roßkopf, & Linder, 2014), and carbonates systems (e.g. $PbCO_3/PbO$, $CaCO_3/CaO$) (Barker, 2007, Kato, Harada, & Yoshizawa, 1999) can be used to store heat at 450 °C and above. Redox reactions (e.g. Co_3O_4/CoO, BaO_2/BaO, Mn_2O_3/Mn_3O_4) (Tescari et al., 2013; Carrillo et al., 2016; Carrillo, Serrano, Pizarro, & Coronado, 2016) have been investigated for high temperature heat storage: 700 – 1100 °C. Nevertheless, interval of operating temperatures of thermal chemical storage materials is unacceptable for systems of district heating and decentralized heat supply because of Sanitary Regulations. Furthermore, reversible exothermic reaction is not always affected by cooling the product mixture the endothermic reaction. Moreover, a wide range of chemical reactions are irreversible, and so, they are suitable only for one-time operation.

Adsorptive materials hold an intermediate place between thermal chemical and phase-change materials as for operating temperatures and energy storage density.

Adsorption is known to be the result of the interaction between a solid (adsorbent) and a liquid (adsorbate) on the basis of physical or chemical reaction (Hauer, 2007). Physical adsorption proceed when the adsorbate molecules at the expense of the Van der Waals forces are fixed on the surface of porous solids (adsorbent), which leads to the accumulation of adsorbate on the surface of the adsorbent (sorption). As heated, the adsorbate molecules are removed from the surface of the adsorbent (desorption). Contrariwise, the result of chemical adsorption is ionic or covalent bonds formed between adsorbate and adsorbent molecules. The bonds strength is much greater than for physical adsorption and more heat is released. However, since this type of compound contributes to the chemical change of the adsorbed substance, it is difficult to carry out the reversible chemical process and therefore the molecules of adsorbate and adsorbent will not return to its original state after desorption (Wang, L.W., Wang, R.Z., & Oliveira, 2009; Gordeeva & Aristov, 2018). Therefore, most of developed adsorptive heat converters (AHTs) are mainly based on physical adsorption.

The basic principle of adsorptive thermal energy storage in closed system is presented on Figure 2 (Schreiber, Lanzerath, & Bardow, 2018). When charged, the heat input Q_{des} leads to an increase of the temperature in the zeolite and to desorption, i.e. to the release of water vapor. The released water vapor can be condensed at a lower temperature. The enthalpy of condensation can be used as low temperature heat Q_{cond} or be dismissed to the environment. During the storage period, zeolite and water have to be stored separately until the stored heat energy is needed. During discharging of the storage unit, the charging process is reversed: low-grade heat Q_{vap} is used to vaporize water at low temperatures. The water vapor is adsorbed by the zeolite. Hereby, the enthalpy of adsorption is released and can be used as process heat Q_{ads}

Figure 2. Working principle of adsorption thermal energy storage with zeolite and water: during charging, the heat input Q_{des} is used to desorb the zeolite. Released water vapor is condensed. During discharging, water vaporizes and is adsorbed by the zeolite, releasing the heat output Q_{ads}
(Schreiber, Lanzerath, & Bardow, 2018)

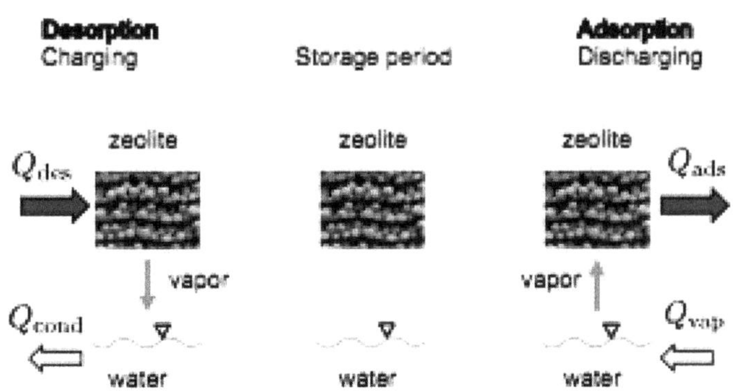

at a higher temperature. The low grade heat in the discharging process can be waste heat or come from the environment. Thus, adsorption TES allows to integrate low-grade heat into the process heat supply to enhance energy efficiency.

In adsorption heat amplification cycle during desorption stage (adsorbent regeneration) the heat is consumed at middle temperature TM from the external heat source, which is used to drive the cycle (Figure 3a) (Gordeeva & Aristov, 2018). That results in the vapor desorption from the adsorbent, and then water collects in the condenser. The condensation heat is dissipated to the ambient media, which is used as heat sink at low temperature TL. Then the adsorption stage is carried out, during which the heat is released at high temperature TH and transferred to a consumer (Figure 3b). The evaporation heat is consumed from the driving heat source at temperature TM. So, the heat of the source at middle temperature is partially upgraded to higher temperature level suitable for heating, due to rejecting the remaining fraction to a lower temperature level (Figure 3b).

Figure 3. The principle scheme of AHA cycle: (a) adsorption/heat release stage; (b) desorption/adsorbent regeneration stage
(Gordeeva & Aristov, 2018).

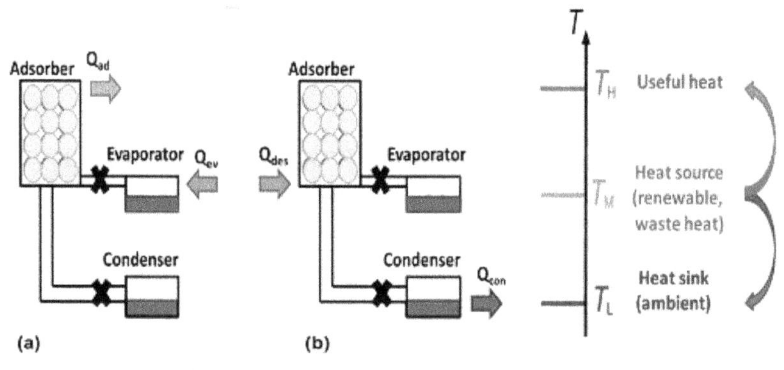

The main advantages of the adsorptive system of energy storage and conversion are concerned with their high energy density and their insignificant loss of heat and the repetition of accumulation operations (Demir, Mobedi, & UË lkËu, 2008). Few of existing solutions are capable to useful transformation such a low-potential heat that is therefore often wasted.

Existing technologies of heat energy storage during heating of buildings are based mainly on the sensible mechanism (Lanahan & Tabares-Velasco,

2017). At the same time, the sensible materials have the lowest energy storage density and as a result, the largest mass and volume for the accumulation of 1 GJ of heat: stones 67 tones and 30 m^3 and water 16 tones and 16 m^3 (Kolomiyets & Sukhyy, 2016).

As compared with sensible, phase-change and thermochemical mechanism of heat storage, adsorptive mechanism is one of the most promising due to optimal operating temperatures interval and environmental friendliness of applied materials. Nevertheless, the main challenge to adsorptive heat storage technology is focused on increasing of heat storage density up to values which are commensurable with thermochemical heat energy storage materials.

ADSORPTIVE HEAT STORAGE MATERIALS

The main characteristics determining the efficiency of the adsorption heat storage system involve high storage density, high temperature rise, low charging temperature and high thermal conductivity. Among the various systems, chemical reactions between salts ($MgCl_2$, Na_2S, $MgSO_4$, $SrBr_2$ etc.) and water vapor with hydrate formation are characterized by high sorption capacity and therefore high energy storage density (Mugnier & Goetz, 2001). Though, the corrosion activity of salts, their low heat transfer, as a consequence, the low specific power of the accumulation unit strongly prevents their practical application (Visscher & Veldhuis, 2005).

Salts are allowed to realize the energy storage density of 1.9–2.7 GJ/m^3 of crystalline hydrate. However, their application is limited not only by physical and chemical instability along with the corrosive activity of these salts at high temperatures, but instability in repeated cycling, degradation in time, and also an underdeveloped heat exchange surface. For example, energy storage density of magnesium sulfate heptahydrate is 1 GJ/m^3, but the dehydration rate of $MgSO_4 \cdot 7H_2O$ at low pressures occur to be low (Ferchaud, Scherpenborg, Zondag, & de Boer, 2014).

Ideally, pure crystals reach a stable state automatically after several successive cycles of hydration and dehydration, as for instance observed for $BaCl_2$ (Andersson, Azoulay, & De Pablo J., 1983). What may have happened in this case is illustrated by Figure 4, showing an example of a macro-crystal, which is cracked during one of cycling experiments. If such a cracked structure would be found to hold stable, with sufficient vapor and heat transport, no additional stabilizing material would be needed (Jong, Trausela, Fincka, van Vlieta, & Cuypers, 2014). As it is, during massive salts hydratation in steady-state condition water access anhydrous salts is generally disabled by hydrate films formation, deceleration of sorption and cooling being resulted from these processes. Massive salt cycling is impossible without mechanical dispersion during operating process.

Apparently, similar problems are typical.

Figure 4. Macro-crystal with cracks due to cycles of hydration and dehydration (Jong et al., 2014).

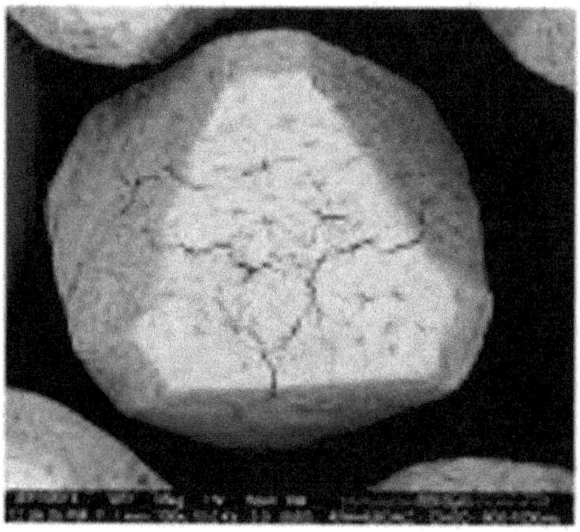

LiCl according to (Bales, 2008) has an energy storage density of 253 kWh/ m^3 but the use of this salt on an industrial scale is not possible due to the high price. Na_2S has a high heat energy storage density 1.1 kWh/kg (1980 kWh/ m^3) for heat and 0.7 kWh/kg (1300 kWh/m^3) for cooling with coefficient of energy performance 0.84 and 0.57, respectively, but its operating processes being rather complicated by high vacuum (Boer, Haije, Veldhuis, Smeding, 2004).

The heat energy storage density of $Na_2S{\cdot}5H_2O$ is almost 2.79 GJ/ m^3 for heating and 1.58 GJ/m^3 for open and closed system respectively (Donkers, Sögütoglu, Huinink, Fischer, & Adan, 2017). Theoretically, magnesium sulphate heptahydrate according to Donkers et al. (2017) has a very good potential in terms of accumulation of solar energy with a storage density of almost 2.27 GJ/m^3. It is non-toxic and non-aggressive, but it being expensive and has a low bulk density.

Porous materials such as silica gels, aluminum oxide, activated carbon and zeolites have a high sorption capacity and high specific surface area.

These materials occur to be chemically inert, non-toxic and available in large quantities. So, porous materials can be used as effective adsorbents for heat storage because of negligible regeneration temperatures.

System based on synthetic zeolite 13X was used to equalize the loads within the thermal network of the district (Hauer, 2002), an experimental heat energy storage density being 124 kWh/m^3 for heating and 100 kWh/m^3 for cooling with an efficiency of 0.9 and 0.86, but the desorption of moisture at a temperature of about 180°C makes its application economically unviable (N'Tsoukpoe et al., 2009). Austrian scientists in several studies (Jaehnig, Hausner, Wagner, & Isaksson, 2006) used silica gel in closed loops, and experimentally achieved density storage of about 50 kWh/m^3, whereas the theoretical estimates corresponded to 200 – 300 kWh/m^3.

Activated carbon is shown to offer a good compromise between high adsorption and desorption capacity. Natural zeolites have to be used in large quantities, because only a small amount of adsorbate is desorbed when the temperature increased. However, zeolite adsorption isotherms reveal a very non-linear dependence on pressure, which is relevant for solar refrigeration systems. On the contrary, activated carbon and silica gel have almost linear dependence of the isotherm on pressure. Silica gel is corresponded with most of the criteria for adsorbents, but it is expensive and cannot be available in most countries. Additionally, the phenomenon of decrease of the adsorption potential and aging of silica gel is another urgent problem (Alghoul, Sulaiman, Azmi, & Wahab, 2007; Pantos et al., 1994; Wang, Zhang, Xia, Han, & Wang, 2012).

There are several promising options to improve the properties of this adsorbent. Basic adsorbents can be doped to present better properties when applied in adsorptive heat conversion systems. At the end of the 20th century, activated carbon fiber was developed for adsorptive refrigeration devices. An example was described by Vasiliev et al. (1996), who reported the development of a prototype solar refrigerator based on activated carbon fiber/ethanol, activated carbon fiber and methanol working pairs. According to Wang et al. (1997) a specially treated activated carbon fiber and concluded should have a two to three times higher capacity to adsorb methanol than standard activated carbon, and a shorter adsorption cycle.

Modification of traditional adsorbents with salt was proposed in the early XXIth century to increase their sorption capacity. Further progress in the adsorptive heat energy conversion was shown to be enhanced only by the use of innovative adsorbents with additional properties, i.e. composites 'salt in the porous matrix' (SIPM) (Aristov, Tokarev, Cacciola, & Restuccia, 1996,

Aristov, Restuccia, Cacciola, & Parmon, 2002, Gordeeva, Freni, Restuccia & Aristov, 2007; Sharonov, Veselovskaya, & Aristov, 2006). They are a two-component system: one component is a matrix, and the other is an inorganic salt placed inside the pores of the matrix. Salt S reacts with the vapours of adsorbed substances (water, methanol, ethanol and ammonia), which leads to the formation of a complex S·NV (hydrate, ethanolate, etc.) of the molecules of salt and adsorbate by the reaction of $S + NV = S \times NV$. Further adsorption leads to complex dissolution and transformation of the salt – adsorbate system (Gordeeva, Restuccia, Cacciola, & Aristov, 1998; Petit, Karwacki, Peterson, & Bandosz, 2007). The equilibrium between sorbent and sorbate can be fundamentally changed by the salt content in the pores of the matrix. In particular, the bi-variant sorption equilibrium which is typical for conventional adsorbents turns into a mono-variant of the equilibrium between salt and vapour. The porous matrix is also important: it helps to disperse salt particles, affects the phase composition and adsorptive properties of the salt.

Due to modification by salt the specific adsorptive capacity of the composites exceeds the common adsorbents of the main working fluids (water, methanol, ethanol and ammonia). The sorption capacity of the composite $CaCl_2$ –silica gel (SWS-1l) on water is shown to reach 0.6 g/g, which surpasses the adsorptive capacity of the base silica gel by five to six times (Sharonov, et al., 2006). A similar effect was observed for LiBr –silica composites (Gordeeva et al., 1998). A sharp increase in the adsorption of ethanol on aluminum oxide and silica gel was stated, when modified by various inorganic salts (Cui, Tao, Chen, Gui, & Yao, 2005; Gordeeva & Aristov, 2010). The rise of sorption of ammonia after impregnation of the porous matrix (activated carbon, exfoliated vermiculite and aluminium) chlorides of calcium, copper and zinc was stated to be in 4 – 70 times (Sharonov et al., 2006; Bandosz & Petit, 2009). The adsorptive capacity of the composites LiCl – silica gel and LiBr – silica gel relative to methanol is five times higher than for the base silica gel (Gordeeva, Freni, & Krieger, 2008; Aristov & Gordeeva, 2009).

The change of adsorption properties of γ-alumina, vermiculite, sibunite and carbon tissue was stated when chloride, calcium, strontium and barium are introduced into their pores (Veselovskaya, 2011). The modification of the matrix with salt is shown to increase the adsorption capacity of the material with respect to ammonia to 0.7 g NH_3 per 1 g of the composite. The distribution of the salt in the pores was observed to result in a change the adsorption equilibrium of the salt with ammonia as compared to massive salts. The equilibrium in micro - and mesopores of aluminum oxide and carbon matrix occur to be bi-variant. The values of heat and entropy of adsorption decrease by 2 - 20 kJ/mole and 7 – 50 kJ/mole·K, respectively, when the amount of sorbed ammonia increased. For composites based on macroporous vermiculite monovariant equilibrium is observed, and the values of heat and entropy of adsorption are close to those of the corresponding massive salts.

Obviously, the adsorption capacity of SIPM is strongly affected by the salts content in the composite (Sukhyy et al., 2018). It was shown in (Oliveira, Wang, & Li, 2010) that the increase in sorption capacities of LiCl (13.7%) composites – activated carbon and LiCl (21.3%) –activated carbon relative to methanol, respectively, exceed the initial capacity by 56 and 86%. According to Fortier et al. (2008), the linear growth of the sorption capacitance with increasing salt content broke up by 3.5 mmol (or Cs = 32% by weight) for $ZnCl_2$ per 1 g of activated carbon. This behavior war explained by the fact that when the salt content is less than 3.5 mmol/g, a salt monolayer is formed on the surface of activated carbon. At Cs > 3.5 mmol/g a decrease in sorption occurs, since relatively large aggregates of salt particles block the pore space.

A similar dependence was also found for water sorption on a composite of $CaCl_2$-microporous silica gel (Gordeeva et al., 1998). Adsorption capacity increases with increasing salt content up to Cs = 31 mass% and then remained practically constant. Continuous increase in the sorption of water with composites '$CaCl_2$-activated carbon' is observed up to Cs = 70% by weight (Okada et al., 2010), and the composite '$CaCl_2$ –silica gel' up to Cs = 51% by weight. The salt content affects not only the sorption capacity but also the type of sorption equilibrium (Gordeeva et al., 2006). For adsorbents with high content of salts (Cs > 10 – 20% by weight) with the crystalline phase of salt forms, sorption of water occurs by the formation of crystalline hydrates of salts. The dispersion of such a system is 1, and sorption isobars are staged, which is desirable for the AHT. With low salt content, an X-ray-amorphous phase of saline forms is observed. Isobars of sorption of such composites are smooth curves, typical for a bi-variant system.

The influence of the chemical nature of the salt on the sorption properties of the composites was studied in (Xin et al., 2010, Wu, Wang, & Zhu, 2007). Xin et al. (2010) suggested that an increase in the sorption capacitance of $MeCl_n$ composites - mesoporous silica gel (Me = Li, K, Na, Ca, Mg, Zn) correlates with the ionic radius of the cation.

Most of technologies of obtaining of composite adsorbents are based technologies of dry or wet impregnation of porous matrix with salts (Grekova, Gordeeva, & Aristov, 2017; Hiremath & Kadoli, 2013).

Conditions on the impregnation stage affect the sorption properties (Wu et al., 2007). The increase of saline solution, temperature and time of impregnation were stated to enhance the sorption capacity of the composite '$CaCl_2$ – silica gel'. Perhaps the decrease of the viscosity of the saline solution and the increase in temperature result in a greater volume of salt solution that impregnates into silicon dioxide grains at a fixed point in time. Consequently, the salt content in the composites increases and contribute to an increase in the sorption capacity. The pH of the salt solution and the drying temperature affect the phase composition of the salt and the sorption equilibrium between water or methanol vapor (Xin et al., 2010). Thus, composite sorbents have intermediate properties between solid adsorbents, salts, hydrates and liquid absorbers. A sorption capacity and energy storage density are increased the combination of high density of salt storage and a good mass transfer of the porous matrix. Properties of the composites can be influenced by changing the salt content in pores of the matrix, the choice of salt according to its chemical nature, the conditions of synthesis (impregnation), the pore size of the matrix of the starting material. This enables the synthesis of composites with desired properties.

Parallel to possibility of task-oriented modification of properties of composite adsorbents, their main advantages of composite adsorbent are low charging temperature and high heat storage capacity. For example, regeneration temperature of the composite $LiNO_3$/vermiculite (SWS-9V) is 65 – 70ºC, it being demonstrated heat storage capacity of 900 kJ/kg (Sapienza, Glaznev, Santamaria, Freni, & Aristov, 2012). The composite 'expanded vermiculite – 40$SrBr_2$' presents water uptake of 0.53 g/g, mass energy storage density of 0.46 kWh/kg and volume energy storage density of 105.36 kWh/m^3 (Zhang et al., 2016). Through this method, a composites 'silica gel – $CaCl_2$ content of 43 wt.%' provides a high cycle loading lift of 0.40 g/g and an unprecedented energy storage density for this type of material of 211 kW h/m^3 of packed bed composite, in conditions of a solar heat storage system (adsorption at 30 °C, desorption at 80 °C, and water vapor pressure of 12.5 mbar) can be reached (Courbon et al., 2017). New composite based on silica gel and $SrBr_2$ provides a high cycle loading lift of 0.22 g/g corresponding to an energy storage capacity of 230Wh/kg and an energy storage density of 203 kW h/m^3 of packed bed composite (between 30 °C and 80 °C at 12.5 mbar) with an excellent stability over 14 sorption/desorption cycles (Courbon et al., 2017). The adsorption kinetics of this composite is enhanced as compared with bulk salt.

However, their industrial application is strongly inhibited by their quite complex technology production based on rather expensive dry and wet impregnation of porous media by crystalline hydrate solutions (Courbon et al., 2017). One of alternative solutions is synthesis of composite 'silica gel – crystalline hydrate' with sol – gel method (Mrowiec-Bialon, Jarzebski, & Pajak, 1999; Sukhyy, Belyanovskaya, Kozlov, Kolomiyets, & Sukhyy, 2014) https://serials.unibo.it/cgi-ser/start/it/spogli/df-s.tcl?prog_art=6535176&language=ITALIANO&view=articoli. Sorption capacities sol – gel sorbents $CaCl_2$/silica gel can reach 1 kg of water/kg of adsorbent, and regeneration temperature of these materials is 200°C (Mrowiec-Bialon et al., 1999).

The adsorption properties of the composites 'silica gel – sodium sulphate' and 'silica gel – sodium acetate' and mechanical mixture of bulk salt and silica gel are in a strong juxtaposition because of the characteristics of this disperse system (Sukhyy et al., 2014). This can be shown by comparing the adsorption capacities of these substances. The actual values of the adsorption capacities of the studied composites are states to surpass by approximately 30% the linear superposition of the limit adsorption of mechanical mixtures of silica gel and salts. As a result, the heat of adsorption is enhanced by 30%.

Sol-gel synthesis method allows to obtain crystals of salts distributed in the silicon-oxygen matrix at the molecular level. So, the adsorption properties of the synthesized composite adsorbents 'silica gel – crystalline hydrate' are not a linear combination of characteristics of silica gel and massive salt. The synthesized composite adsorbents reveal a unique structure that enhances the rate of reactions between salt crystals and water vapor in the developed pores of the silicon-oxygen matrix, which is confirmed by the kinetic curves of sorption.

The kinetic curves of water adsorption by composites 'silica gel – sodium sulphate' and 'silica gel – sodium acetate' in comparison with massive salts Na_2SO_4 and CH_3COONa are shown in Figure 5 (Sukhyy et al., 2018).

Qualitative difference of the mechanism of adsorption of water vapor for the composite adsorbents "silica gel – crystalline hydrate" and massive salts is clearly demonstrated by the adsorption kinetic curves.

So, for massive salts, hydration occurs in a kinetic regime, as shown by the S-shaped form of the kinetic curve (Fig. 1.5, *a*, *b*). For the composite adsorbents, the initial section of the kinetic curve is practically linear in the coordinates $a - \tau^{0.5}$ (Fig. 1.5, *c*, *d*). This indicates the diffusion character of the process, i.e. the adsorption of water by the crystalline hydrate is rapid, and the limiting factor being the transport of water through the silica gel pore system. Kinetic equilibrium for massive salts is stated to reach much slower than for the composites. The increase in temperature obviously increases the rate both of diffusion and of the chemical reaction. As temperature increased, the initial section of the kinetic curve of the adsorption of water by massive salts is

Figure 5. Kinetic curves of water sorption with: a – massive salt sodium sulphate; b massive salt sodium acetate; c – composite sorbent 'silica gel – sodium sulphate'; d – composite sorbent 'silica gel – sodium acetate'. Water vapour pressure 1010.8 mbar. Temperature, °C: 1 – 20; 2 – 40; 3 – 60
(Sukhyy et al., 2018)

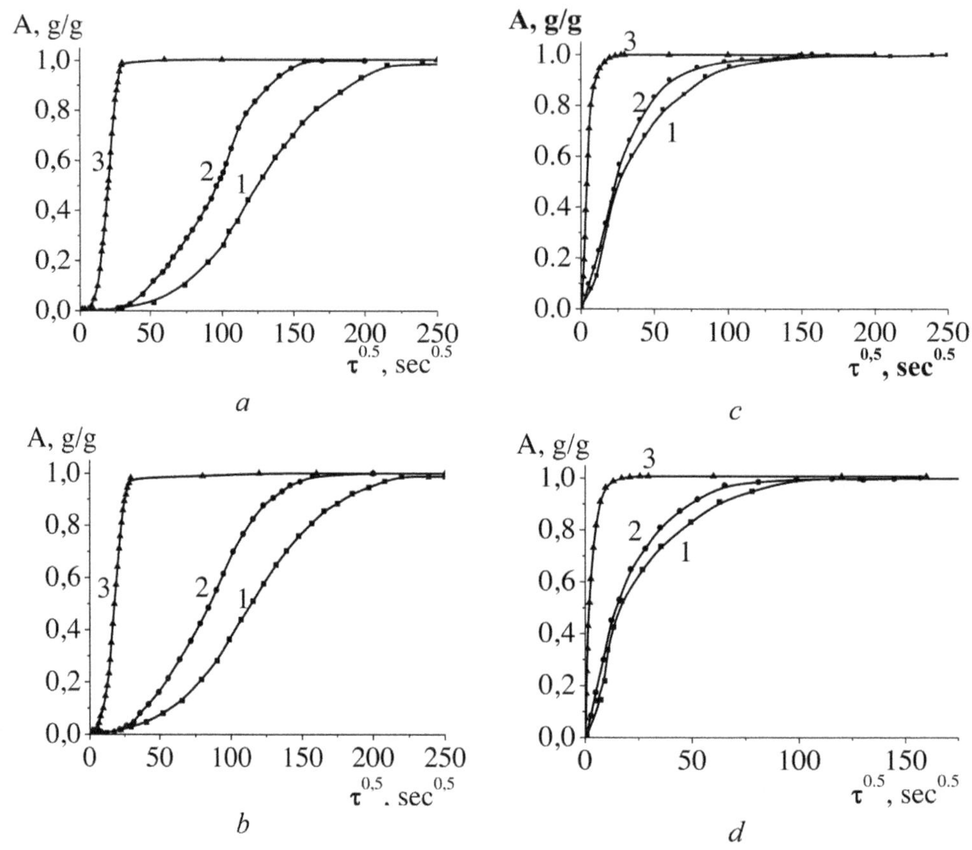

observed to be straightened (Fig. 1.5, *a*, *b*), it being probably a consequence of a change in the limiting stage. At 60 °C, the shape of the kinetic curves of bulk salts and composites becomes similar (curve 3).

So, the synthesized composite adsorbents can be called a highly porous three-dimensional silicon-oxygen matrix, in the pores of which molecules of crystalline hydrate are located.

The adsorption of water with massive crystalline hydrates is complicated by kinetic limitations, and water transport through the pore system of the composite adsorbents is much easier as compared with massive salts. Negative changes in the appearance, granulometric composition and adsorption properties of the composite adsorbent were not revealed during 2–3 cycles per day for 4–6 months.

So, water vapor diffusion coefficient in the pore system as function of temperature is stated to be linear in semilog coordinate for both composite sorbents. For sorbents 'silica gel sodium sulphate' and 'silica gel – sodium acetate':

$$\ln D = -2.95 - 6282.02 \cdot \frac{1}{T} \tag{2}$$

$$\ln D = 2.52 - 7733.59 \cdot \frac{1}{T} \tag{3}$$

The activation energy of this process is determined from the slope of the dependences (1.2) and (1.3) to be about 52 kJ/mole for the 'silica gel-sodium sulphate' composite, and 64 kJ/mole for the 'silica gel-sodium acetate' composite.

Therefore, synthesized composite sorbents can be said as a highly porous three-dimensional silicon-oxygen matrix, in the pores of which molecules of crystalline hydrate are located.

So, the most promising adsorptive heat storage media are composite adsorbents 'salt inside porous matrix' obtained by sol gel method.

REFERENCES

Alghoul, M. A., Sulaiman, M. Y., Azmi, B. Z., & Wahab, M. A. (2007). Advances on multi/purpose solar adsorption systems for domestic refrigeration and water heating. *Applied Thermal Engineering*, *27*(5/6), 813–822. doi:10.1016/j.applthermaleng.2006.09.008

Andersson, J., Azoulay, M., & De Pablo, J. (1983). Kinetic investigation of the sorption of water by barium chloride monohydrate. *Thermochimica Acta*, *70*(1-3), 291–302. doi:10.1016/0040-6031(83)80203-2

Aristov, Yu. I., & Gordeeva, L. G. (2009). Salt in a porous host/matrix adsorbents: Design of the phase composition and sorption properties. *Russ Kinet Catal.*, *50*(1), 65–72. doi:10.1134/S0023158409010091

Aristov, Yu. I., Restuccia, G., Cacciola, G., & Parmon, V. N. (2002). A family of new working materials for solid sorption air conditioning systems. *Applied Thermal Engineering*, *22*(2), 191–204. doi:10.1016/S1359-4311(01)00072-2

Aristov, Yu. I., Tokarev, M. M., Cacciola, G., & Restuccia, G. (1996). Selective water sorbents for multiple application, 1. $CaCl_2$ confined inmesopores of silica gel: Sorption properties. *Reaction Kinetics and Catalysis Letters*, *59*(2), 325–335. doi:10.1007/BF02068130

Bales, C. (2008). Final report of Subtask B Chemical and Sorption Storage' ' the overview. *Renewable & Sustainable Energy Reviews*, 23–28.

Bandosz, T. J., & Petit, C. (2009). On the reactive adsorption of ammonia on activated carbons modified by impregnation with inorganic compounds. *Journal of Colloid and Interface Science*, *338*(2), 329–345. doi:10.1016/j.jcis.2009.06.039 PMID:19615690

Barker, R. (2007). The reactivity of calcium oxide towards carbon dioxide and its use for energy storage. *Journal of Applied Chemistry & Biotechnology*, *24*(4-5), 221–227. doi:10.1002/jctb.2720240405

Boer, R. D., Haije, W., Veldhuis, J., & Smeding, S. (2004). Solid sorption cooling with integrated storage: the SWEAT prototype. *3rd international heat powered cycles conference – HPC 2004,* 687 – 689.

Carrillo, A. J., Sastre, D., Serrano, D. P., Pizarro, P., & Coronado, J. M. (2016). Revisiting the BaO_2/ Ba-redox cycle for solar thermochemical energy storage. *Physical Chemistry Chemical Physics*, *18*(11), 0168039–0168048. doi:10.1039/C5CP07777J PMID:26922970

Carrillo, A. J., Serrano, D. P., Pizarro, P., & Coronado, J. M. (2016). Manganese oxide-based thermochemical energy storage: Modulating temperatures of redox cycles by Fe-Cu codoping. *J. Energy Storage.*, *5*, 169–176. doi:10.1016/j.est.2015.12.005

Courbon, E., D'Ans, P., Permyakova, A., Skrylnyk, O., Steunou, N., Degrez, M., & Frère, M. (2017). Further improvement of the synthesis of silica gel and $CaCl_2$ composites: Enhancement of energy storage density and stability over cycles for solar heat storage coupled with space heating applications. *Solar Energy*, *157*, 532–541. doi:10.1016/j.solener.2017.08.034

Cui, Q., Tao, G., Chen, H., Gui, X., & Yao, H. (2005). Environmentally benign working pairs for adsorption refrigeration. *Energy, 30*(2-4), 261–271. doi:10.1016/j.energy.2004.05.005

Demir, H., Mobedi, M., & Uëlkëu, S. (2008). A review on adsorption heat pump: Problems and solutions. *Renewable & Sustainable Energy Reviews, 12*(9), 2381–2403. doi:10.1016/j.rser.2007.06.005

Donkers, P. A. J., Sögütoglu, L. C., Huinink, H. P., Fischer, H. R., & Adan, O. C. G. (2017). A review of salt hydrates for seasonal heat storage in domest applications. *Applied Energy, 199*, 45–68. doi:10.1016/j.apenergy.2017.04.080

Ferchaud, C. J., Scherpenborg, R. A. A., Zondag, H. A., & de Boer, R. (2014). Thermochemical seasonal solar heat storage in salt hydrates for residential applications - Influence of the water vapor pressure on the desorption kinetics of $MgSO_4 \cdot 7H_2O$. *Energy Procedia, 57*, 2436–2440. doi:10.1016/j.egypro.2014.10.252

Fortier, H., Westreich, P., Selig, S., Zelenietz, C., & Dahn, J. R. (2008). Ammonia, cyclohexane, nitrogen and water adsorption capacities of an activated carbon impregnated with increasing amounts of $ZnCl_2$, and designed to chemisorb gaseous NH_3 from an air stream. *Journal of Colloid and Interface Science, 320*(2), 423–435. doi:10.1016/j.jcis.2008.01.018 PMID:18261740

Gordeeva, L., & Aristov, Yu. (2010). Novel sorbents of ethanol—Salt confined to porous matrix for adsorptive cooling. *Energy, 35*(6), 2703–2708. doi:10.1016/j.energy.2009.04.001

Gordeeva, L., Freni, A., Krieger, T., Restuccia, G., & Aristov, Y. I. (2008). Composites lithium halides in silica gel pores: Methanol sorption equilibrium. *Microporous and Mesoporous Materials, 112*(1-3), 264–271. doi:10.1016/j.micromeso.2007.09.040

Gordeeva, L. G., & Aristov, Yu. I. (2018). Adsorptive heat storage and amplification: New cycles and adsorbents. *Energy, 167*, 440–453. doi:10.1016/j.energy.2018.10.132

Gordeeva, L. G., Freni, A., Restuccia, G., & Aristov, Yu. I. (2007). Influence of characteristics of methanol sorbents – salt in mesoporous silica on the performance of adsorptive air conditioning cycle. *Industrial & Engineering Chemistry Research, 46*(9), 2747–2752. doi:10.1021/ie060666n

Gordeeva, L. G., Glaznev, I. S., Savchenko, E. V., Malakhov, V. V., & Aristov, Y. I. (2006). Impact of phase composition on water adsorption on inorganic hybrids – salt/silica. *Journal of Colloid and Interface Science*, *301*(2), 685–691. doi:10.1016/j.jcis.2006.05.009 PMID:16765372

Gordeeva, L. G., Restuccia, G., Cacciola, G., & Aristov, Yu. I. (1998). Selective water sorbents for multiple applications: 5. LiBr confined in mesopores of silica gel: sorption properties. *Reaction Kinetics and Catalysis Letters*, *63*(1), 81–88. doi:10.1007/BF02475434

Grekova, A. D., Gordeeva, L. G., & Aristov, Yu. I. (2017). Composite "LiCl/vermiculite" as advanced water sorbent for thermal energy storage. *Applied Thermal Engineering*, *124*, 1401–1408. doi:10.1016/j.applthermaleng.2017.06.122

Hadj Ammar, M. A., Benhaoua, B., & Bouras, F. (2016). Thermodynamic analysis and performance of an adsorption refrigeration system driven by solar collector. *Applied Thermal Engineering*, *112*, 1289–1296. doi:10.1016/j.applthermaleng.2016.09.119

Hauer, A. (2002). Thermal energy storage with zeolite for heating and cooling applications, *Proceedings of 3rd workshop of annex 17 ECES IA/IEA*, 258 – 267.

Hauer, A. (2007). *Sorption theory for thermal energy storage. Thermal energy storage for sustainable energy consumption*. Springer.

Hiremath, C. R., & Kadoli, R. (2013). Experimental studies on heat and mass transfer in a packed bed of burnt clay impregnated with $CaCl_2$ liquid desiccant and exploring the use of gas side resistance model. *Applied Thermal Engineering*, *50*(1), 1299–1310. doi:10.1016/j.applthermaleng.2012.08.002

Jaehnig, D., Hausner, R., Wagner, W., Isaksson, C. (2006). Thermo-chemical storage for solar space heating in single-family house. *ECOSTOCK Conference*, *4*, 58 – 63.

Jong, A. J., Trausela, F., Fincka, C., van Vlieta, L., & Cuypers, R. (2014). Thermochemical heat storage – system design issues. *Energy Procedia*, *48*, 309–319. doi:10.1016/j.egypro.2014.02.036

Kato, Y. (2007). *Chemical energy conversion technologies for efficient energy use. In Thermal energy storage for sustainable energy consumption*. Springer.

Kato, Y., Harada, N., & Yoshizawa, Y. (1999). Kinetic feasibility of a chemical heat pump for heat utilization of high-temperature processes. *Applied Thermal Engineering*, *19*(3), 239–254. doi:10.1016/S1359-4311(98)00049-0

Kolomiyets, O.V., Sukhyi, K.M. (2016). Akumuliuvannia tepla, Vseukr. nauk./prakt. konf. «Problemy enerhoresursozberezhennia promyslovomu rehioni. Nauka i praktyka»: Zb.tez dopov. *Mariupol, DVNZ «PDTU»*, 48 – 49.

Koohi-Fayegh, S., & Rosen, M. A. (2020). A review of energy storage types, applications and recent developments. *Journal of Energy Storage*, *27*, 101047. doi:10.1016/j.est.2019.101047

Lanahan, M., & Tabares-Velasco, P. C. (2017). Seasonal Thermal-Energy Storage: A Critical Review on BTES Systems, Modeling, and System Design for Higher System Efficiency. *Energies*, *10*(6), 743–767. doi:10.3390/en10060743

Mastronardo, E., Bonaccorsi, L., Kato, Y., Piperopoulos, E., Lanza, M., & Milone, C. (2017). Strategies for the enhancement of heat storage materials performances for $MgO/H_2O/Mg(OH)_2$ thermochemical storage system. *Applied Thermal Engineering*, *120*, 626–634. doi:10.1016/j.applthermaleng.2017.04.004

Mrowiec-Bialon, J., Jarzebski, A. B., & Pajak, L. (1999). Water vapor adsorption on the SiO_2-$CaCl_2$ sol-gel composites. *Langmuir*, *15*(19), 6505–6509. doi:10.1021/la990231n

Mugnier, D., & Goetz, V. (2001). Energy storage comparison of sorption systems for cooling and refrigeration. *Solar Energy*, *71*(1), 47–56. doi:10.1016/S0038-092X(01)00013-5

N'Tsoukpoe, K. E., Liu, H., Le Pierres, N., & Luo, L. (2009). A review on long/term sorption solar energy storage. *Renewable & Sustainable Energy Reviews*, *13*(23), 85–96. doi:10.1016/j.rser.2009.05.008

Nielsen, K. (2003). Thermal energy storage—a state-of-the-art. A report within the research program Smart Energy-Efficient Buildings at the Norwegian University of Science and Technology and SINTEF, Norwegian University of Science and Technology and SINTEF, 123

Okada, K., Nakanome, M., Kameshima, Y., Isobe, T., & Nakajima, A. (2010). Water vapor adsorption of $CaCl_2$/impregnated activated carbon. *Materials Research Bulletin*, *45*(11), 1549–1553. doi:10.1016/j.materresbull.2010.07.027

Oliveira, R. G., Wang, R. Z., & Li, T. X. (2010). Adsorption characteristics of methanol in activated carbon impregnated with lithium chloride. *Chemical Engineering & Technology*, *33*(10), 1679–1686. doi:10.1002/ceat.201000075

Pantos, E. J., West, B., Dokter, W. H., Van Garderen, H. F., Beelen, T. P. M., & Van Santen, R. A. (1994). Growth and aging phenomena in silica gels. *Journal of Sol-Gel Science and Technology*, *2*(1-3), 273–276. doi:10.1007/BF00486255

Pardo, P., Deydier, A., Anxionnaz-Minvielle, Z., Rougé, S., Cabassud, M., & Cognet, P. (2014). A review on high temperature thermochemical heat energy storage. *Renewable & Sustainable Energy Reviews*, *32*, 591–610. doi:10.1016/j.rser.2013.12.014

Petit, C., Karwacki, C., Peterson, G., & Bandosz, T. J. (2007). Interactions of ammonia with the surface of microporous carbon impregnated with transition metal chlorides. *The Journal of Physical Chemistry C*, *111*(34), 12705–12714. doi:10.1021/jp072066n

Sapienza, A., Glaznev, I. S., Santamaria, S., Freni, A., & Aristov, Yu. I. (2012). Adsorption chilling driven by low temperature heat: New adsorbent and cycle optimization. *Applied Thermal Engineering*, *32*, 141–146. doi:10.1016/j.applthermaleng.2011.09.014

Sarbu, I., & Sebarchievici, C. (2018). A Comprehensive Review of Thermal Energy Storage. *Sustainability*, *10*(2), 191. doi:10.3390u10010191

Scapino, L., Zondag, H. A., Van Bael, J., Diriken, J., & Rindt, C. C. M. (2017). Sorption heat storage for long-term low-temperature applications: A review on the advancements at material and prototype scale. *Applied Energy*, *190*, 920–948. doi:10.1016/j.apenergy.2016.12.148

Schmidt, M., Szczukowski, C., Roßkopf, C., Linder, M., & Wörner, A. (2014). Experimental results of a 10 kW high temperature thermochemical storage reactor based on calcium hydroxide. *Applied Thermal Engineering*, *62*(2), 553–559. doi:10.1016/j.applthermaleng.2013.09.020

Schreiber, H., Lanzerath, F., & Bardow, A. (2018). Predicting performance of adsorption thermal energy storage: From experiments to validated dynamic models. *Applied Thermal Engineering*, *141*, 548–557. doi:10.1016/j.applthermaleng.2018.05.094

Sharma, A., Tyagi, V. V., Chen, C. R., & Buddhi, D. (2009). Review on thermal energy storage with phase change materials and applications. *Renewable & Sustainable Energy Reviews*, *13*(2), 318–345. doi:10.1016/j.rser.2007.10.005

Sharonov, V., Veselovskaya, J. V., & Aristov, Yu. I. (2006). Ammonia sorption on composites —$CaCl_2$ in inorganic hostmatrix: Isosteric chart and its performance. *The International Journal of Low Carbon Technologies*, *1*(3), 191–200. doi:10.1093/ijlct/1.3.191

Sukhyy, K., Belyanovskaya, E., Kovalenko, V., Kotok, V., Sukhyy, M., Kolomiyets, E., Gubynskyi, M., Yeromin, O., & Prokopenko, O. (2018). The study of properties of composite adsorptive materials "silica gel – crystalline hydrate" for heat storage devices. *Eastern-European Journal of Enterprise Technologies*, *91*(1), 52–58. doi:10.15587/1729-4061.2018.123896

Sukhyy, K. M., Belyanovskaya, E. A., Kozlov, Ya. N., Kolomiyets, E. V., & Sukhyy, M. P. (2014). Structure and Adsorption Properties of the Composites 'Silica Gel – Sodium Sulphate', obtained by Sol – Gel Method. *Applied Thermal Engineering*, *64*(1-2), 408–412. doi:10.1016/j.applthermaleng.2013.12.013

Tescari, S., Agrafiotis, C., Breuer, S., De Oliveira, L., Neises-Von Puttkamer, M., Roeb, M., & Sattler, C. (2013). Thermochemical solar energy storage via redox oxides: Materials and reactor/heat exchanger concepts. *Energy Procedia*, *49*, 1034–1043. doi:10.1016/j.egypro.2014.03.111

Vasiliev, L. L., Gulko, N. V., & Khaustov, V. M. (1996). Solid adsorption refrigerators with active carbon/acetone carbon ethanol pairs. *Proceedings of the International Sorption Heat Pump Conference*, *1*, 3–6.

Veselovskaya, J.V. (2011). *Sorbtsionnuye svoistva kompositnykh poglotyteley na osnove dyspersnykh khloridov shelochnozemelnykh metallov: avtoref. dis. na soiskaniye nauch. zvaniya kand. khim. nauk: 02.00.04.* Moskva, 21.

Visscher, K., & Veldhuis, J. B. J. (2005). Comparison of candidate materials for seasonal storage of solarheat through dynamic simulation of building and renewable energy system, *Proceedings of the ninth international building performance simulation association.*

Wang, D., Zhang, J., Xia, Y., Han, Y., & Wang, S. (2012). Investigation of adsorption performance deterioration in silica gel–water adsorption refrigeration. *Energy Conversion and Management*, *58*, 157–162. doi:10.1016/j.enconman.2012.01.013

Wang, L. W., Wang, R. Z., & Oliveira, R. G. (2009). A review on adsorption working pairs for refrigeration. *Renewable & Sustainable Energy Reviews*, *13*(3), 518–534. doi:10.1016/j.rser.2007.12.002

Wang, R. Z., Jia, J. P., Zhu, Y. H., Teng, Y., Wu, J. Y., Cheng, J., & Wang, Q. B. (1997). Study on a new solid adsorption refrigeration pair: Active carbon ðber – methanol. *Sol. Energy Eng.*, *119*(3), 214–218. doi:10.1115/1.2888021

Wu, H., Wang, S., & Zhu, D. (2007). Effects of impregnating variables on dynamic sorption characteristics and storage properties of composite sorbent for solar heat storage. *Solar Energy*, *81*(7), 864–871. doi:10.1016/j.solener.2006.11.013

Xin, L., Huiling, L., & Siqi, H. (2010). Dynamics and isotherms of water vapor sorption on mesoporous silica gels modified by different salts. *Kinetics and Catalysis*, *51*(5), 754–761. doi:10.1134/S0023158410050186

Zalba, B., Marrın, J. M., Cabeza, L. F., & Mehling, H. (2003). Review on thermal energy storage with phase change: Materials, heat transfer analysis and applications. *Applied Thermal Engineering*, *23*(3), 251–283. doi:10.1016/S1359-4311(02)00192-8

Zhang, Y. N., Wang, R. Z., Zhao, Y. J., Li, T. X., Riffat, S. B., & Wajid, N. M. (2016). Development and thermochemical characterizations of vermiculite/$SrBr_2$ composite sorbents for low-temperature heat storage. *Energy*, *115*, 120–128. doi:10.1016/j.energy.2016.08.108

Chapter 2
Basic Principles for Substantiation of Working Pair Choice

ABSTRACT

The chapter is devoted to basic principles for substantiation of working pair choice. Principles for substantiation for selection of adsorbate and adsorbent are considered. The main requirements to adsorbate are formulated as follows: low cost, easy of obtaining, small molecular size to facilitate the adsorption effect; high latent heat of evaporation and small volume in liquid state; high thermal conductivity; low viscosity, thermal stability with adsorbent in the operating temperature range; chemical stability in the working temperature range; non-toxicity for animals and human, non-aggressiveness and incombustibility; low pressure saturation (slightly above atmospheric pressure) at normal operating temperature; the absence of environmental problems. Water is shown to conform to these requirements. The crucial requirements to adsorbent are the ability to adsorb large amounts of adsorbate when cooled to environment temperature and give a high cooling effect; high values of maximal adsorption; desorption of the major portion of adsorbate (ideally all) when heated by an accessible source of heat; low heat capacity; good

DOI: 10.4018/978-1-7998-4432-7.ch002

heat conductivity, short cycle time; no deterioration and loss of adsorption capacity over time or use; non-toxicity, non-aggressiveness; chemical physical compatibility with the selected adsorbate; low cost and wide availability. Properties of various types of adsorbents were compared. Composites 'salt inside porous matrix' are shown to be promising media for heat storage and transformation. Characteristics of thermodynamic cycles of heat conversion were analysed. The ways to improve the coefficient of performance were analysed and shown to be affected by a proper choice of an adsorbent.

NOMENCLATURE

COP_c^{id} – coefficient of performance for cooling;
COP_h^{id} – coefficient of performance for heating;
COP_c^{Carnot} – Carnot coefficient of performance for cooling;
COP_h^{Carnot} – Carnot coefficient of performance for heating;
M_{ads} – adsorbent mass, kg;
P – pressure, Pa;
P_{H_2O} – partial pressure of water vapor, Pa;
P_o – saturated vapor pressure;
ΔF – Dubinin adsorption potential, kJ/mole;
Q_a – heat of adsorption, kJ;
Q_c –heat of condensation, kJ;
Q_d –heat of desorption, kJ;
Q_e – heat of evaporation, kJ;
$\Delta_r H$ – enthalpy during the reaction;
$\Delta_r S$ – change of ethropy during the reaction;
m – amount of removed water;
R – absolute gas constant;
SCP – specific adsorptive cooling/heating power;
T_c – temperature of heat rejection, K;
T_{cond} – water temperature in the condenser, K;
T_{ev} – water temperature in the evaporator;

T_g – temperature of external heat source, K;
$T_3 = T_{des}$ – temperature of the dehydration of the sorbent, K;
w – equilibrium water uptake;
τ_c – cycle time.

Adsorption is the process of redistribution of molecules between volume of a gas or liquid medium and a surface (external and internal). A solid substance on the surface of which uptakes another substance is called an adsorbent. A substance removed from the gas or liquid phase and concentrated on the surface of adsorbent is called an adsorbate.

'Adsorbent – adsorbate' interaction is mainly co-operative and the most important contribution of various adsorption is known to determine the overall process of distribution between phases of substances. It depends on the chemical nature of the interacting components of the adsorption system.

Adsorption of substances when interphase equilibrium reached depends on: temperature; the chemical nature (affinity to the sorbent) and the concentration or pressure of the gas of the adsorbate; the nature of the chemical, the structure and surface area of the adsorbent. Generally, adsorption is an exothermic process, therefore, as temperature increased, the amount of adsorbed substance decreases. Adsorption is intensified by increasing surface area of the adsorbent, as well as the concentration of adsorbate.

SELECTION OF ADSORBATE

For technologies of heat energy storage, the adsorption capacity (or maximal adsorption) of the substances used (adsorbents) is crucial for both quantitative estimation of the adsorption value and transfer of adsorbate from the bulk phase to the adsorbent surface. This value is affected by the nature of the adsorbent, its structure, dispersion, the nature of adsorbate, pressure and temperature.

For heat storage technologies, adsorbate have to meet the next requirements (Sur, & Das, 2010, Riffel, Belo, & Leite, 2007): low price, easy of obtaining, small molecular size to facilitate the adsorption effect; high latent heat of evaporation and small volume in liquid state; high thermal conductivity; low viscosity, thermal stability with adsorbent in the operating temperature range; chemical stability in the working temperature range; non-toxicity for animals and human, non-aggressiveness and incombustibility; low pressure saturation (slightly above atmospheric pressure) at normal operating temperature; the absence of environmental problems. Water conforms to these requirements.

For cooling purposes or heat pumps, limitation of evaporation temperature down to 0^0C ought to be added to above-mentioned requirements (Riffel et al., 2007).

Mainly, a natural refrigerant used in adsorption refrigeration systems does not affect the environment. The most common refrigerants applied in adsorptive chilling devices are ammonia, methanol and water, which have relatively high values of latent heat (1368, 1160 and 2258 kJ/kg, respectively) and low levels of specific volumes (about 10 m^3/kg). To work with water and methanol at operational temperatures, it is necessary to maintain sub-atmospheric saturation pressure, and any penetration of the air immediately leads to a malfunction in the system. As for ammonia, for a while small leaks can be allowed, but its saturation pressure is 13 bar at a condensation temperature of 35°C. Ammonia is toxic and aggressive, and methanol is flammable. Water is the most thermostable adsorbate, followed by methanol and ammonia. However, even the cheapest adsorbate, water, cannot be used for cooling purposes below 0°C (Odesola & Adebayo, 2010).

SELECTION OF ADSORBENT

The effectiveness of adsorptive heat conversion processes is strongly affected by properties of applied adsorbent. The most important characteristics to select an adsorbent are (Alghoul, Sulaiman, Azmi, & Wahab, 2007; Leite, Grilo, Andrade, & Belo, 2004): the ability to adsorb large amounts of adsorbate when cooled to environment temperature, and give a high cooling effect; high values of maximal adsorption; desorption of the major portion of adsorbate (ideally all) when heated by an accessible source of heat; low heat capacity; good heat conductivity, short cycle time; no deterioration and loss of adsorption capacity over time or use; non-toxicity, non-aggressiveness; chemical physical compatibility with the selected adsorbate; low cost and wide availability.

The most suitable adsorbent should be sufficiently porous (with a specific surface of about 600 m^2/g or above (Ruthven, 1984). To ensure the adsorption of a large amount of refrigerant, this leads to low thermal conductivity, which limits the performance of the cooling system. Therefore, there must be a compromise between the high porosity required for rapid steam diffusion and the high density required for good thermal conductivity (Tather, Tantekin, Ersolmaz, & Senatalar, 1999).

Activated carbon, zeolites and silica gel are most often used as adsorbents. Several studies have been carried out in (Wang, Zhang, Tian, Liu, & Sumathy, 2014, Al Alili, Hwang, & Radermacher, 2014) to improve the productivity of some adsorbents by combining with other substances.

As considered the sanitary rules, regeneration temperature of adsorbent ought to be limited up to 110°C. Nevertheless, regeneration temperatures for silica gels are varied between 100 – 150°C (Publishing House of standards, 1976; Ng et al., 2001). The minimal regeneration temperatures of 100°C are stated for nano-porous silica gels produced by Fuiji Silicia Ltd, Japan (Ng et al., 2001). However, its adsorptive capacity is 0.35 – 0.4 kg/kg.

TGA results of samples of $CaCl_2$ and zeolite cycled 4 times reveals similar profile. The final result is the same: around 38% of anhydrous $CaCl_2$ remains inside the crucible. So, thermal measurements of the $CaCl_2$ are repeatable (Barreneche, Fernández, Cabeza, & Cuypers, 2015).

DSC results are shown in Figure 2 for both materials under study. The difference between adsorbents is important because DSC curves of zeolite are quite flat but an energy-related process is taking place. Nevertheless, $CaCl_2$ DSC reveals two different peaks corresponded to two energy-related processes. The peak integration resulted from energy involved in those processes.

Figure 1. TGA curves of $CaCl_2$ and zeolite spheres sample cycled once with TGA (Barreneche et al., 2015).

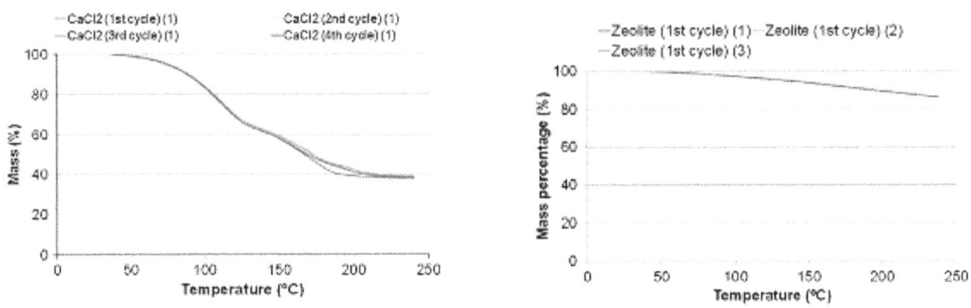

a b

The first peak is supposed to be related to the solid–liquid phase change from $CaCl_2 \cdot 6H_2O$. According to TGA curves, there is not mass loss at 50 °C (Barreneche et al., 2015). The obtained phase change temperature is around 50

°C which is higher, as compared to expected 45.3 °C. This result is explained by the experimental conditions used: the heating rate is not slow enough to achieve the thermal equilibrium in the sample. Moreover, the second peak corresponds to dehydration water loss which takes place applying heat to $CaCl_2$. Contrariwise, DSC curves for zeolite show only one broad peak which resulted from the zeolite dehydration. This energy-related process is almost finished at 400 °C. The effect of the heating rate was estimated performing DSC analyses using three different heating rates (2 K·min^{-1}, 5 K·min^{-1} and 10 K·min^{-1}) between 40 °C and 240 °C for $CaCl_2$ and between 50 and 400 °C for zeolite. Resultant DSC curves (Cp vs. temperature) are given in Fig. 2, respectively. As it is stated for the $CaCl_2$, the peak reaction temperature applying 10 K·min^{-1} heating rate is 186 °C. However, this reaction temperature for analyses performed using slower heating rate is decreased to 165 °C. So, the measured temperature of reaction is affected by the heating/cooling rate applied. The calculated energy density is varied from 1.47 GJ·m^{-3} applying 10 K·min^{-1} to 1.31 GJ·m^{-3} applying 2 K·min^{-1}. Therethrough, the energy density is also decreased when the heating rate is slower. On the other hand, DSC curves for zeolite revealed that the highest heating rate corresponds with the lowest process temperature.

The process of uptake and release of water is a crucial point for the selection of adsorbent because final dehydration temperature will determine the total energy stored by a certain application.

According to thermophysical data (Figure 1), thermochemical materials are more energy efficient than the sorptive materials as shown by the much higher calculated energy density. Salt hydrates are more promising than the sorption TCM as shown by the much higher calculated energy density (Table 1).

Figure 2. $CaCl_2$ and zeolite DSC curves performed applying three different heating rate (10 K· min^{-1}, 5 K·min^{-1} and 2 K·min^{-1})
(Barreneche et al., 2015).

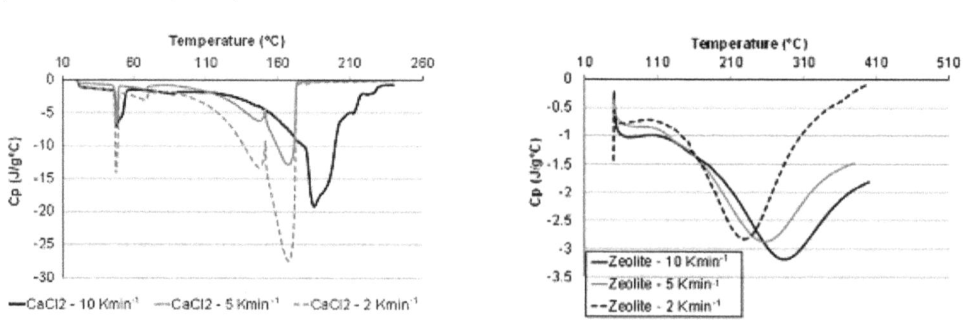

However, other parameters as the corrosion behavior as well as degradation process ought to study for this substances in order to reveal the best candidate to be implemented in a storage system.

Moreover, thermal stability of salts in the projected temparature interval ought to take into account. According to data by Donkers et al. (2017), $KAl(SO_4)_2$, $(NH4)_2Zn(SO_4)_2$, $NiSO_4$, $NH_4Co(SO_4)_2$ can be considered as promising for thermal energy storage not only due to high energy storage density but low dehydration temperatures 21 – 61°C. Despite these facts, above-mentioned salts can be cycled only one-time because of hydrolysis.

Table 1. DSC results and calculated energy density obtained for $CaCl_2$ and zeolite

Material	Energy, J/g	T°C	Energy, J/g	T_{peak} °C	Energy Density, GJ/m^3
$CaCl_2$	-32	50.1	801	190.0	1.47
Zeolite	–	–	141	274.0	0.18

(Barreneche et al., 2015)

Attempts to improve adsorptive capacity of adsorptive heat storage media are generally focused on the choice among existing adsorbents (Jong et al., 2014) or synthesis and investigation of composite sorbents ‘porous matrix-salt’ due to increased sorption capacity to ordinary working fluids (Cabeza, Solé, & Barreneche, 2017; Gordeeva, Grekova, Krieger, & Aristov, 2013).

Furthemore, the distribution in the porous matrix results in stabilization of the salt during repeated hydration cycles-dehydration (Jong et al., 2014). Silica gel (Gordeeva et al., 2013), zeolites (Scapino et al., 2017), vermiculite (Grekova et al., 2017), graphite (Zamengo & Kato, 2017), and salts, ie. magnesium chloride and lithium bromide (Tanashev, Krainov, & Aristov, 2013), lithium chloride (Grekova et al., 2017), magnesium oxide (Zamengo & Kato, 2017), calcium chloride (Hiremath & Kadoli, 2013) are used as a matrix. The adsorbtive is water vapor (Grekova et al., 2017; Frazzica, & Freni, 2017), ammonia (Hiremath & Kadoli, 2013), methanol (Zalba et al., 2003).

THE THERMODYNAMIC CYCLE OF ADSORPTIVE HEAT TRANSFORMATION DEVICE

The thermodynamic cycle is the main criterion for substantiation of a working pair 'adsorbent – adsorbate' and the hardware design of an AHT device, because it determines the temperature of sorption-desorption and the amount of adsorbed substance.

Sorption of water vapor with inorganic salts or adsorbents can be described as follows:

$$S_s + nH_2O_g = S\cdot nH_2Os \tag{1}$$

where S is inorganic salt or adsorbent; n is amount of water molecules.

For obtaining and decomposition of hydrates dependence equilibrium pressure of water vapour is described by equation (Bajin, Ivatshenko, & Parmon, 2002):

$$\ln P_{H_2O} = \frac{\Delta_r S}{m \cdot R} - \frac{\Delta_r H}{m \cdot R \cdot T} \tag{2}$$

where $\Delta_r S$ and $\Delta_r H$ are change of entropy and enthalpy during the reaction; R is absolute gas constant; T is reaction temperature; m refers to amount of removed water.

According to the Clapeyron-Clausius equation, the equilibrium is described by a straight line in the coordinates lnp vs. T^{-1}. The pressure of water vapor is described by the following equations (Rabinovich & Khavin, 1978):

as temperatures are lower than 0°C $$\ln P_{H_2O} = -\frac{6135.2}{T} + 17.362 \tag{3}$$

as temperatures are higher than 0°C $$\ln P_{H_2O} = -\frac{3806.8}{T} + \frac{218171}{T^2} + 11.768 \tag{4}$$

The thermodynamic cycle of the adsorption heat converter, which work is based on the process of adsorption-desorption of water vapour with solutions of hygroscopic salts or porous adsorbents, is presented in Figure 3.

Figure 3. Thermodynamic cycle of adsorptive heat energy transformer (Aristov, (2012))

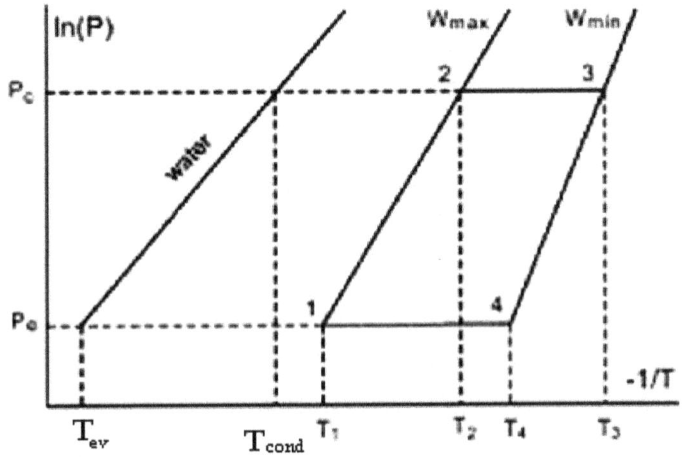

The cycle consists of two isobars (1-4 and 2-3) and two isosteres (1-2 and 4-3). The cycle is characterized by three temperatures: the water temperature in the condenser T_{cond}, in the evaporator T_{ev} and the temperature of the dehydration of the sorbent $T_3 = T_{des}$. The temperature of condensation usually depends on the temperature of the environment, as there dissipates heat in the condenser, and in the summer time is 25 – 50°C. The temperature of the evaporator is set at a cooling temperature and is usually in the range from 0 to 10°C. The temperature of the dehydration of the sorbent is determined by the available heat source.

For every particular case the thermodynamic cycle of AHT has to be considered, because it determines the choice of AHT equipment and its efficiency. In the relevant chapters the thermodynamic APT cycles will be presented for each field of application of AHT.

A three temperature (3T) adsorptive heat/cooling (ACH) cycle consists of two isosters and two isobars and is commonly presented on the Clapeyron diagram ln(P) vs. (1/T) (Figure 4a).

Other usable ways to present this cycle are w(T) and T(S) diagrams, where S is the entropy change. Additionally, the ACH cycle is considered by the diagram w(ΔF) that is rarely used.

The Dubinin adsorption potential $\Delta F_{min} = -RT \ln(P_e/P_o)$ was applied as a universal measure of the adsorbent affinity to the adsorptive (here P_o is the saturated vapour pressure at temperature T). In particular, it is valid for

Figure 4. Common Clapeyron diagram of 3T AHT cycle (left) and the same cycle presented in the w vs. ΔF co-ordinates (right)
(Aristov, 2014).

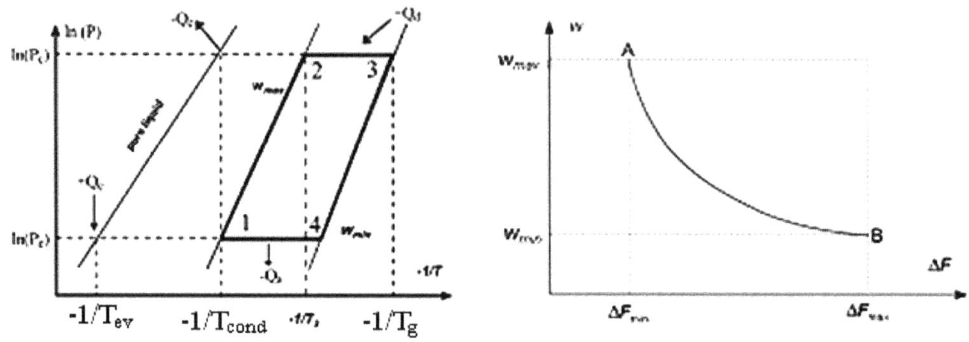

adsorbent – adsorbate pairs described by the Dubinin-Astakhov or Dubinine-Radushkevich equations.

The ACH cycle in the w(ΔF) presentation is very simple: one line represents both isobaric adsorption and desorption stages of a 3T ACH cycle (Figure 4b). If hysteresis is present, two different curves correspond with the ad/desorption steps. Point A in Figure 4b states the boundary conditions at the end of the adsorption step (point 1 on Figure 4a). It is concerned with the adsorption potential $\Delta F_{min} = -RT_1 \cdot \ln[P_e/P_o(T_1)]$ that characterizes the minimal adsorbent affinity necessary to bind an adsorptive at temperature $T_1 = T_c$ and pressure $P(T_e)$. This potential is affected by the adsorptive nature.

Five basic applications, such as air-conditioning, heat pumping, and deep freezing can be applied.

The boundary conditions at the end of the desorption stage (point 3, Figure 4a) are presented by point B in Fig. 2.4b. They are corresponded to the desorption potential $\Delta F_{max} = -RT_3 \ln[P_c/P_o(T_3)]$ which is defined by the conditions of heat rejection (T_c) and adsorbent regeneration T_3). Adsorbate molecules bound to the adsorbent with an affinity lower than $\Delta F_{max} = -RT_g \ln[P_c/P_o(T_g)]$ can be desorbed during the regeneration stage 2 – 3. Those bound more strongly adsorbed and are not involved in the heat transformation cycle.

For an ideal 3T cycle (with zero thermal masses) the energy balance is given as

$$Q_c - Q_e - Q_d + Q_a = 0 \quad (5)$$

(see definition of heats on Fig. 4a). The first law efficiency or the Coefficient Of Performance (COP) is then defined as

$$COP_c^{id} = \frac{Q_e}{Q_d} \tag{6}$$

$$COP_h^{id} = \frac{Q_c + Q_a}{Q_d} \tag{6a}$$

for cooling and heating, respectively. For real systems, the COP-value is obviously also dependent on the adsorbate mass Δw exchanged in the cycle

$$COP_c\left(\Delta w\right) = \frac{Q_e \Delta w}{Q_d \Delta w + C\left(T_g - T_{cond}\right)} = COP_c^{id} \cdot \frac{\Delta w}{\Delta w + B} \tag{7}$$

where $B = \frac{C\left(T_g - T_{cond}\right)}{Q}$ is the ratio of the sensible heat of all inert masses to the latent heat necessary for desorption. The first law efficiency can be increased by maximizing of exchanged mass

$$\Delta w = w_{max} - w_{min} = w\left(\Delta F_{min}\right) - w\left(\Delta F_{max}\right).$$

The important highlights for increasing the first low COP are the next:

1. searching for adsorbents with very large Δw values (>0.5 g/g) is the ultimate priority for heat storage rather than for cooling/heating;
2. for any cooling application, the efficiency increases at lower heats of desorption. E.g. for small-pore aluminophosphates the heat of water desorption is Q_d = 55 – 60 kJ/mol (Ristic, Zabukovec, Henninger, & Kaucic, 2012; Kakiuchi et al., 2005), therefore the cooling COP_c can not exceed 0.65 – 0.75. For silica gels, that have lower affinity for water vapour, the COP_c can reach about of 0.9;
3. the maximum COP_c (0.7 – 0.9) for “single effect” AC cycles, is significantly lower than predicted for a reversible Carnot cycle with the same boundary temperatures.

For an ideal 3T system, the second law efficiency could be should be supplemented with the entropy balance ensued by the energy balance (the first law) of Eq. (2.5) and given as:

$$-\frac{Q_c}{T_{cond}}+\frac{Q_e}{T_{ev}}+\frac{Q_d}{T_g}-\frac{Q_a}{T_a}=\Delta S\geq 0 \tag{8}$$

As all the processes are completely reversible, the entropy generation is equal to zero ($\Delta S = 0$), and the Carnot COP can be calculated for cooling and heating (Aristov, 2014):

$$COP_c^{Carnot}=\frac{\frac{1}{T_{cond}}-\frac{1}{T_g}}{\frac{1}{T_{ev}}-\frac{1}{T_{cond}}} \tag{9}$$

$$COP_h^{Carnot}=\frac{\frac{1}{T_{ev}}-\frac{1}{T_g}}{\frac{1}{T_{ev}}-\frac{1}{T_{cond}}} \tag{9a}$$

No entropy is generated, when heat for regeneration supplied directly at the minimal desorption temperature ($T_g = T_{min}$). As a result, that the maximum(Carnot) efficiency is reached (Sharonov & Aristov, 2008). Moreover, the COP-values calculated from the first and second laws are equal. This imposes an appropriate requirement on the shape of the sorption isobar for an optimum adsorbent: the rich and weak isosters must coincide, so that the cycle occurs to be degenerated. The isobaric adsorption 4 – 1 and desorption 2 – 3 (Fig.2.4a) occur immediately and completely at the temperatures T_c and T_g (Figure 5a), respectively (Aristov, 2014).

The degenerated cycle becomes possible with an adsorbent characterized by a mono-variant equilibrium. The adsorbent has to sorb a large amount of adsorbate in a stepwise manner directly at T_{cond} and P_e and completely desorb it at T_g and P_c (Fig. 2.3b). Theoretically, an optimal adsorbent for ACH should have a step-wise adsorption isobar. Real adsorbents, however, exhibit S-shaped adsorption isobars (Fig. 5c) instead of strictly step-wise ones. Such adsorption equilibrium corresponds to isotherms of types V and VI n the IUPAC classification. The harmonization of the adsorbent and the cycle by second law means that the position of the step have to coincide with the boundary temperatures $T_{cond} = T_4$ at $P = P_e$ and $T_2 = T_g$ at $P = P_c$ as shown in Figure 5b.

If the adsorbate is removed at $T_g > T_{min}$, the efficiency decreases due to the thermal entropy production caused by the external thermal coupling $\Delta T = T_g - T_{min}$ (Aristov, 2014). So, a high second law efficiency may be attained by a proper choice of adsorbent.

As opposed to the thermodynamic efficiencies, the adsorbent dynamic efficiency depends less on adsorbent inherent properties and more on a number of external factors (Aristov, 2014).

Among these are the designs of the "adsorber – heat exchanger", efficiency of heat supply/rejection, pressure and temperature levels, etc. The optimization criterion of the dynamic analysis is the specific ACH power which can be defined for cooling as:

$$SCP = \frac{Q_e}{M_{ads}\tau_c} \tag{10}$$

where M_{ads} is the adsorbent mass and τ_c is the cycle time.

The step-like-shaped adsorption isobars are promising from both thermodynamic and dynamic points of view. The dynamic requirement to maximize the average temperature difference between the adsorbent and the fin is in direct contradiction with the thermodynamic requirement to minimize this difference (Aristov, 2014). So, proper choice of the isobar step position greatly depends on which output parameter is to be maximized – COP or SCP.

If both parameters are important, a reasonable compromise ought to be reached by intelligent design of adsorber – heat exchanger unit.

Figure 5. Degenerated AHT cycle characterized by a single ad-/desorption isoster (a); corresponding step-like (b, d) and S-shaped (c) isobars of adsorption (at P = P_e) and desorption (at P ¼ Pc) optimum from the thermodynamic (b, c) and dynamic (d) points of view
(Aristov, 2014).

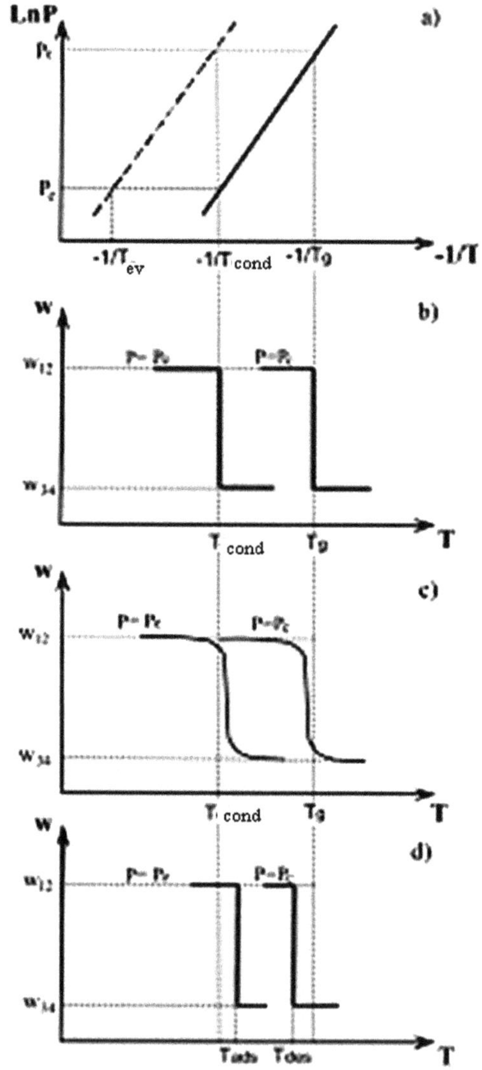

REFERENCES

Al Alili, A., Hwang, Y., & Radermacher, R. (2014). Review of solar thermal air conditioning technologies. *J. Refrig.*, *39*, 4–22. doi:10.1016/j.ijrefrig.2013.11.028

Alghoul, M. A., Sulaiman, M. Y., Azmi, B. Z., & Wahab, M. A. (2007). Advances on multi/purpose solar adsorption systems for domestic refrigeration and water heating. *Applied Thermal Engineering, N°*, *27*(5/6), 813–822. doi:10.1016/j.applthermaleng.2006.09.008

Aristov, Y. I. (2012). An optimal sorbent for adsorption heat pump nhermodynamik requirements and molecular design, *VI Minsk International Seminar —Heat Pipes, Heat Pumps, Refrigerators,* 342 – 353.

Aristov, Yu. I. (2014). Concept of adsorbent optimal for adsorptive cooling/heating. *Applied Thermal Engineering*, *2014*(72), 166–175. doi:10.1016/j.applthermaleng.2014.04.077

Bajin, N. M., Ivatshenko, V. A., & Parmon, V. N. (2002). Thermodynamic for chemists, Moskaw. *Chemistry (Weinheim an der Bergstrasse, Germany)*, 407.

Barreneche, C., Fernández, A. I., Cabeza, L. F., & Cuypers, R. (2015). Thermophysical characterization and thermal cycling stability of two TCM: $CaCl_2$ and zeolite. *Applied Energy, 137*, 726–730. doi:10.1016/j.apenergy.2014.09.025

Cabeza, L. F., Solé, A., & Barreneche, C. (2017). Review on sorption materials and technologies for heat pumps and thermal energy storage. *Renewable Energy*, *110*, 1–37. doi:10.1016/j.renene.2017.04.054

Donkers, P. A. J., Sögütoglu, L. C., Huinink, H. P., Fischer, H. R., & Adan, O. C. G. (2017). A review of salt hydrates for seasonal heat storage in domestic applications. *Applied Energy*, *199*, 45–68. doi:10.1016/j.apenergy.2017.04.080

Frazzica, A., & Freni, A. (2017). Adsorbent working pairs for solar thermal energy storage in buildings. *Renewable Energy*, *110*, 87–94. doi:10.1016/j.renene.2016.09.047

Gordeeva, L., Grekova, A., Krieger, T., & Aristov, Yu. (2013). Composites “binary salts in porous matrix” for adsorption heat transformation. *Applied Thermal Engineering*, *50*(2), 1633–1638. doi:10.1016/j.applthermaleng.2011.07.040

Grekova, A. D., Gordeeva, L. G., & Aristov, Yu. I. (2017). Composite "LiCl/vermiculite" as advanced water sorbent for thermal energy storage. *Applied Thermal Engineering, 124,* 1401–1408. doi:10.1016/j.applthermaleng.2017.06.122

Hiremath, C. R., & Kadoli, R. (2013). Experimental studies on heat and mass transfer in a packed bed of burnt clay impregnated with $CaCl_2$ liquid desiccant and exploring the use of gas side resistance model. *Applied Thermal Engineering, 50*(1), 1299–1310. doi:10.1016/j.applthermaleng.2012.08.002

Jong, A. J., Trausela, F., Fincka, C., van Vlieta, L., & Cuypers, R. (2014). Thermochemical heat storage – system design issues. *Energy Procedia, 48,* 309–319. doi:10.1016/j.egypro.2014.02.036

Kakiuchi, H. M., Iwade, S., Shimooka, K., Ooshima, K., Yamazaki, M., & Takewaki, T. (2005). Water vapour adsorbent FAM-Z02 and its applicability to adsorption heat pump. *Kagaku Kogaku Ronbunshu, 31*(4), 273–277. doi:10.1252/kakoronbunshu.31.273

Leite, A.P.F., Grilo, M.B., Andrade, R.R.D., & Belo, F.A. (2004). Dimensioning, thermal analysis and experimental heat loss coefficients of an adsorptive solar icemaker. *Renewable Energy, 29,* 1643 – 1663.

Ng, K. C., Chua, N. T., Chung, C. Y., Loke, C. H., Kasiwagi, T., Akisawa, A., & Saha, B. B. (2001). Experimental investigation of silica gel water adsorption isotherm characteristics. *Applied Thermal Engineering, 21*(16), 1631–1642. doi:10.1016/S1359-4311(01)00039-4

Odesola, I. F., & Adebayo, J. (2010). Solar adsorption technologies for ice/making and recent developments in solar technologies: A review. *J. Adv Eng Technol., 1*(3), 284–303.

Rabinovich, V. A., & Khavin, Z. Ya. (1978). Short chemical guide, Leningrad. *Chemistry,* 392.

Riffel, D.B., Riffel, D.B., Belo, F.A., & Leite, A.P.F. (2007). Ar condicionado solar por adsorção: Fundamentos e estado da arte. *I CBENS – I Congresso Brasileiro de Energia. Solar,* 257 – 262.

Ristic, A., Zabukovec, N. L., Henninger, S., & Kaucic, V. (2012). The performance of small-pore micro-porous aluminophosphates in low-temperature solar energy storage: The structure – property relationship. *Advanced Functional Materials*, *22*(9), 1952–1957. doi:10.1002/adfm.201102734

Ruthven, D. M. (1984). *Principles of adsorption and adsorption processes*. John Wiley & Sons.

Scapino, L., Zondag, H. A., Van Bael, J., Diriken, J., & Rindt, C. C. M. (2017). Sorption heat storage for long-term low-temperature applications: A review on the advancements at material and prototype scale. *Applied Energy*, *190*, 920–948. doi:10.1016/j.apenergy.2016.12.148

Sharonov, V. E., & Aristov, Yu. I. (2008). Chemical and adsorption heat pumps: Comments on the second law efficiency. *Chemical Engineering Journal*, *136*(2-3), 419–424. doi:10.1016/j.cej.2007.07.026

State standard 3956-76. Technical Silica gel. Technical conditions, Moskaw, Publishing House of standards, 13

Sukhyy, K., Belyanovskaya, E., Kovalenko, V., Kotok, V., Sukhyy, M., Kolomiyets, E., Gubynskyi, M., Yeromin, O., & Prokopenko, O. (2018). The study of properties of composite adsorptive materials “silica gel – crystalline hydrate” for heat storage devices. *Eastern-European Journal of Enterprise Technologies*, *91*(1), 52–58. doi:10.15587/1729-4061.2018.123896

Sur, A., & Das, R. K. (2010). Review on solar adsorption refrigeration cycle. *J. Mech Eng.Technol.*, *1*(1), 190–226.

Tanashev, Yu. Yu., Krainov, A. V., & Aristov, Yu. I. (2013). Thermal conductivity of composite sorbents “salt in porous matrix” for heat storage and transformation. *Applied Thermal Engineering*, *61*(2), 401–407. doi:10.1016/j.applthermaleng.2013.08.022

Tather, M., Tantekin, B., Ersolmaz, A., & Senatalar, E. (1999). A novel approach to enhance heat and mass transfer in adsorption heat pumps using the zeolite–water pair. *Microporous and Mesoporous Materials*, *27*(1), 1–10. doi:10.1016/S1387-1811(98)00174-7

Wang, J. D., Zhang, X., Tian, D., Liu, K., & Sumathy, K. (2014). Progress in silica gel–water adsorption refrigeration technology. *Renewable & Sustainable Energy Reviews*, *30*, 85–104. doi:10.1016/j.rser.2013.09.023

Zalba, B., Marrın, J. M., Cabeza, L. F., & Mehling, H. (2003). Review on thermal energy storage with phase change: Materials, heat transfer analysis and applications. *Applied Thermal Engineering*, *23*(3), 251–283. doi:10.1016/S1359-4311(02)00192-8

Zamengo, M., & Kato, Yu. (2017). Comparison of magnesium hydroxide/expanded graphite composites for thermal energy storage in cogeneration nuclear power plants. *Energy Procedia*, *131*, 119–126. doi:10.1016/j.egypro.2017.09.463

Chapter 3
Structure and Properties of Composite Adsorbents Salt Inside Porous Matrix

ABSTRACT

The chapter is devoted to structure and properties of composite adsorbents 'salt inside porous matrix'. Characteristics of adsorbents 'salt inside porous matrix', such as 'zeolite – crystalline hydrate', 'vermiculite – crystalline hydrate', 'silica gel – crystalline hydrate' were analysed. Main advantages of composite adsorbents are shown to be higher adsorptive capacity and lower regeneration temperature as compared with host matrix. Adsorptive capacities of composite materials are shown to be significantly enhanced by introduction of salts in host matrix such as zeolite, vermiculite, or silica gel. Water uptake by composite adsorbent is shown to be increased by rising the salt content in it. The drawback of most of existing impregnation technologies is shown to be impossibility of obtaining composite with salt content more than 40 – 60% along with complexity. Sol gel method is shown to be an alternative for conventional impregnation methods. Properties of adsorbents 'silica gel – sodium sulphate' synthesized according to sol gel method developed by authors were considered. The composite 'silica gel – sodium sulphate' composition and structure were studied by IR-spectroscopy and wide-angle x-ray scattering. Adsorptive properties of crystalline Na_2SO_4 when allocated in silicon oxygen matrix are shown to result from dispersion up to nanoscale. Adsorptive capacities and heat of adsorption of composites

DOI: 10.4018/978-1-7998-4432-7.ch003

'silica gel – sodium sulphate' and 'silica gel – sodium acetate' surpass almost by 30% the value calculated from the linear superposition of the sorption capacities of the sorbent and massive salt. Their adsorption properties are shown to be not a linear combination of properties of silica gel and salt. The formation of a unique structure promoting an increase in the rate of reaction between crystalline hydrates and water vapor in the developed pores of the silicon-oxygen matrix is confirmed. It leads to increasing the heat of adsorption and the heat energy storage density. Strong difference of water sorption kinetic curves of composite 'silica gel – sodium sulphate' and massive sodium sulphate is revealed. The correlation of their composition, structure, water adsorption kinetic, and operating characteristic as heat storage material is stated.

NOMENCLATURE

a – mass content of sodium sulphate in the composite a=0.44;

$m_{Na_2SO_4}$, $m_{CH3COONa}$, m_{SiO_2} – mass of water (g), adsorbed by 1 g of massive salts Na_2SO_4 or CH_3COONa and 1 g of silica gel;

Q_m – mass energy storage density, kWh/ kg adsorbent;

SCP – specific cooling power;

x_w – summarized water adsorption, g/g adsorbent.

During the last decade, a growing interest is focused on adsorptive heat energy conversion and adsorptive heat storage materials, with the aim of obtaining materials with high heat storage density and cycle stability for long-term operation. Traditional adsorptive materials such as zeolites (Jänchen, Ackermann, Stach, & Brösicke, 2004; Jochen & Helmut, 2012), zeotypes (Yu, Wang, R.Z., & Wang, L.W., 2013) and silica gels (Wang, L.W., Wang, R.Z., & Oliveira, 2009), but among them, salt hydrates present outstanding properties, in terms of theoretical storage capacity in the range of temperatures of interest (<100 °C), which can reach 3 GJ/m^3 (Donkers, Sögütoglu, Huinink, Fischer, & Adan, 2017; Rammelberg, Osterland, Priehs, Opel, & Ruck, 2016). Nevertheless, their practical application still requires significant research efforts to overcome different problems arising in the realization of storage system. The most critical of them are the low thermal conductivity of the materials (Kleiner, Posern, & Osburg, 2017) and the agglomeration and swelling phenomena, that limit vapor permeation, inducing degradation after cycling (De Jong, Trausel, Finck, Van Vliet, & Cuypers, 2014; Donkers, Sögütoglu, Huinink, Fischer, & Adan, 2017).

As a thoroughgoing solution of this problem the use of composite adsorbents can be considered. It currently represents a topic of great interest (Aydin, Casey, & Riffat, 2015; Gordeeva & Aristov, 2012). According to Aydin et al. (2015), these classes of adsorptive materials represent rather high heat storage density and wider interval of operating temperatures as compared with silica gels and most of zeolites.

This class of materials is based on two components: the host matrix and the inorganic salt placed inside the matrix pores.

The salt S reacts with a sorbate vapor V (water, methanol/ethanol and ammonia) that results in the formation of a complex S·NV (hydrate, methanolate etc.) of the salt and sorbate molecules according to the reaction

Figure 1. Energy storage densities of sorption materials
(Aydin et al., 2015)

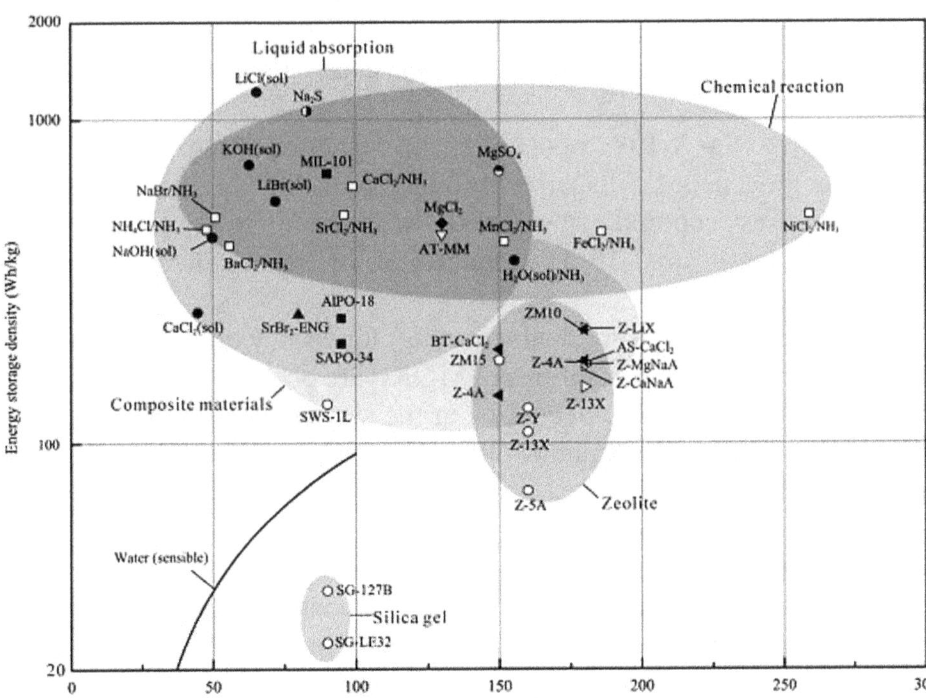

$$S + NV = S \times NV \tag{1}$$

Further sorption leads to the complex dissolution and transformation to the 'salt–sorbate' solution (Gordeeva & Aristov, 2012). The distribution of the salt into the matrix pores resulted in fundamental change the equilibrium 'adsorbent–adsorbate'.

For application with salt crystalline hydrates several types of matrices and/or additives were proposed, such as mesoporous materials, clays and carbonaceous structures (Scapino, Zondag, Van Bael, Diriken, & Rindt, 2017; Hiremath & Kadoli, 2013; Zamengo & Kato, 2017; Xu, Wang, & Li, 2014; Zhao, Wang, Zhang, & Yu, 2016).

Obviously, operating parameters of heat energy conversion unit are strongly determined by adsorbent properties.

COMPOSITES 'ZEOLITE – CRYSTALLINE HYDRATE'

Zeolites of various grades can be appertained to the typical porous matrixes of composite adsorbents.

Whiting et al. (2013) used the wet method to impregnate $MgSO_4$ into different types of zeolites, such as Faujasite Na–X (Na) (Alfa Aesar, 13X, beads 1.6 – 2.5 mm), Mordenite (Na) (Alfa Aesar), Faujasite (Na and H) Y (Alfa Aesar).

Except for the Na–X/15 composite, the impregnation of $MgSO_4$ is indicated to be non-homogeneously dispersed throughout the host zeolite, however, 17 wt% $MgSO_4$ was recorded using for Na–X/15, with slightly lower $MgSO_4$ content (than the desired 15 wt%) measured for the zeolite Y composites (12.5–13.8 wt%).

BET surface area measurements (in m^2/g) for each zeolite and composite series containing varying quantities of $MgSO_4$ are observed in Figure 2. The surface area of each zeolite decreases in the succession: Na–Y4H–Y4Na–X4MOR. For zeolite Na–X and MOR, when the quantity of $MgSO_4$ in the matrix increased, a large decrease in surface area is stated, it being agreed with Hongois et al. (2011).

The internal micropore surface area of each zeolite is far greater as compared with the external surface area. The substantial decline in surface area with increasing $MgSO_4$ concentration can be resulted from blocking of the zeolite pores with salt particles, which inhibits the adsorption of N_2 molecules. Pore volume measurements for both zeolite Na–X and MOR

Figure 2. Plot of BET surface area in each composite material, with increasing magnesium sulphate quantity present
(Whiting et al., 2013).

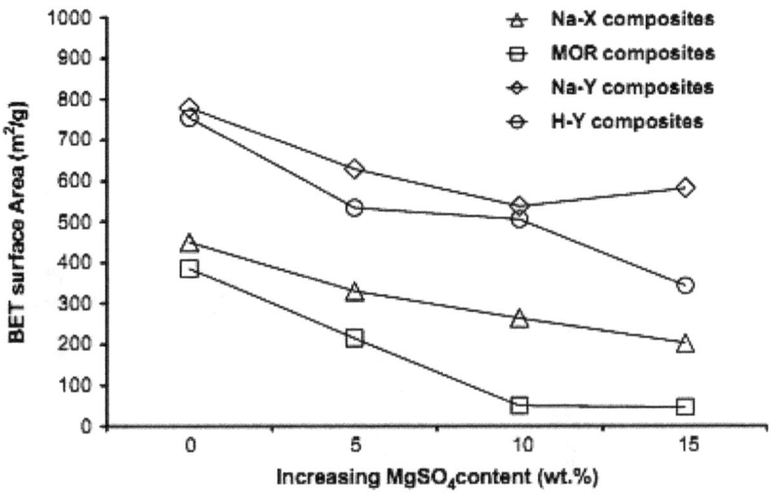

(0.20 and 0.17 $cm^3 \cdot g^{-1}$), respectively) reveal a sufficiently large volume for salt impregnation. Nevertheless, upon the impregnation of 15 wt% MgSO4, an extensive reduction in the volume of the pores for both composites (0.08 and 0.01 $cm^3 \cdot g^{-1}$, respectively).

For zeolite Na–Y composite materials, increasing $MgSO_4$ concentration also results in a decrease in surface area. Nevertheless, a high concentration of $MgSO_4$ (Na–Y/15) reveals a higher surface area (620 $m^2 \cdot g^{-1}$) than that of Na–Y/10 (488 $m^2 \cdot g^{-1}$). It can be caused by a higher distribution of the salt throughout the zeolite. Pore volume measurements of both zeolite Na–Y (0.32 $cm^3 \cdot g^{-1}$) and Na–Y/15 (0.27 $cm^3 \cdot g^{-1}$) confirm a successful incorporation of salt into the zeolite pores, without blocking.

Despite the fact that the sorption capacities of composite materials such as silica, zeolites and activated carbons are be greatly enhanced by introduction of inorganic salts as compared with compared to the host porous matrix alone (Hongois et al., 2011; Bandosz & Petit, 2009; Sharonov, Veselovskaya, & Aristov, 2006), the zeolite sorption capacity is decreased by the presence of $MgSO_4$ decreases the zeolite sorption capacity (Figure 3), suggesting the salt is not utilizing its full storage potential, and is in fact inhibiting the adsorption capabilities of the host matrix. According to Hongois et al. (2011), it results from replacing Na^+-cations in Na–X by Mg^{2+} ions in the salt ($MgSO_4$).

Figure 3. Chart comparing the percentage mass loss (of water) of fully hydrated zeolites Na–X, MOR, Na–Y and H–Y and their corresponding composites containing 5,10 and 15 wt% $MgSO_4$, when undergoing dehydration at an isothermal temperature of 150 °C for 3 h
(Whiting et al., 2013).

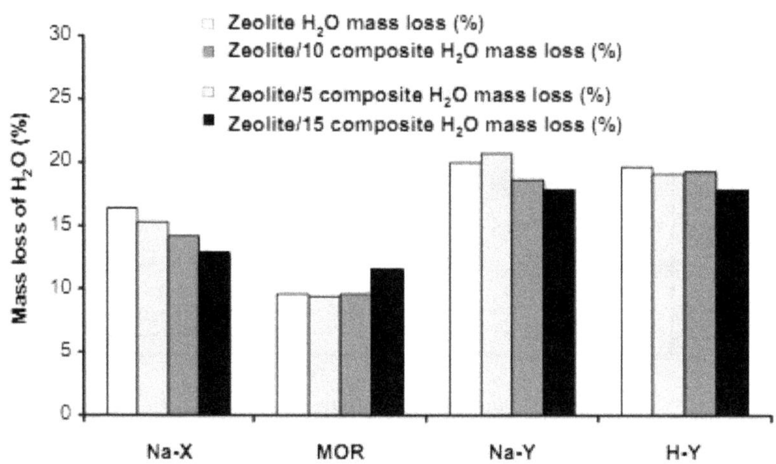

Thus magnesium sulphate was unable to utilize its full sorption capabilities, zeolites Na–Y and H–Y containing 15 wt% MgSO4 are shown to provide the highest heats upon hydration/dehydration (1090 and 876 $J \cdot g^{-1}$, respectively) as compared to those of Na–X and MOR composites (731 and 507 $J \cdot g^{-1}$, respectively).

Zettle B. et al. studied composites which prepared by wet impregnation of the natural zeolite clinoptilolite with solutions of varying salt mixtures (Zettl, Englmair, & Somitsch, 2015). Two series of composite materials were prepared, both with particle sizes 0.5 – 1 mm. In series A, the salt loading reached (on a dry weight basis at 150°C) 1.5% LiCl and 13.5% $MgSO_4$ (Cpt-A1); 4.5% LiCl and 10.5% $MgSO_4$ (Cpt-A2); 7.5% LiCl and 7.5% $MgSO_4$ (Cpt-A3). In series B, the salt loading yielded 7.5% $MgSO_4$ and 7.5% $MgCl_2$ (Cpt-B1); 11% $MgSO_4$ and 11% $MgCl_2$ (Cpt-B2).

As compared to untreated clinoptilolite, the composite materials are stated to require longer time to reach full adsorption capacity. After one hour exposure to humidity of series A composites only 32 – 38% of the capacities at 6 hours have been adsorbed (Fig. 3). In series B, the water uptake after one hour reached 43% (Cpt-B1) and 37% (Cpt-B2), see Figure 4 (Zettl et al., 2015). Due to the impregnation, the total moisture adsorption capacities increased and have more than doubled. Series B composites are located between Cpt-A1 and Cpt-A2 and are therefore advantageous compared to low LiCl-doping because of their low material costs. Composites with more than 4.5% LiCl reveal higher water vapor uptake capacity, up to three times the value of natural clinoptilolite (Cpt-A3, Figure 3).

Figure 4. Time-dependent moisture adsorption in Cpt – LiCl – $MgSO_4$ – composites (Zettl et al., 2015).

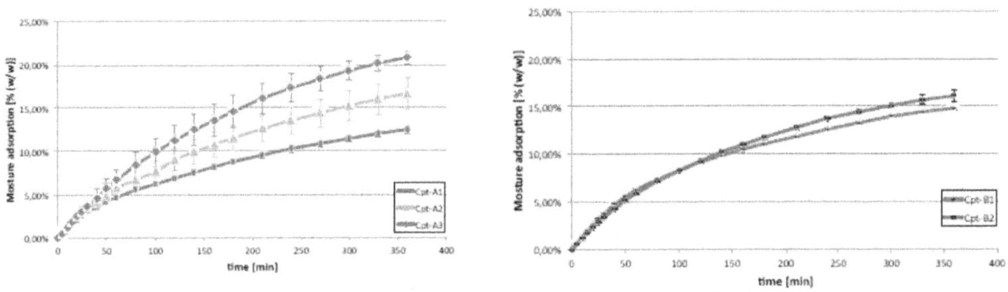

Series A Series B

According to data by Xue et al. (2019), composites 'zeolite-13X – $CaCl_2$' prepared by wet impregnation reveal notably decline due to the embedding of $CaCl_2$ molecular upon zeolite frame (Table 1). After reimpregnation the pore volume pore volumes and surface areas for CA40% are decreased by 52.63% and 48.60%, it correlates with more adhering of $CaCl_2$ in zeolite as a higher $CaCl_2$ solution concentration.

Table 1. Pore structure of different adsorbents

Zeolite	Pore Volume ($cm^3 \cdot g^{-1}$)	Surface Area ($m^2 \cdot g^{-1}$)
13X beads	0.36	663.04
zeolite13X –20%$CaCl_2$	0.30	570.71
zeolite13X –40%$CaCl_2$	0.18	346.96

(Xue et al., 2019)

Despite this fact, equilibrium water uptakes for 'zeolite13X –20%$CaCl_2$' and 'zeolite13X –40%$CaCl_2$' of 0.350 and 0.449, respectively, as compared with 0.289 for zeolite13X. Furthermore, the values of adsorption heat for

13X beads, 'zeolite13X–20%$CaCl_2$' and 'zeolite13X–40%$CaCl_2$' are 272.11, 285.41 and 307.86 kW·h·m^{-3}, respectively.

This features correlate with the temperature variation during the operation, i.e. steam generation (Figure 5).

Figure 5. Temperature vs. time during generation process. (a) zeolite 13X; (b) 'zeolite13X –40%$CaCl_2$'
(Xue et al., 2019).

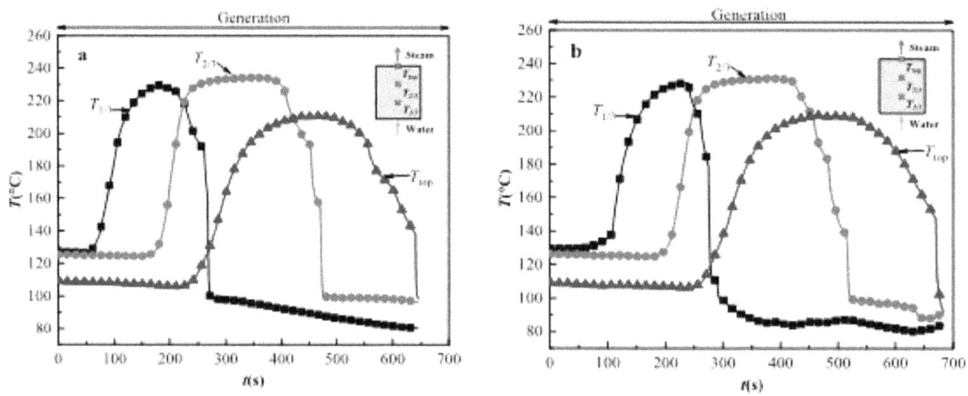

Temperature-time dependences were registered inside the packed bed ($T_{1/3}$ and $T_{2/3}$) and at the outlet (T_{top}).

For 'zeolite13X –40%$CaCl_2$' in Figure .4 (b), the maximum temperature inside packed bed is almost the same as that for 13X shown in Figure 4(a) (Sharonov et al., 2006). The same maximum temperature confirms a fact that only physical adsorption occurs at high temperature. However, peak of generated steam temperature at 209.8 °C for 'zeolite13X –40%$CaCl_2$' lasts longer as compared with that for zeolite 13X due to more heat stored and released for steam generation during the ascent of liquid water interface. Adsorption equilibrium for the whole range can be obtained easier with larger overall released adsorption heat for 'zeolite13X –40%$CaCl_2$'.

C.Y. Tso et al. (2015) compared performance of an adsorption cooling system with a novel composite material (zeolite 13X – $CaCl_2$) and silica gel as the adsorbents and water as the adsorbate under various working conditions. It is shown in Fig. 6 that both coefficients of performance COP for silica gel and composite adsorbent increase with desorption temperatures lower than 70 °C, but decrease for desorption temperatures higher than 70 °C. It is particularly perceptible for the COP value when an adsorptive cooling system used silica gel as the adsorbent. This can be resulted from an increase in heat loss at higher desorption temperatures. For the lower desorption temperatures, the heat loss is less as compared with the higher desorption temperatures. The result of specific cooling powers SCP for both silica gel and composite 'zeolite13X – calcium chloride' good conforms with the prediction, increasing monotonically when temperature of desorption ranged from 55°C to 95°C. Since water vapor is desorbed faster at the higher desorption temperature, the degree of dryness of adsorbent and the driving force for adsorption are increased simultaneously, more water vapor being allowed to adsorb during the next adsorption process. In other words, the increasing the desorption temperature enhances the specific cooling power of the adsorption cooling system. Nevertheless, it should be noticed that the temperature of desorption is affected by the real situation.

As the intensity of sun light is not steady, the achievable hot water temperature may not be high. So, standard hot water inlet temperature was selected of 85 °C.

Figure 6. Effect of desorption temperature on the specific cooling power SCP and coefficient of performance COP of adsorptive cooling systems based on silica gel and .composite 'zeolite 13X – calcium chloride' [Remark: 31°C cooling water inlet temperature; 15 min adsorption/desorption phase time for both adsorbents; 8 kg/min hot and cooling water mass flow rate; 3.6 kg/min chilled water mass flow rate; 14 °C chilled water inlet temperature; 50 s heat recovery time and 10 s mass recovery time.)
(Tso et al. (2015)

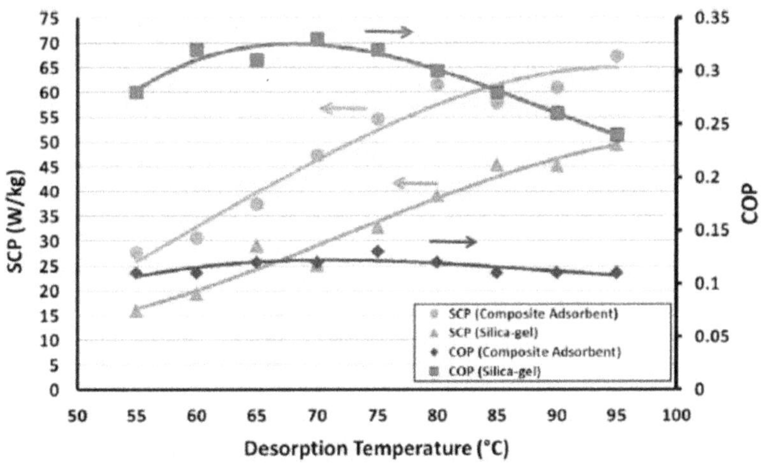

SCP of an adsorption cooling system using the composite adsorbent 'zeolite 13X – $CaCl_2$ is stated to be on average 30% higher than that of the silica gel adsorbent because of the adsorption capacity of the composite adsorbent is larger than that of the silica gel adsorbent, allowing more water vapor to be adsorbed during the adsorption process. However, the COP of the adsorption cooling system based on the composite adsorbent is about 60% lower than that of the silica gel adsorbent because the composite adsorbent includes calcium chloride, a chemical absorbent. The chemical absorbent (e.g. $CaCl_2$) requires a larger amount of thermal energy for desorption than the physical adsorbent (e.g. silica gel) because the absorbate (water vapor) is absorbed inside the absorbent ($CaCl_2$) while the adsorbate (water vapor) is adsorbed on the surface of the adsorbent (silica gel). The intermolecular force between the chemical absorbent ($CaCl_2$) and absorbate (water vapor) is much larger as compared with that between the physical adsorbent (silica gel) and adsorbate (water vapor). As a result, the COP of the adsorption cooling system using the composite adsorbent is smaller than that using the silica gel adsorbent. However, it is expediently to note that the thermal energy input should actually come from renewable energy (e.g. solar energy or waste heat) which is free of charge or from natural resources. So, the thermal COP values shown in this study are given just for reference and comparison with other studies. SCP is indeed a very important parameter which indicates how large the adsorption cooling system and the cooling power of the adsorption cooling system are. A much lower desorption temperature can be used to desorb the composite adsorbent when compared with the zeolite 13X adsorbent. When the adsorption cooling system equipped with the composite adsorbent, a longer time of adsorption/desorption phase is required. The adsorption cooling system equipped with the composite adsorbent shows a better cooling performance than that of the silica gel if a high cooling water temperature, above 29°C, or low chilled water inlet temperature, below 17 °C, is utilized.

As alternative to composite adsorbents, novel adsorbtive materials "AQSOA®", belonging to the class of (Silico) aluminophosphates, can be considered (Kakiuchi, Takewaki, Takumi, Yamazaki, & Watanabe, 2003). They have been developed by Mitsubishi Plastic Inc. (MPI) for desiccant and adsorptive heat conversion system. Adsorption equilibrium and durability studies of adsorbent AQSOA-FAM-Z01 and AQSOA-FAM-Z02 are presented in (Kakiuchi et al., 2005a; Kakiuchi et al., 2005b; Sapienza, Freni, Vasta, Restuccia, & Cacciola, 2007; Kayal, Baichuan, & Saha, 2016). Along with conveniently low desorption temperature (60–90 °C), both materials are able to exchange the amount of water up to 0.25 g/g) within narrow temperature and humidity ranges. Obviously, low values of water uptake lead to large volumes of adsorptive units which impede their involvement in heat supply systems. However, a general issue of (silico)aluminophosphate materials still remains the high capital cost, as the synthesis process is more expensive than standard commercial zeolites and any industrial mass-production has not yet been established.

A cheaper alternative to AlPO and SAPO may be Y-type zeolites. These materials developed by UOP Llc., a Honeywell Company. A novel zeolite "DDZ-70" is a high silica content Y zeolite exchanged with cerium to reduce its hydrophilicity (Freni et al., 2016; Dunne & Bradley, 2005). Application of DDZ-70 can be advantageous for low temperature adsorption cycles, due to the lower regeneration temperature (<150 °C) (Dawoud, 2014). Nevertheless, regeneration temperatures of these materials are still too high to involve them in domestic heat supply systems. On the other hand, hydrophobic zeolites can be employed for methanol or ethanol adsorption. With this regard a very high silica content Y-type zeolite CBV-901 from Zeolyst Int., has been proposed as methanol adsorbent for cooling applications driven by low temperature heat sources (Restuccia, Freni, Russo, & Vasta, 2005).

COMPOSITE MATERIALS BASED ON VERMICULITE HOST MATRIX

Vermiculite is one of the most attractive matrixes to be used for composite sorbents preparation, thanks to the high specific pore volume and pore size distribution, which perfectly fits with the integration of salt inside the porous structure, leaving sufficient space for water vapour transfer (Palomba & Frazzica, 2018).

According to data by Zhang et al. (2016) composite 'vermiculite – $SrBr_2$' was synthesized by impregnation of expanded vermiculate (EVM) with $SrBr_2$ solutions with mass concentration of 10%, 20%, 30% and 40% $SrBr_2$. Samples with salt content ranging from 10 to 40 wt% were prepared by the wet impregnation method which includes the next steps: drying the EVM in an electric furnace for more than 4 h at 200°C to remove the adsorbed water, impregnation the dried EVM into the $SrBr_2$ solutions for 48 h, respectively, separation the wet EVM with $SrBr_2$ solution in its pores from the solution by a sieve and washing by distilled water to remove the salt adhered to the surface of EVM, drying the wet samples at 120 °C in an electric furnace.

As a result $SrBr_2$ crystals were adhered to the surface of pores inside EVM. According to the mass concentration of impregnating $SrBr_2$ solution, the four 'EVM – $SrBr_2$' adsorbents were denoted to as EVMSr$Br_2$10, EVMSr$Br_2$20, EVMSr$Br_2$30 and EVMSr$Br_2$40, respectively. A homogenous distribution of the salt inside the pores was experimentally verified.

Water sorption kinetics diagram of EVM, $SrBr_2 \cdot H_2O$ and 'EVM – $SrBr_2$' adsorbents are shown in Figure 7. $SrBr_2$ exits in the form of $SrBr_2 \cdot H_2O$ in the original state. The composite adsorbents are stated to have a much larger amount of water uptake than raw EVM. Adsorption equilibrium for EVM occur in 30min with small water uptake (adsorption) of only 0.03 g/g. $SrBr_2 \cdot H_2O$ reaches to adsorption equilibrium at about 150 min, equilibrium water uptake being 0.34 g/g. It's verified that at 30 °C and 60% RH, only hydration reaction occur and deliquescence phenomenon can't appear. For 'EVM – $SrBr_2$' adsorbents, the rate of water uptaking gradually decreases. A higher salt content results in a longer time to obtain adsorption equilibrium for composite sorbents. It shows that mass transfer barrier becomes stronger as more salt content is filled into the space of pores in EVM. Because of the larger mass transfer barrier, EVMSr$Br_2$40 exhibits lower speed of water uptake than EVMSr$Br_2$30 before 120 min. The water uptake of EVMSr$Br_2$10, EVMSr$Br_2$20, EVMSr$Br_2$30 and EVMSr$Br_2$40 reaches 0.29, 0.42, 0.48 and 0.49 g/g dry adsorbent at 420 min, respectively.

Thermochemical characterizations of composites 'EVM – $SrBr_2$' including desorption performance and desorption heat were studied by simultaneous

Figure 7. Water sorption kinetics of EVM, $SrBr_2 \cdot H_2O$ and EVM/$SrBr_2$ sorbents at 30 °C and 60% RH
(Zhang et al., 2016).

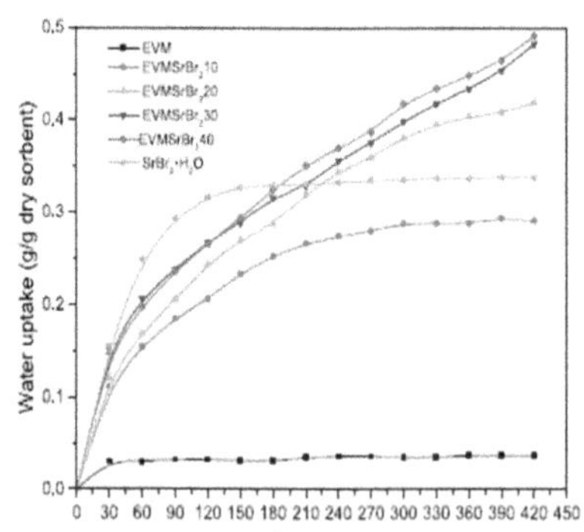

thermal analyze STA. STA measurement results of SrBr2·6H_2O, EVM and composite 'EVM– $SrBr_2$' are given in Fig. 8. The green lines and blue lines represent TGA and DSC curves. Peak value and turning point are marked in Figure 8.

According to the adsorption kinetic result, the water adsorption with EVM (x_w, EVM) is only 0.03 g/g in TGA result. In TGA/DSC curve of $SrBr_2$·6H_2O,

Figure 8. Results of STA measurement: (a) $SrBr_2$·6H_2O; (b) EVM; (c) EVMSrBr$_2$10; (d) EVMSrBr$_2$20; (e) EVMSrBr$_2$30; (f) EVMSrBr$_2$40
(Zhang et al., 2016).

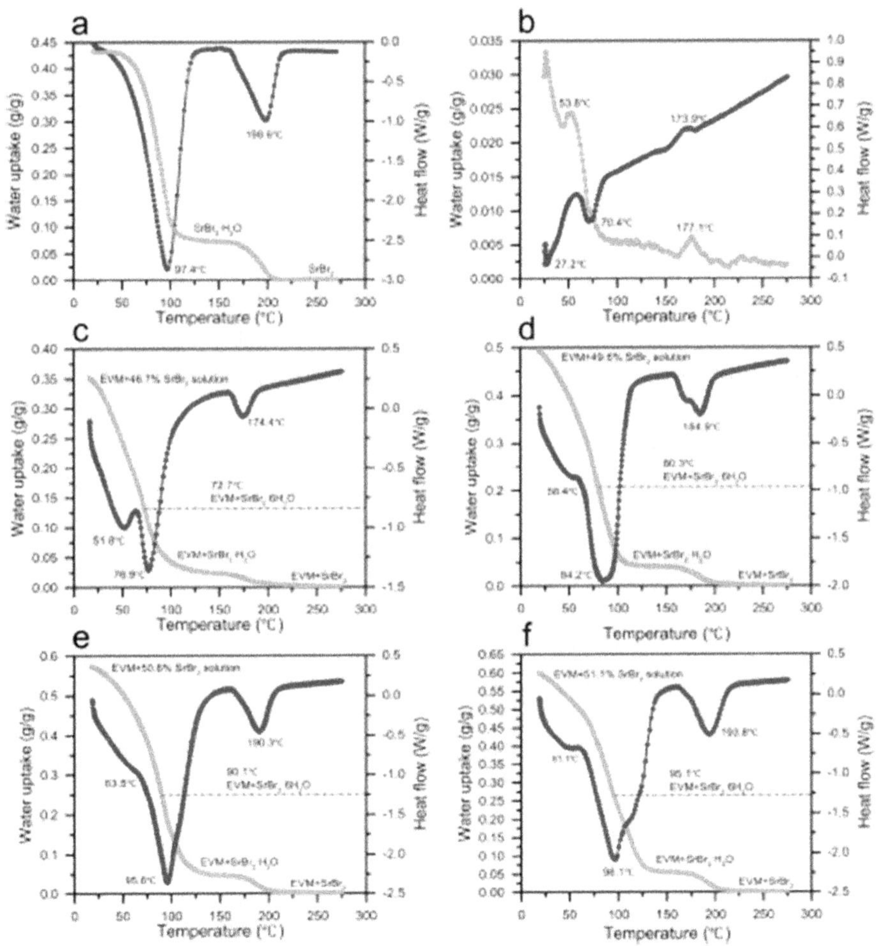

two weight plateaus and related desorption peaks are observed, corresponding with two dehydration reactions respectively. Theoretically, dehydration reaction of $SrBr_2 \cdot 6H_2O$ is divided into two procedures: the formation of $SrBr_2 \cdot H_2O$ and the formation of $SrBr_2$. Two reactions are as follows (Zhang et al., 2016):

Reaction 1:

$$SrBr_2 \cdot 6H_2O_{(s)} + HEAT \rightarrow SrBr_2 \cdot H_2O_{(s)} + 5H_2O_{(g)}$$

$$\Delta H_r = +346.90 kJ / mol \tag{1}$$

Reaction 2:

$$SrBr_2 \cdot H_2O_{(s)} + HEAT \rightarrow SrBr_{2(s)} + 5H_2O_{(g)}$$

$$\Delta H_r = +77.27 kJ / mol \tag{2}$$

TGA/DSC curves for $EVMSrBr_2 10$, $EVMSrBr_2 20$, $EVMSrBr_2 30$ and $EVMSrBr_2 40$ reveals similar shapes with $SrBr_2 \cdot 6H_2O$. Two obvious plateaus occur in TGA curve. The second plateau indicates the end of desorption process and formation of $SrBr_2$ in EVM pores. Considering that water adsorbed by EVM loses below 100°C, desorption process between the two plateaus results from the interaction force between $SrBr_2$ and water vapor. According to calculation based on the water uptake related to the first plateau and salt content of composite adsorbent, it is stated that the first plateau denotes to the formation of $SrBr_2 \cdot H_2O$. The phase between the original state and the first plateau includes concentration of $SrBr_2$ solution, dehydration of $SrBr_2 \cdot 6H_2O$ and desorption of EVM. Neglecting the desorption process of EVM based on different mechanics of desorption process, desorption process of composite adsorbent is divided into two steps correlated with dehydration of strontium bromide (the content included in the brackets means the composition of sorbents) (Zhang et al., 2016):

Process 1:

$$\left(EVM + SrBr_2 Solution\right) + HEAT \rightarrow \left(EVM + SrBr_2 \cdot 6H_2O\right)_{(s)} + nH_2O_{(g)} \tag{3}$$

Process 2:

$$\begin{cases} \left(EVM + SrBr_2 \cdot 6H_2O\right)_{(s)} + HEAT \rightarrow \left(EVM + SrBr_2 \cdot H_2O\right)_{(s)} + 5H_2O_{(g)} \\ \left(EVM + SrBr_2 \cdot H_2O\right)_{(s)} + HEAT \rightarrow \left(EVM + SrBr_2\right)_{(s)} + H_2O_{(g)} \end{cases} \tag{4}$$

Similarly, the sorption process of composite sorbents 'EVM – $SrBr_2$' below 100 °C can be presented by three steps: water adsorption of EVM, water adsorption of $SrBr_2{\cdot}H_2O$ crystal, and liquid-gas absorption of $SrBr_2$ solution.

Further the adsorption behavior of EVM, $SrBr_2{\cdot}H_2O$ and 'EVM – $SrBr_2$' adsorbents is estimated and the result is shown in Figure 9a (Zhang et al., 2016). Summarized water adsorption (x_w) is divided into water adsorbed by EVM ($x_{ad'EVM}$), water adsorbed by $SrBr_2{\cdot}H_2O$ ($x_{ad'salt}$) and water absorbed by $SrBr_2$ solution ($x_{ab'salt}$). Values of $x_{ad'EVM}$, $x_{ad'\ salt}$ and $x_{ab'\ salt}$ are normalized to the final mass of the first plateau in TGA curve. Value of x_w is obtained from TGA curve while that of $x_{ad'\ EVM}$, $x_{ad'\ salt}$ and $x_{ab'\ salt}$ of whole adsorbent are calculated as follows:

$$x_{ad,EVM} = \left(1 - \eta\right) \cdot x_{w,EVM} \left(\frac{g}{g\,adsorbent}\right) \tag{5}$$

$$x_{ad,salt} = \eta \cdot x_{w,salt} \left(\frac{g}{g\,adsorbent}\right) \tag{6}$$

$$x_{ab,salt} = x_w - x_{ad,EVM} - x_{ad,salt} \left(\frac{g}{g\,adsorbent}\right) \tag{7}$$

where x_w, $x_{w,\ EVM}$ (0.03 g/g) and x_w,salt (0.34 g/g) are total water uptake of composite sorbents, raw EVM and bulk $SrBr_2{\cdot}H_2O$, respectively, they are obtained from TGA curve in Figure 9 (Zhang et al., 2016).

For $SrBr_2{\cdot}6H_2O$ only hydration reaction is considered. Total water uptake with EVM is only 0.03 g/g, while that of $SrBr_2{\cdot}6H_2O$ is 0.34 g/g. However, the composites 'EVM – $SrBr_2{\cdot}6H_2O$' represent promising potential of water uptake, as compared with that of $SrBr_2{\cdot}6H_2O$ (Zhang et al., 2016). The water uptake by composite sorbents is greatly increased. $EVMSrBr_220$, $EVMSrBr_230$ and $EVMSrBr_240$ reveal a higher total water uptake than $SrBr_2{\cdot}6H_2O$. The highest x_w of 0.53 g/g is stated for $EVMSrBr_240$, and the value is 1.6 times that of $SrBr_2{\cdot}6H_2O$. Values of x_w, $x_{ad,salt}$ and $x_{ab,salt}$ of composite sorbents are observed to increase with salt content. The ratio of $x_{ab,salt}/x_{ad,salt}$ decreases from 1.9 to 1.4 with the rising of salt content.

Figure 9. Evaluation of water uptake (a) and energy storage density (b) of EVM, $SrBr_2{\cdot}6H_2O$ and EVM/$SrBr_2$ adsorbents
(Zhang et al., 2016).

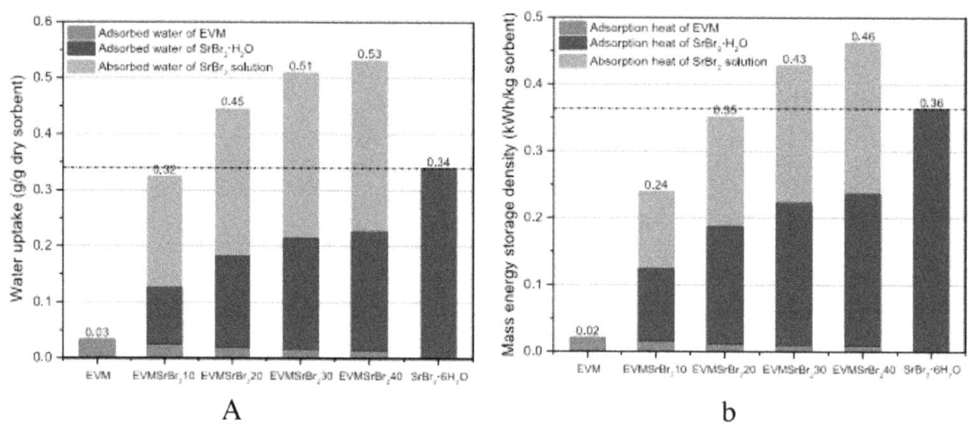

The decreasing amount of absorbed water is corresponded with the salt concentration of salt solution in EVM. It is possibly resulted from the complex influence of adsorption kinetics and pore structures of EVM.

Mass energy storage density (Q_m) is a vital parameter to estimate the energy storage efficiency of TES system. Ordinarily, sorption process is regarded as the reversible process of desorption process. So storage density is equal to desorption heat. Estimation of mass energy storage density of EVM, $SrBr_2{\cdot}H_2O$ and EVM/$SrBr_2$ adsorbents is shown in Figure 9b (Zhang et al., 2016). Mass energy storage density is obtained from DSC curve based on weight at the end of first weight plateau in TGA curve, Q_m consists of three parts: adsorption heat of EVM ($Q_{m,EVM}$), adsorption heat of $SrBr_2{\cdot}H_2O$ crystal ($Q_{ad,\,salt}$) and absorption heat of $SrBr_2$ solution ($Q_{ab,\,salt}$), the calculation formula are as follows:

$$Q_{ad,EVM} = (1-\eta)\cdot Q_{m,EVM}\left(\frac{kWh}{kg\,adsorbent}\right) \quad (8)$$

$$Q_{ad,salt} = \eta\cdot Q_{m,salt}\left(\frac{kWh}{kg\,adsorbent}\right) \quad (9)$$

$$Q_{ab,salt} = Q_m - Q_{ad,EVM} - Q_{ad,salt}\left(\frac{kWh}{kg\,adsorbent}\right) \quad (10)$$

where Q_m, $Q_{w,EVM}$ (0.02 kWh/kg) and $Q_{w,salt}$ (0.36 kWh/kg) are total mass storage density of composite adsorbents, raw EVM and bulk $SrBr_2{\cdot}H_2O$.

The Q_m of EVM is an extremely small of 0.02 kWh/kg, it being only one eighteenth that of $SrBr_2{\cdot}6H_2O$ (0.36 kWh/kg) (Zhang et al., 2016). Q_m of 'EVM/$SrBr_2$' composites increases from 0.24 kWh/kg to 0.46 kWh/kg as the salt concentration ranges from 30.19% to 63.02%. The largest Q_m of 0.46 kWh/kg is shown by EVM$SrBr_2$40, which is 1.3 times that of $SrBr_2{\cdot}6H_2O$. The rate of $Q_{ab,\,salt}/Q_{ad,\,salt}$ ranges from 0.93 to 1.05. The value of $Q_{ab,\,salt}/Q_{ad,\,salt}$ is not correlated with salt concentration change. The amounts of adsorption heat of salt crystal and the absorption heat of salt solution are almost equal as shown in Figure 8b. So, they have the same contributions to the mass energy storage of composite sorbent. Moreover, the total two parts of sorption heats take an account for more than 94% of the overall mass energy storage.

So, the composites 'EVM – $SrBr_2$' present more significant potential for low-temperature TES as compared with bulk $SrBr_2{\cdot}6H_2O$. EVM$SrBr_2$40 is suggested to be an the optimal adsorbent among different EVM/$SrBr_2$ composites, with water adsorption of 0.53 g/g, mass energy storage density of 0.46 kWh/kg and volume energy storage density of 105.36 kWh/m^3 (Zhang et al., 2016). Moreover, for the composite with 40 wt % of salt, a water vapour desorption of 0.53 g/g^{-1} was measured, representing a storage density of 1650 kJ/kg.

The similar matrix of expanded vermiculite was impregnated by dry method with an aqueous $BaCl_2$ solution or mixed aqueous solution of $BaCl_2$ and $BaBr_2$. Using composite 'vermiculite – $BaCl_2$ –$BaBr_2$' as ammonia adsorbent for air conditioner operating in the steady state mode can allow to obtain Specific Cooling Power related to unit mass of the adsorbent, higher than 500W/kg, Specific Cooling Power related to unit area of heat exchanger over 270 W/m^2 and coefficient of performance more than 0.45 (Veselovskaya & Tokarev, 2011).

Another composite based on vermiculate matrix is presented by Sapienza et al. (2012). As a salt lithium nitrate was used. This composite 'vermiculite – $LiNO_3$' (SWS-9V) has been specifically developed at Boreskov Institute of Catalysis of Russian Academy of Science, Novosibirsk, Russia (BIC-RAS) for adsorption chilling. This composite was prepared with a dry impregnation of the expanded vermiculite by a saturated solution of $LiNO_3$ with consequent drying at 150 °C until the weight of composite remained constant. The salt content in the composite was found to be 63 mass %.

The experimental isobars were used to plot a thermodynamic cycle of adsorption chiller for both the adsorbents as given in Figure 10 (Sapienza et al., 2012). The isobars of SWS-9V are more steeper and shifted to lower temperatures than those of SWS-9L: e.g. the new composite exchanges app.

Figure 10. Equilibrium isobars of water adsorption at 12.6 mbar (solid symbols) and desorption at 56.2 mbar (empty symbols) and appropriate AC cycle for the composites SWS-9V (A-B-C-D) and SWS-9L (A0-B0-C0-D)
(Sapienza et al., 2012).

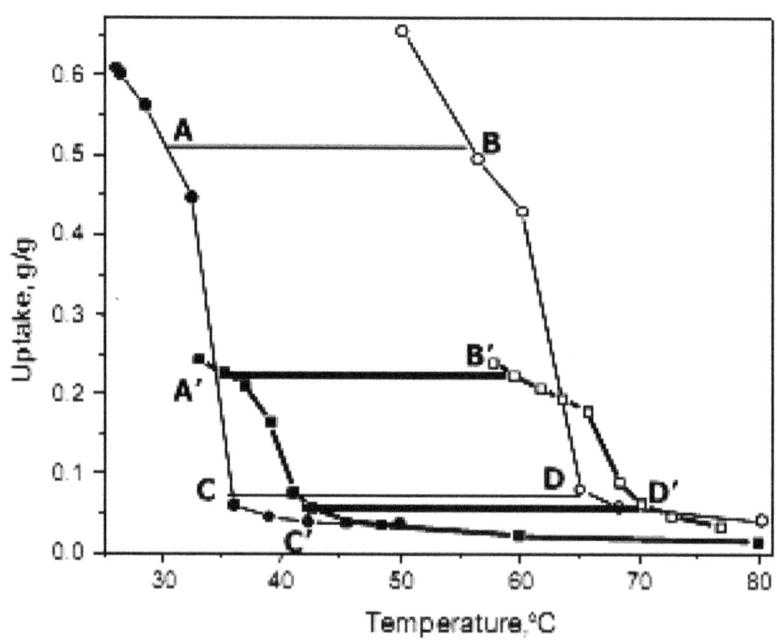

0.4 g H_2O/g in a very narrow temperature range 33 – 36 °C (adsorption) and 62 – 65 °C (desorption).

The measurement of the equilibrium properties showed that this composite is able to exchange a large amount of water (0.4 – 0.5 g H_2O/g) under operating conditions typical for air conditioning applications as well as to be driven by an exceptionally low grade thermal source (<70 °C) (Sapienza et al., 2012). Composite sorbent 'vermiculite – $LiNO_3$' (SWS-9V) can be a promising solution for heat storage applications. Under the tested boundary conditions it exhibited a remark- able heat storage capacity (1.15 MJ/kg) while if relate the stored heat to the total volume of the unit 'adsorbent – heat exchanger', it gives 0.45 MJ/dm^3. This is far superior to common materials suggested for storage of heat with temperature potential of 60 – 80 °C.

Recently, Grekova et al. (2017) developed a composite sorbent based on vermiculite and LiCl. 59 wt% of salt were loaded up to inside the porous structure by means of the wet impregnation method. The selection of the amount of salt was performed taking into consideration the free volume space left available for the water vapour diffusion during sorption as well as to allow the expansion of the salt solution avoiding any leakage from the pores. The prepared samples were then characterized by means of XRD and N_2 porosimetry to check crystalline as well as porous structure before and after the synthesis process. The composite sorbent was then characterized both from the kinetic and thermodynamic point of view, applying two reference storage cycles, namely, daily storage (DS) and seasonal storage (SS), which operating conditions are summarized in Table.2 (Grekova et al., 2017).

Table 2. Operating conditions of the SS and DS cycles

Cycle	T_{dis}, °C	T_{ev}, °C	T_{ch}, °C	T_c, °C	ΔF_{dis}, kJ/mol	ΔF_{ch}, kJ/mol
SS	35	10	75 – 85	30	3.9	6.4 – 7.7
DS	35	5	75	15	4.8	9.0

(Grekova et al., 2017)

The operating conditions of heat storage cycles (Figure 11) are determined by four temperatures, namely, the temperatures of released heat (T_{dis}), evaporation (T_{ev}) during the discharge stage, available heat source (T_{ch}) and heat sink (T_c) during the charge stage (Table 2). These four temperatures

Figure 11. The SS (blue line) and DS cycles (red line)
(Grekova et al., 2017).

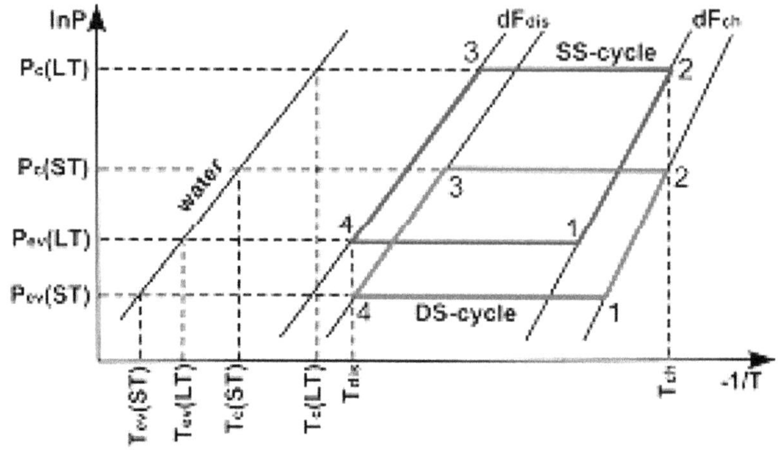

were selected according to the climatic data in the regions where the selected cycles can be realized (Grekova et al., 2017).

The tested sample exhibit ability to exchange up to 0.60 – 0.85 g/g for both DS and SS even at low charging temperature (i.e. 75 °C). This is reflected by a very high heat energy storage density, up to 2.6 kJ/g and 910 MJ/m^3 (Grekova et al., 2017). High charging and discharging powers were estimated out of the kinetic tests, with achievable discharging power of 3.9 and 2.4 kW/kg for DS and SS cycle, and charging power up to 11.8 and 15.6 kW/kg for DS and SS cycle, respectively. It ought to notice that high charging power is achieved, probably due to the more favourable working conditions. The driving heat transfer force represented by the temperature difference between adsorbent material and heat source is higher for desorption than for adsorption, thus enhancing the overall heat transfer efficiency. Moreover, the absolute water vapor pressure during the phase of desorption is always higher than in the adsorption one, thus limiting the mass transfer resistance offered by the sorbent material towards the vapour transfer. So, it can be concluded that the heat and mass transfer efficiency is higher during desorption (charging) than adsorption (discharging) runs.

COMPOSITE ADSORBENTS BASED ON SILICA GEL HOST MATRIX

Composites 'silica gel – salt' are among of the most popular representatives of adsorbents 'salt in porous matrix'. They are used for water, methanol, ethanol and ammonia adsorption (Dunne & Bradley, 2005). When water used as adsorbate, such composites are subsumed to Selective Water Sorbents or SWSs. The studies by Aristov et al. reported in (Gordeeva & Aristov, 2018; Aristov et al., 1996; Aristov et al., 2002; Gordeeva & Aristov, 2012) exhibit that the composite materials based on meso-porous silica gel impregnated with $CaCl_2$ (SWS-1L) and LiBr (SWS-2L) showed adsorption of water up to 0.75 g/g, with desorption being practically completed at relatively low temperatures (70 – 120°C).

Adsorptive properties of similar materials can be significantly enhanced by introduction of salts forming crystalline hydrate into host matrix. According to A. Jabbari-Hichri et al. (2015) as such material can be considered silica–alumina (SA, from LA LPV 13% Al_2O_3, KETJEN) because of a high specific surface area (628 m^2/g), pore volume (0.8 cm^3/g), and a pore diameter(4.7 nm) in the mesoporosity range. Hygroscopic salts ($CaCl_2$, $Ba(OH)_2$ or $LiNO_3$)

Figure 12. Dehydration heats as a function of the salt content for the three series of composites
(Simonova et al., 2009).

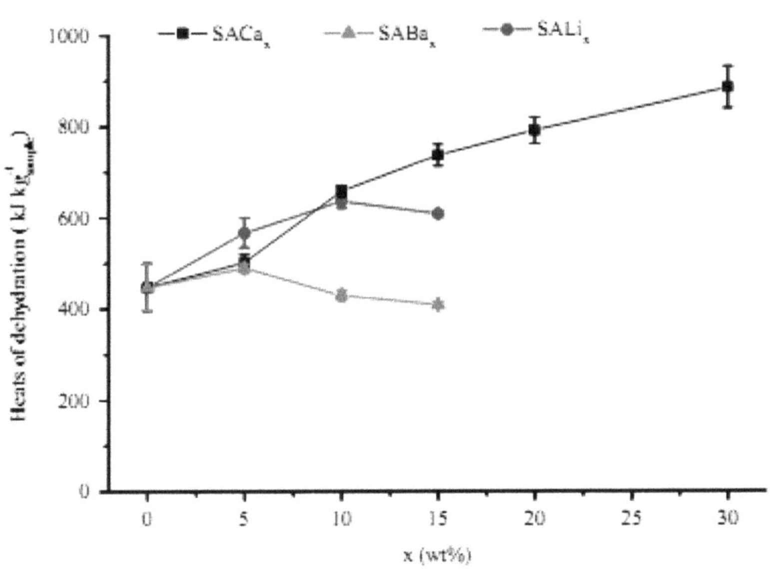

were inserted into host material by an incipient wetness impregnation starting from the aqueous solution of the three salts. They were named $SACa_x$, $SABa_x$, and $SALi_x$, x being the wt% of anhydrous salt theoretically added.

The dehydration heats as a function of the amount of deposited salt for the three series of composites are presented on the Figure 12 (Simonova, Freni, Restuccia, & Aristov, 2009).

The three curves occur to pass through a maximum value depending on the amount of deposited salt. $SACa_x$ composites gave the best results, exhibiting a continuous increase of the measured dehydration heat with the amount of $CaCl_2$ deposited up to 30 wt. % (Simonova et al., 2009). The content of salt is not sufficient to reach the maximum point of the curve $SABa_x$ and $SALi_x$ following the same trend of $SACa_x$ respectively up to 5 wt% $Ba(OH)_2$ and 10 wt. % $LiNO_3$, while for higher amounts of deposited salt the heat values decrease. $SACa_x$ composites show an experimental dehydration heat (Figure 12) higher as compared with that expected by simple addition of the two contributions (contribution of the salt, considering $CaCl_2 \cdot 6H_2O$, and contribution of the support, derived by the experiment performed on the pure salt and SA), but lower than the heat value calculated considering the $CaCl_2 \cdot 6H_2O$ form. The hydration extent of $CaCl_2$ is clearly improved when deposited on the SA surface. Due to its high adsorption heat of water and low cost, the composite 'silica–alumina – 15 wt% $CaCl_2$' (738 kJ/kg) occurs to be the most promising as compared to lithium nitrate or barium hydroxide composites (609 and 408 kJ/kg, respectively) (Simonova et al., 2009).

The characteristic curves of water adsorption on several CSPMs are given in Figure 13.

All of composites 'silica gel – salt' are S-shaped curves with steep rise of uptake, which correspond to the formation of the complexes $S \cdot nH_2O$ (Gordeeva & Aristov, 2018). The position of the step on the isotherm is strongly affected by the salt. The variation of salts allows shift the step in order to match the conditions of the specific adsorptive heat storage cycle. Generally, nitrates possess low affinity, they can be applied for cycles with driving temperature of 60°C at the condensation temperature 30°C, but provide the low temperature lift $\Delta T = 20$°C during adsorption. Chlorides, and particularly bromides, reveal a higher affinity. Consequently they provide a larger temperature lift at the expense of higher charging temperature.

Another composite sorbent 'meso-porous silica gel impregnated with $Ca(NO_3)_2$', SWS-8L, revealed sufficiently high water sorption capacity (up to 0.2 g/g) for very low desorption temperature (75–80 °C) (Simonova et al., 2009). As reported in (Gordeeva, Freni, & Krieger, 2008; Aristov et al., 2007),

Figure 13. The characteristic curves of water adsorption on composites 'silica gel – LiCl' (1), 'silica gel – $MgSO_4$' (2), 'silica gel – LiBr' (3), 'silica gel – $LiNO_3$' (4), 'silica gel – $Ca(NO_3)_2$' (5)
(Gordeeva & Aristov, 2018).

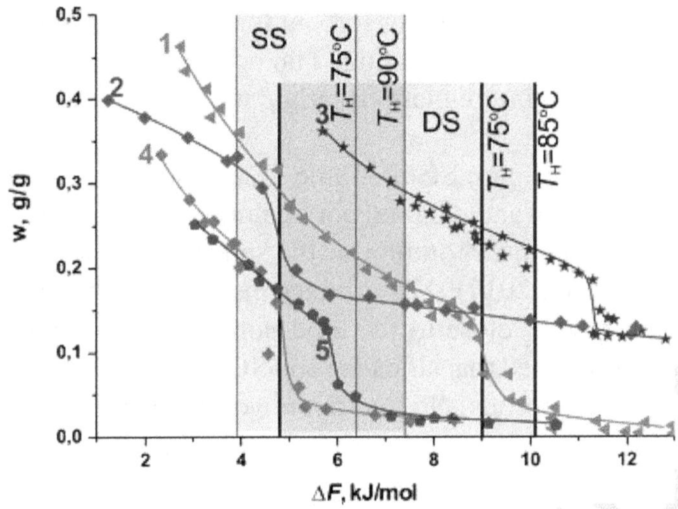

the composite materials based on meso-porous silica gel impregnated with $CaCl_2$, LiBr and LiCl presented outstanding methanol sorption capacity (up to 0.8 g/g), which is attractive for cooling applications with subzero evaporation temperatures. For example, in the 'silica gel – LiCl'-methanol chiller under nominal operating conditions the evaporator temperature, coefficient of performance and specific cooling power can reach 15 °C, 0.41 and 244 W/kg (Lu & Wang, 2013).

On the other hand, the composite mesoporous silica gel impregnated with LiBr, showed very high ethanol adsorption capacity (up to 0.65 g/g), which is less toxic than methanol (Gordeeva & Aristov, 2010). A possible issue related to utilization of composite materials in adsorptive heat conversion cycles is the risk of solution leakage because of refrigerant over-saturation. Several means how to avoid or reduce this risk are considered in (Aydin et al., 2015). Hydrothermal stability is another important issue that has never been systematically tested yet.

In works Courbon et al. (2017) stronthium bromide is suggested to impregnate into a silica gel matrix. As the porous matrix, silica gel Davisil grade 62 from Grace, labelled SG62, is used. This silica gel contains mainly mesoporosity with disperse pore sizes. The hexahydrated strontium bromide is chosen because of the excellent matching between the sorption and desorption temperatures of hexahydrated $SrBr_2$ and those required for solar heat storage for building applications, the high heat of reaction (67.5 kJ/mol of water) coupled with the gain/loss of 5 mol of water per mole of salt. All of these features make this salt an appealing sorbent for solar thermal energy storage applications coupled to space heating. The composite was synthesized by using incipient wetness impregnation method with successive impregnation/drying steps.

In comparison to the pure SG62 (Table 3), a significant decrease of both the specific surface area and the total pore volume of the composite is reveal in agreement with the incorporation of the salt into the pores of the silica matrix (Courbon et al., 2017). The primary mesopore volumes obtained by as-plot method are very close to the total pore volumes obtained by BET analysis, which confirmed that silica gel and SG62/$SrBr_2$ 58 wt.% composite are both mainly mesoporous. No micropore volume is stated for both silica gel and composite 'SG62 – 58 wt.% $SrBr_2$'. As compared to the pure silica gel, composite exhibits a small decrease of the external surface area, the possible presence of a small amount of salt into the macropores or onto the surface of the matrix being indicated. This could result in a problem of stability of the composite.

Table 3. Structural characteristics of silica gel SG62 and composite silica gel/$SrBr_2$: specific surface area, total pore volume and average pore diameter obtained from BET analysis and primary mesopore volume, micropore volume and external surface area calculated from as-plot analysis of the nitrogen adsorption isotherm at 77 K

Adsorbent	Specific Surface Area (m^2/g)	Total Pore Volume (cm^3/g)	Average pore Diameter (nm)	Primary Mesopore Volume (cm^3/g)	Micropore Volume (cm^3/g)	External Surface Area (m^2/g)
SG62	320	1.14	14	1.13	0	1.93
SG62 – 58 wt.% $SrBr_2$	62	0.33	21	0.32	0	0.39

(Courbon et al., 2017)

The water vapor sorption behavior of the composite was studied at the temperatures from 20 °C to 80 °C (Courbon et al., 2017). According to a given isotherm, a first mass uptake is revealed as pressure increased (Figure 14). Then, after a plateau, a second lift of the adsorbed mass occurs. Depending on the temperature, it is followed by a plateau or by a low slope curve. At 20°C and 30°C for high water vapor partial pressures, another lift of the adsorption is observed.

Figure 14. Water vapor sorption isotherms of 'SG62 – $SrBr_2$' at 58 wt.%. Adsorbed mass of water per 1 g anhydrous composite vs. water pressure (a); Moles of water adsorbed per mol of $SrBr_2$ vs. water pressure (b).
(Courbon et al., 2017).

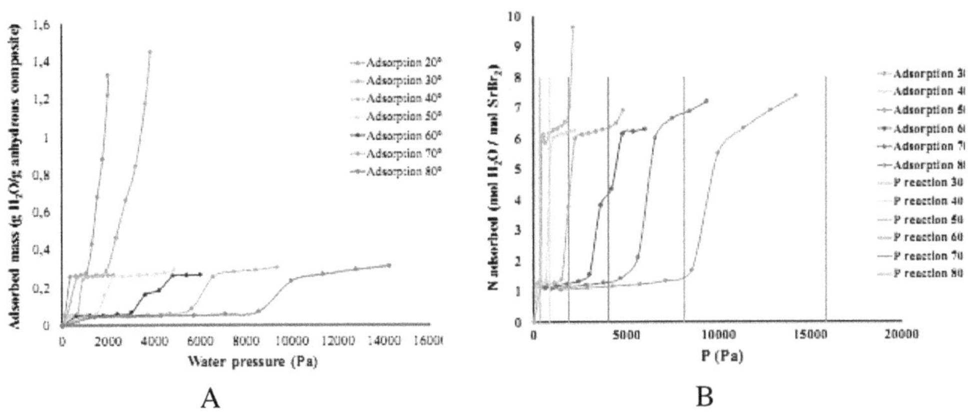

So, changing of the structure of 'SG62 – $SrBr_2$' at the highest water contents was studied by XRD-method (Zhang et al., 2016). X-ray diffraction patterns were registered when the temperature varied (Figure 15). The material was first hydrated until 0.69 g/g before conducting these X-ray diffraction experiments.

As temperature increased, the peaks corresponded with the hexahydrate form of $SrBr_2$ gradually alternate by peaks correlated with monohydrate and anhydrous forms of $SrBr_2$ (Courbon et al., 2017). For example, at 30 °C, the X-ray diffraction pattern of the composite reveal XRD peaks, the assignment of which is consistent with the presence of the hexahydrate form of $SrBr_2$. The low intensity of those XRD peaks confirms that $SrBr_2$ hexahydrate is present in the composite in small amount, the main part of the salt being confined into the pores of silica as amorphous or partially solubilized salt. The high intensity of the peak at 2h =40° in comparison to the pattern of the $SrBr_2$ hexahydrate indicates coexisting the different hydrates of $SrBr_2$ coexist in the sample. Upon increasing the temperature, a dehydration process of $SrBr_2$ is confirmed by XRD. The XRD peaks of the hexahydrate form of $SrBr_2$ are clearly observed in the range 40-50 °C, the XRD pattern of the monohydrate

Figure 15. X-ray diffraction patterns of composite 'SG62/SrBr$_2$' from 30 °C to 170 °C, the composite is initially hydrated at 0.69 g/g. "0", "1" and "6" represent the XRD pattern of the anhydrous, monohydrate and hexahydrate forms of SrBr2, respectively
(Courbon et al., 2017).

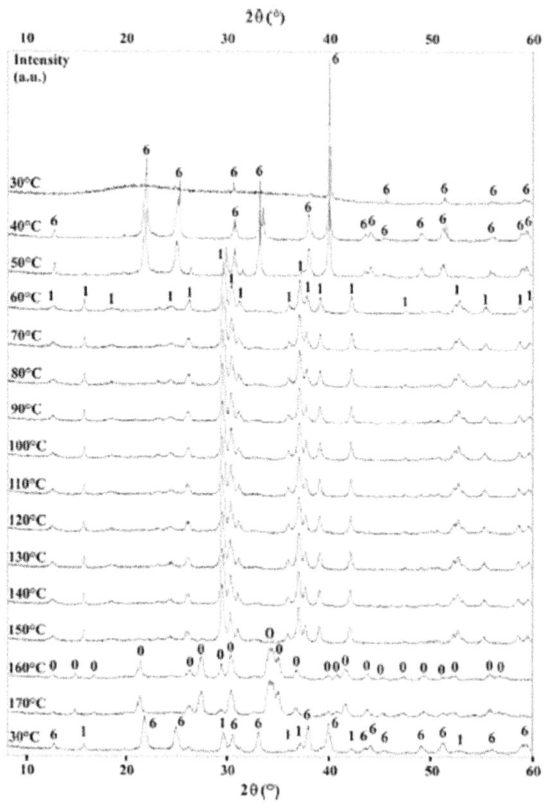

form of SrBr$_2$ being recorded from 60 to 150°C. Finally, the anhydrous form of SrBr$_2$ is revealed by XRD at 160 °C. Upon decreasing the temperature back to 30 °C, inverted changes are observed. So, the XRD pattern exhibits mainly the XRD peaks of the hexahydrate form of SrBr$_2$ with minor peaks of the monohydrate and anhydrous forms of SrBr$_2$.

So, the sorption mechanism of water vapor on composite 'SG62 – $SrBr_2$' can be presented by two steps (Courbon et al., 2017). The first step is the chemical reaction of $SrBr_2$ from the anhydrous to the monohydrate:

$$SrBr_{2(s)} + H_2O_{(v)} = SrBr_2{\cdot}H_2O_{(s)}. \quad (11)$$

The second step corresponded to the hydration of monohydrate of $SrBr_2$:

$$SrBr_2{\cdot}H_2O_{(s)} + 5H_2O_{(v)} = SrBr_2{\cdot}6H_2O_{(s)}. \quad (12)$$

Both plateaus with a very small mass uptake are resulted from a slight sorption of water on composite (Courbon et al., 2017). The last water uptake observed at 20 °C and 30 °C should be representative of absorption in solution of $SrBr_2$. These conclusions are conformed to the water vapor isotherms of composite 'silica gel – $SrBr_2$' represented as a function of the number of moles of water per mole of $SrBr_2$ in the composite (Figure 14b).

The first plateau correlates with the adsorption of one mole of water per mole of $SrBr_2$, and the second plateau being corresponded to the adsorption of six moles of water per mole of $SrBr_2$ (Courbon et al., 2017). The matching of the water vapor pressures at which the reaction (12) occurs in the composite and the pressures at which this reaction occurs in the pure salt (named P reaction in Figure 14b) exhibits that for temperatures lower than 60 °C, the reaction in

Figure 16. Comparison of hydration kinetics of SrBr2 (red) and 'SG62 – 58 wt.% SrBr2' (blue), initial conditions: 80°C and 12.5 mbar, final conditions: 30°C and 12.5 mbar
(Courbon et al., 2017).

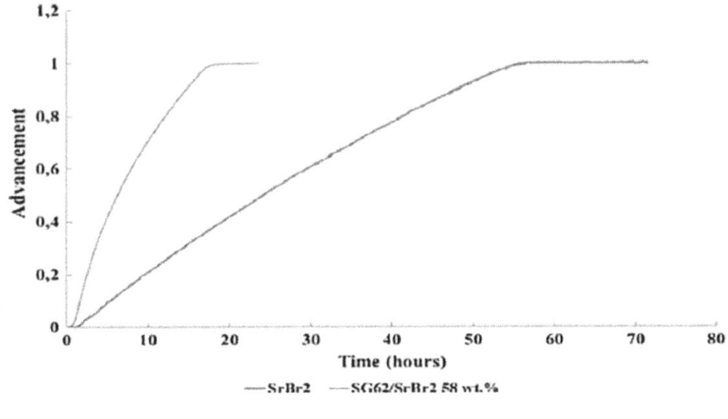

the composite occurs at the same pressure as in the pure salt. Nevertheless, for temperature higher than 60 °C, the reaction in the composite occurs at lower pressure as compared with the pure salt. So, even if the chemical behavior of the salt is preserved in the composite, the thermodynamic equilibrium of the reaction is modified because the incorporation of the salt into the pores of the silica gel. For this reason, the thermodynamic equilibrium data of pure salt cannot describe the adsorption behavior of the composite.

Kinetics of adsorption of water vapor by composite is also different as compared with pure salt, which is strontium bromide (Courbon et al., 2017). Hydration tests between 80°C and 30°C at 12.5 mbar of pure $SrBr_2$ and composite 'SG62 – 58 wt.% $SrBr_2$' (Figure 16) reveal that the hydration step occur to be faster for the composite than for $SrBr_2$. Diffusion is not inhibited by the incorporation of the salt into the matrix pores, the access to the salt being even enhanced as compared to the pure salt. The kinetics is also affected by particle size. As particles of $SrBr_2$ are larger (around 400 μm) compared to those of 'SG62 – 58 wt.%$SrBr_2$' (around 180 μm). The average sorption rate was $1.83 \cdot 10^{-3}$ mg H_2O/g·s (between 10 and 40 h) for $SrBr_2$ and $2.91 \cdot 10^{-3}$ mg H_2O/g·s (between 5 and 15 h) for 'SG62 – 58 wt.%$SrBr_2$'. Moreover, this composite 'silica gel – $SrBr_2$' appear to be of an excellent stability over 14 successive sorption/desorption cycles (Figure 17) (Courbon et al., 2017).

Figure 17. Stability of SG62/$SrBr_2$ at 58 wt.% under successive sorption/desorption cycles
(Courbon et al., 2017).

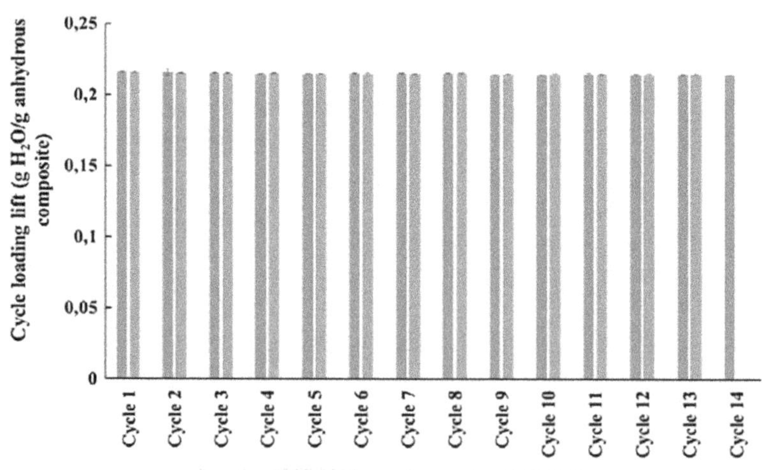

Apparently, the pores of silica gel are able to create a stabilizing media for the salt due to specific interactions and confinement effect.

The experimental value of adsorption per a cycle of this composite between 40°C and 80°C is stated to be of 0.22 g H_2O/g anhydrous composite (Courbon et al., 2017). The integration of the heat flow signal with the correction of the crucible and sample heating is 827 J/g of anhydrous composite which corresponds to an energy storage capacity of 230Wh/kg of anhydrous composite and an energy storage density of 203 kWh/m^3 of packing anhydrous composite (the experimental packing density is 885 kg/m^3).

Providing the adsorption/desorption conditions allow the reaction to proceed. Variation of water vapor pressure does not change the heat storage density.

New type of composite 'silica gel – calcium chloride' was obtained by Courbon et al. (2017b) a new protocol of successive impregnations of a small quantity of salt solution (corresponding to the matrix pore volume) and drying steps. This technique was previously used for obtaining composite 'silica gel – strontium bromide' (Courbon et al., 2017a).

Conforming with all of considered composites, introduction of calcium chloride in porous matrix of silica gel results in significant decreasing specific surface area and total pore volume which confirms the presence of the salt in the matrix, but average pore diameter is increased (Table 4) (Courbon et al., 2017b).

The water vapor sorption isotherms of the composite were recorded in the range of 30 – 80 °C (Figure 18). According to a given isotherm, a first water uptake was observed, followed by a plateau at around 0.14 g/g (Courbon et al., 2017b). A second water uptake was then observed, followed by a linear increase of the mass with pressure. A hysteresis between adsorption and desorption was revealed for every temperature.

Table 4. Structural characteristics of silica gel SG62 and composite 'SG62 – 43 wt% $CaCl_2$' obtained from BET analysis of nitrogen adsorption isotherm at 77 K

Adsorbent	Specific Surface Area (m^2/g)	Total Pore Volume (cm^3/g)	Average Pore Diameter (nm)
SG62	320	1.14	14
SG62 – 44 wt.% $CaCl_2$	75	0.38	20

(Courbon et al., 2017b)

Figure 18. Water vapor sorption isotherms of composite 'SG62 – 43 wt% $CaCl_2$'. Adsorption isotherms are represented by continuous lines whereas desorption isotherms are re- presented by discontinuous lines
(Courbon et al., 2017b).

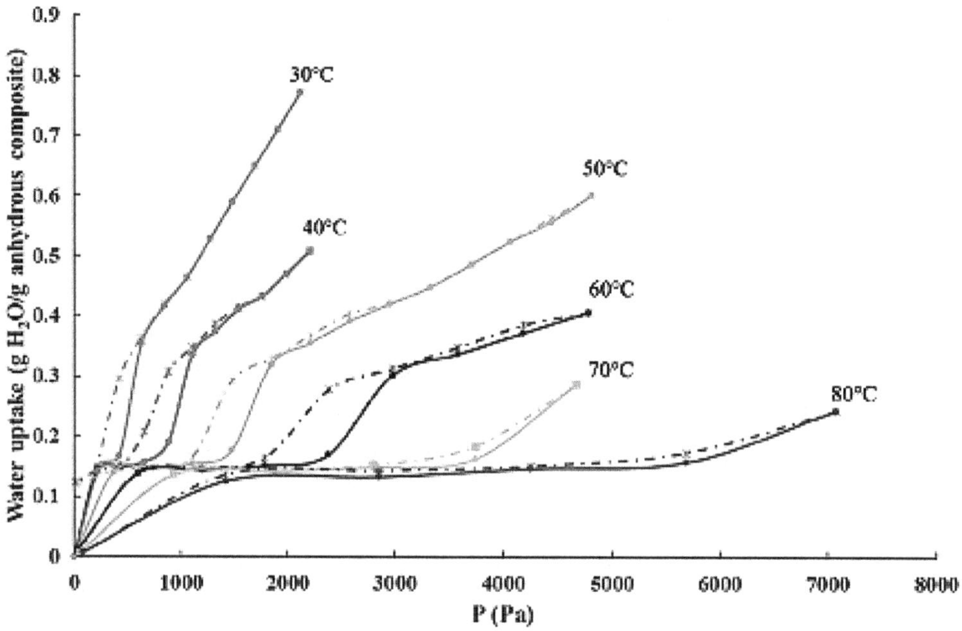

To understand the behavior of the salt in this mechanism of sorption, X-ray diffraction patterns were recorded by varying the temperature on the composite 'SG62 – $CaCl_2$' with high hydration rate (Figure 19).

Between 30 °C and 50 °C, no X-ray diffraction peaks were observed (Courbon et al., 2017b). So, $CaCl_2$ is amorphous or liquid, while crystalline $CaCl_2 \cdot 4H_2O$ was expected. This occurs due to dissolution of $CaCl_2 \cdot 4H_2O$ in the composite. Between 60 °C and 80 °C, the dihydrate form of $CaCl_2$ was stated, and, for temperatures higher than 90 °C, anhydrous $CaCl_2$ was mainly revealed. The monohydrate form of $CaCl_2$ was not observed whatever the temperature.

So, the water vapor sorption by this composite 'SG62 – CaCl2 43 wt%' can be presented by the next steps (Courbon et al., 2017b):

$$CaCl_2 + 2H_2O \rightleftarrows CaCl_2 \cdot 2H_2O \tag{13}$$

Figure 19. Temperature dependent X-ray diffraction patterns of SG62/$CaCl_2$ 43 wt%. Above 110 °C, no more change in the diffraction patterns is observed. At ≥90 °C, small peaks at 2θ = 18.3°, 22.3°, 30.2° and 39.3° cannot be indexed with certainty (Courbon et al., 2017b).

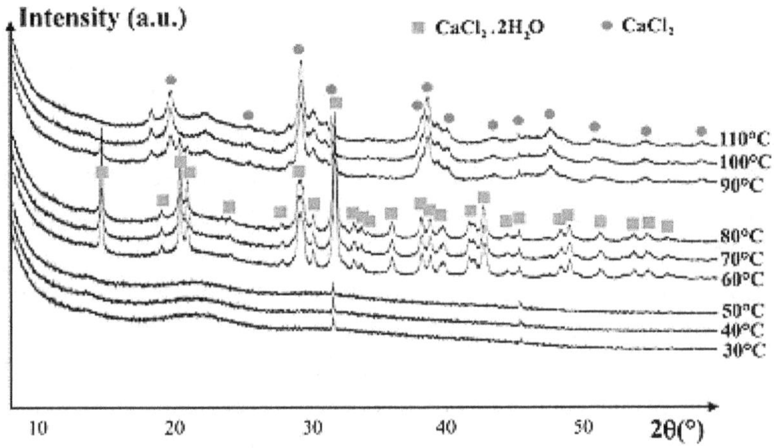

$$CaCl_2 \cdot 2H_2O + 2H_2O \rightleftarrows CaCl_2 \cdot 4H_2O + dissolution\ of\ CaCl_2 \cdot 4H_2O \qquad (14)$$

Then the absorption of water in the $CaCl_2$ solution trapped in the pores of silica gel proceeds.

The hysteresis observed between the sorption and desorption steps can be explained by the recrystallization step.

The composite 'SG62 – 43 wt%$CaCl_2$' can adsorb 0.77 g H_2O/g at 30 °C and 50% relative humidity (Courbon et al., 2017b). At the adsorption temperature 30°C, desorption temperature 80°C and vapor pressure of 12.5 mbar the experimental cycle loading lift is 0.40 g/g, which corresponds to an energy storage capacity of 300 W h/kg, and an energy storage density of 211 kW h/m^3 for the bed of anhydrous composite. The energy storage density is calculated according to the experimental packing density of the anhydrous composite, which is 703 kg/m^3.

Thus, composite adsorbents 'salt inside porous matrix' exhibit low charging temperature and high heat storage capacity. Another their advantage is a possibility of controllable modification of their sorption properties. Nevertheless, commercialization of these promising materials in heat energy conversion devices is strongly inhibited because of their production is concerned with applying of quite complex and rather expensive technologies dry and wet impregnation of porous media by crystalline hydrate solutions.

Promising alternative is sol – gel method. According to Mrowiec-Bialon et al. (1999), sorption capacities sol – gel sorbents 'silica gel – $CaCl_2$' can reach 1 kg of water/kg of adsorbent, regeneration temperature of these materials being 200°C (Mrowiec-Bialon et al. 1999).

Another composites obtained by sol – gel technique are 'silica gel – sodium sulphate' and 'silica gel – sodium acetate' (Sukhyy et al., 2018; Sukhyy, K. M., Belyanovskaya, Kozlov, Kolomiyets, & Sukhyy M. P. 2014).

According to infrared spectrums, bands at 3000 – 3200 cm^{-1} (Figure 20) concerned with hydrogen bond of OH-groups and oxygen atoms of water molecules oscillations attached are typical for synthesized composite 'silica gel – sodium sulphate' (Sukhyy et al., 2014). They are known to be typical for strongly hydrated silica gel surface too (Neymark & Shenfine, 1982). Apparently, such interaction does not occur in the first moment water adsorption. Typical for silica gel bands at 1100 – 900 cm^{-1} corresponded with valency oscillations of Si – O bonds are observed for synthesized composite 'silica gel – sodium sulphate' in SiO tetrahedrons $[Si_4]^{4-}$. Intense bands resulted from deformative oscillations in SiO tetrahedrons (470 – 490 cm^{-1}) and hydroxyls valency and deformative oscillations (1500 – 4000 cm^{-1}) are

Figure 20. IR – absorption spectra: 1 – sodium sulphate; 2 – composite 'silica gel – sodium sulphate' after desorption; 3 – silica gel KSM
(Sukhyy et al., 2014).

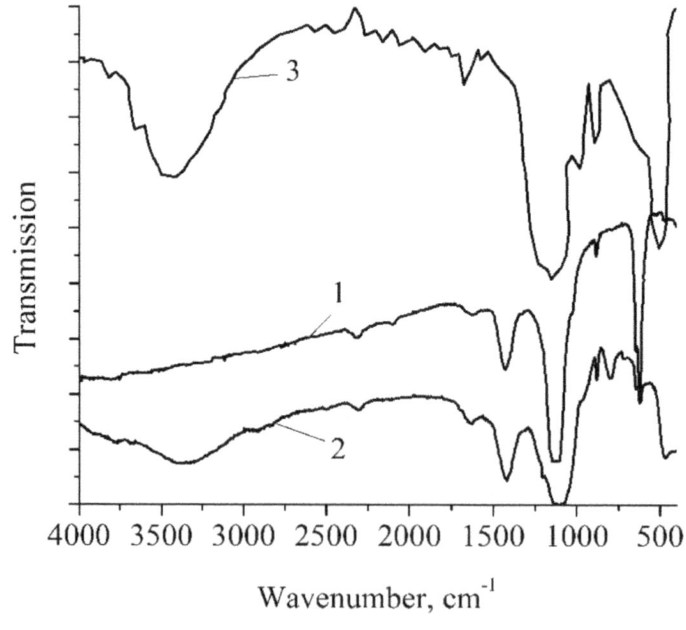

typical for silica gel and composite 'silica gel – sodium sulphate' (Figure 20). Typical bands for silica gel and sodium sulphate are observed in spectrum of composite 'silica gel – sodium sulphate'. However, intense band at 3749 cm^{-1} which is typical for Si – OH groups (Karyakin & Kriventsov, 1972; Slisarenko, 1968) is absent practically. At that depose broad band at 3680 – 3200 cm^{-1} that is typical for silica gel on front-end of water adsorption and strongly hydrated silica gel surface (Neymark & Shenfine, 1982) is observed for composite 'silica gel – sodium sulphate' (Figure 20).

The crystalline hydrates adsorption properties changing at their allocation in silicon oxygen skeleton can be a consequence of salt – matrix interaction or its dispersion up to nanoscale that is size effect. These effects are estimated by X-ray diffraction analysis.

WAXS profiles of the anhydrous silica gel KSM, composite sorbent 'silica gel – sodium sulphate' dehydrated and sodium sulphate Na_2SO_4 are given on Figure 20 (Sukhyy et al., 2014). Reflexes according to crystalline sodium sulphate (Figure 21, profile 3) and silica gel (Figure 21, profile 1) are observed on WAXS profile of composite 'silica gel – sodium sulphate' (Figure 21, profile 2).

The microcrystals of Na_2SO_4 with coherent scattering section dimension on 12 – 17 nm occur in the composite 'silica gel – sodium sulphate'. So, the synthesized composite sorbents can be considered as a three-dimensional

Figure 21. WAXS profiles of the anhydrous samples: 1 – silica gel KSM; 2 – composite 'silica gel – sodium sulphate', 3 – sodium sulphate
(Sukhyy et al., 2014).

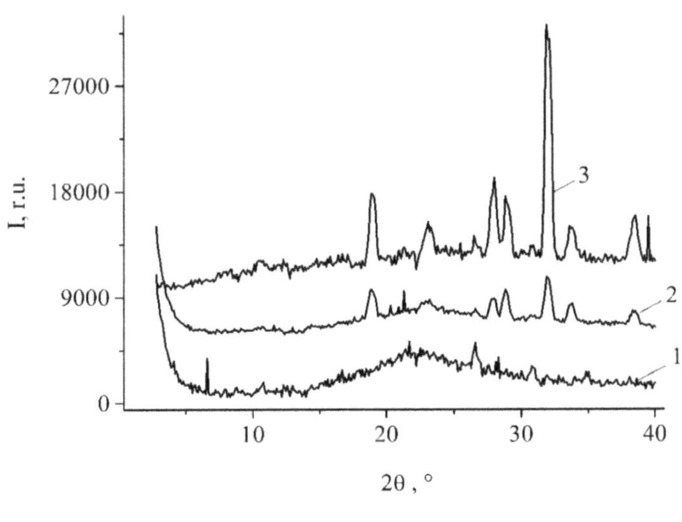

silicon-oxygen frame with developed system of pore which distributed crystalline hydrate molecules in. No crystalline phases of the salt and silicon-oxygen frame interaction are revealed.

In addition to the dimensional peak of sodium sulphate decahydrate, the dimensional reflexes are only observed on WAXS profiles of hydrated samples (Figure 22). As compared to the base salt, these peaks are narrower, their intensity being less. It could be resulted from either the crystalline hydrate dissolution at once after its formation or its X-ray amorphous state. So, size distribution of the nanocrystallites of salt is rather narrow.

Water sorption kinetic curves of composite 'silica gel – sodium sulphate' are strongly differed from massive salts (Figure 23) (Sukhyy et al., 2014). Massive salts kinetic hydration is confirmed by S-shaped curves. On the contrary, curve initial region of composite sorbents is practically linear in coordinate 'm_v/m_s - $\sqrt{\text{time}}$', confirming diffusion nature of process, *i.e.*

Figure 22. WAXS profiles of the hydrated samples: 1 – composite 'silica gel – sodium sulphate', 2 – sodium sulphate
(Sukhyy et al., 2014).

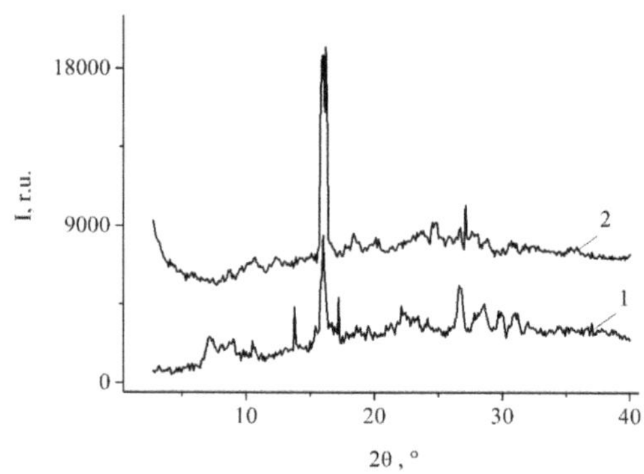

limiting factor is water transport on silica gel pore system. Chemical reaction of water and disperse salt is fast due to salt nanocrystallites narrow distribution by size. Equilibrium for composites reaches well before as compared with massive salts.

To reveal the differences in the thermophysical properties of the investigated composite adsorbents from the properties of massive crystalline hydrates, a differential thermal analysis was carried out for 10-aqueous ($Na_2SO_4 \cdot 10H_2O$)

Figure 23. Kinetic curves of water sorption (P_{H_2O} = 1010,8 mbar), t°C: a – 20; b – 40; c – 60
(Sukhyy et al., 2014).

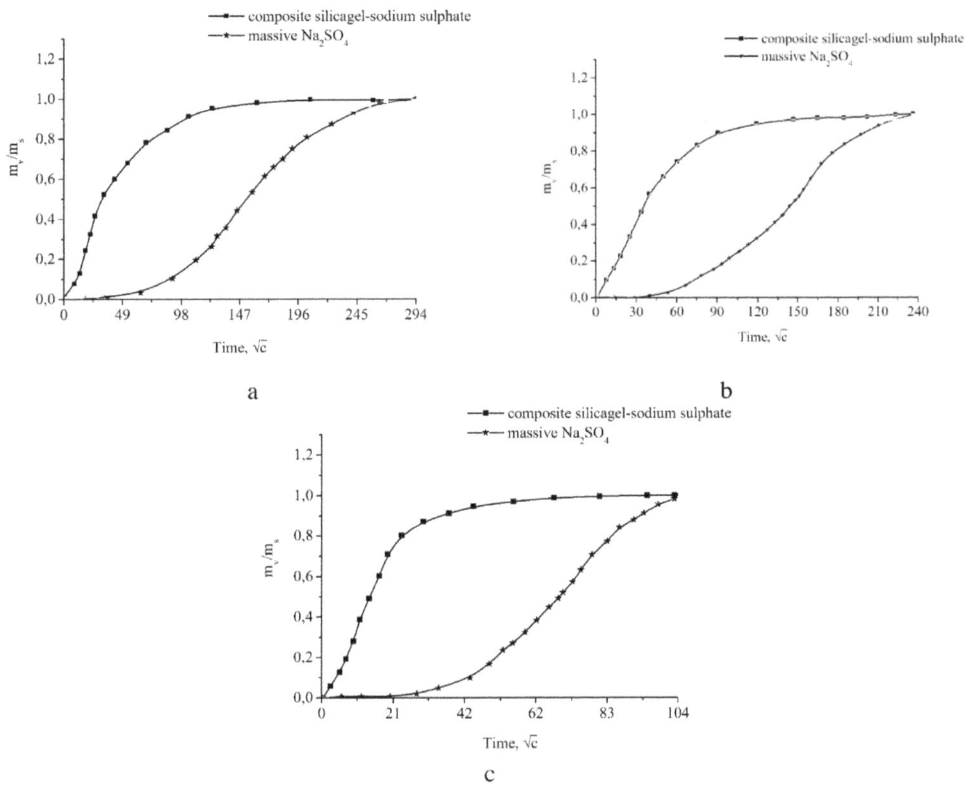

and anhydrous sodium sulphate, as well as for dried KSM branded silica gel at 100 °C.

When the mass loss curves of the composite ‘silica gel – sodium sulphate’ (Figure 24, a) analyzed, it ought to note that the composite after desorption, when heated to 1000 °C, loses only 2% of the mass (curve 2). The mass loss for the composite after adsorption of air moisture is 46%. For the composite ‘silica gel – sodium acetate’ in a fully hydrated state, the mass loss is 42% (Figure 24, b). Silica gel can be considered as an analogue of the silicon-oxygen matrix of the investigated composites (Figure 25).

The sorption properties of the composite ‘silica gel – sodium sulphate’ are shown to rather different as compared with the properties of the bulk salt and silica gel due to the characteristics of this disperse system. This is shown by comparing the sorption capacities of these substances. According to the studies conducted, the sorption capacity of "pure" silica gel is 0.18 g/g, bulk sodium sulfate is 1.27 g/g and bulk sodium acetate is 0.65 g/g (Sukhyy et al., 2018; Sukhyy et al., 2014).

The sorption capacities of the 'silica gel – sodium sulphate' and 'silica gel – sodium acetate' composites are 0.85 g/g and 0.42 g/g. The stoichiometric composition of the composite sorbent, according to the synthesis conditions, is $Na_2SO_4:SiO_2=1:1.27$.

The linear superposition of the sorption abilities of "pure" silica gel and sodium sulphate was calculated by the following formula:

Figure 24. Mass loss curves of the composites: a – 'silica gel – sodium sulphate'; b – 'silica gel – sodium acetate'; 1 – in hydrated and 2 – dehydrated condition *(Sukhyy et al., 2018)*

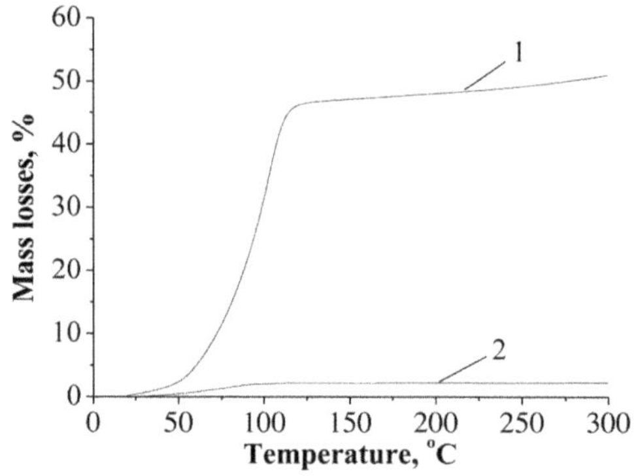

a

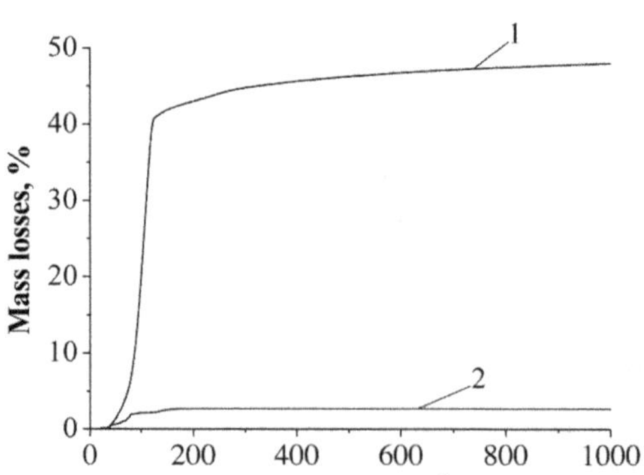

b

Figure 25. Mass loss curves of massive salts and silica gel: a – silica gel – sodium sulphate; b – silica gel – sodium acetate; 1 – salts after adsorption; 2 – silica gel KCM; 3 – salts after desorption
(Sukhyy et al., 2018)

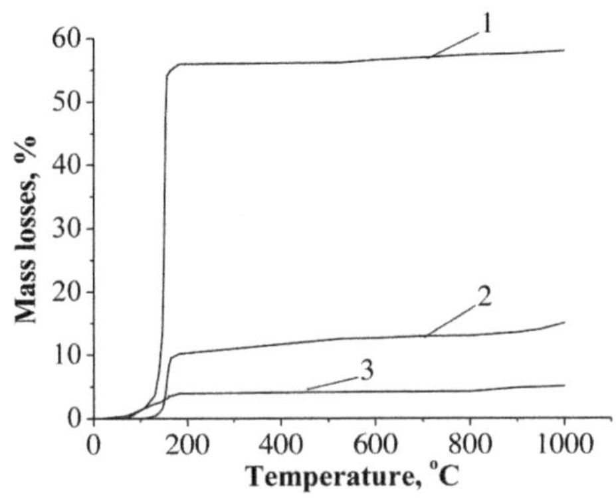

a

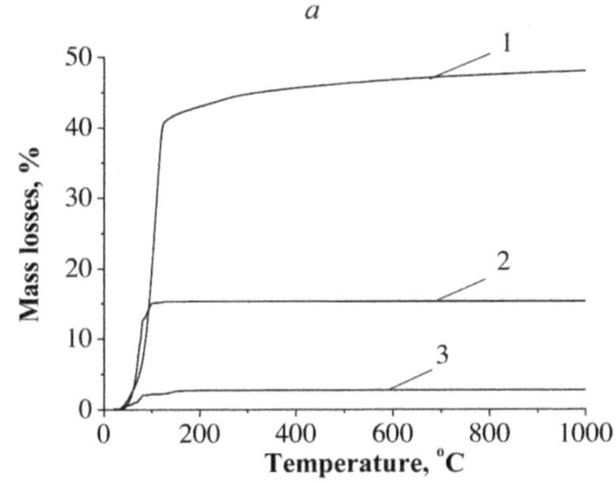

b

Table 5. Adsorptive properties of the composite sorbent 'silica gel – Na_2SO_4'

Content of Silica Gel, %	Content of Na_2SO_4, %	Results of DTA Analysis Mass Loss, %	Maximal Adsorption, kg/kg		Heat of Adsorption, kJ/kg	
			Calculated by Linear Superposition	Actual Value	Calculated by Linear Superposition	Actual Value
20	80	57.4	1.046	1.349	3486.67	4497.81
40	60	51.4	0.822	1.060	2740.00	3534.60
60	40	43.5	0.598	0.771	1993.33	2571.40
80	20	32.5	0.374	0.482	1246.70	1608.24

(Sukhyy et al., 2018)

$$m_{\Sigma} = a \cdot m_{Na_2SO_4} + (1-a) \cdot m_{SiO_2}, \tag{15}$$

where a is the mass content of sodium sulfate in the composite a=0.44; $m_{Na_2SO_4}$, m_{SiO_2} is the mass of water (g), adsorbed by 1 g of massive salts Na_2SO_4 and CH_3COONa and 1 g of silica gel, respectively, $m_{Na_2SO_4} = 1,27$, $m_{CH3COONa}$=0.65 and $m_{SiO_2} = 0,18$.

By substituting the corresponding data into equation (3.15), a linear superposition of the sorption capacity of mechanical mixtures of silica gel and sodium sulphate (Table 5) and also silica gel and sodium acetate was calculated (Table 6).

The actual values of the sorption capacities of the studied composites exceed by approximately 30% the linear superposition of the sorption capacities of mechanical mixtures of silica gel and salts. As a result, the heat of adsorption increases by 30% (Sukhyy et al., 2018).

So, sol – gel method allows to introduce the crystals of sodium sulphate or sodium acetate into the porous silicon dioxide matrix. Obtained composite adsorbents method obviously shows enhanced adsorption capacity which

Table 6. Adsorptive properties of the composite sorbent 'silica gel – CH_3COONa'

Content of Silica Gel, %	Content of CH_3COONa, %	Results of DTA Analysis Mass Loss, %	Maximal Adsorption, kg/kg		Heat of Adsorption, kJ/kg	
			Calculated by Linear Superposition	Actual Value	Calculated by Linear Superposetion	Actual Value
20	80	43.1	0.557	0.756	1856.10	2431.49
40	60	37.3	0.455	0.596	1517.07	1987.36
60	40	31.6	0.353	0.462	1178.05	1543.25
80	20	24.8	0.252	0.330	1099.11	1608.24

(Sukhyy et al., 2018)

exceed by 30% the value calculated from the linear superposition of the sorption capacities of the sorbent and massive salt. Their sorption properties are not a linear combination of properties of silica gel and salt. This confirms the formation of an unique structure that promotes an increase in the rate of reaction between crystalline hydrates and water vapor in the developed pores of the silicon-oxygen matrix. As a consequence, it results in increasing the heat of adsorption and the heat energy storage density.

REFERENCES

Aristov, Y. I., Gordeeva, L. G., Pankratiev, Y. D., Plyasova, L. M., Bikova, I. V., Freni, A., & Restuccia, G. (2007). Sorption equilibrium of methanol on new composite sorbents "$CaCl_2$/silica gel". *Adsorption*, *13*(2), 121–127. doi:10.100710450-007-9012-x

Aristov, Yu. I., Restuccia, G., Cacciola, G., & Parmon, V. N. (2002). A family of new working materials for solid sorption air conditioning systems. *Applied Thermal Engineering*, *22*(2), 191–204. doi:10.1016/S1359-4311(01)00072-2

Aristov, Yu. I., Tokarev, M. M., Cacciola, G., & Restuccia, G. (1996). Selective water sorbents for multiple application, 1. $CaCl_2$ confined inmesopores of silica gel: Sorption properties. *Reaction Kinetics and Catalysis Letters*, *59*(2), 325–335. doi:10.1007/BF02068130

Aydin, D., Casey, S. P., & Riffat, S. (2015). The latest advancements on thermochemical heat storage systems. *Renewable & Sustainable Energy Reviews*, *41*, 356–367. doi:10.1016/j.rser.2014.08.054

Bandosz, T. J., & Petit, C. (2009). On the reactive adsorption of ammonia on activated carbons modified by impregnation with inorganic compounds. *Journal of Colloid and Interface Science*, *338*(2), 329–345. doi:10.1016/j.jcis.2009.06.039 PMID:19615690

Courbon, E., D'Ans, P., Permyakova, A., Skrylnyk, O., Steunou, N., Degrez, M., & Frère, M. (2017). A new composite sorbent based on $SrBr_2$ and silica gel for solar energy storage application with high energy storage density and stability. *Applied Energy*, *190*, 1184–1194. doi:10.1016/j.apenergy.2017.01.041

Courbon, E., D'Ans, P., Permyakova, A., Skrylnyk, O., Steunou, N., Degrez, M., & Frère, M. (2017). Further improvement of the synthesis of silica gel and $CaCl_2$ composites: Enhancement of energy storage density and stability over cycles for solar heat storage coupled with space heating applications. *Solar Energy*, *157*, 532–541. doi:10.1016/j.solener.2017.08.034

Dawoud, B. (2014). On the development of an innovative gas-fired heating appliance based on a zeolite-water adsorption heat pump; System description and seasonal gas utilization efficiency. *Applied Thermal Engineering*, *72*(2), 323–330. doi:10.1016/j.applthermaleng.2014.09.008

De Jong, A.-J., Trausel, F., Finck, C., Van Vliet, L., & Cuypers, R. (2014). Thermochemical heat storage – system design issues. *Energy Procedia*, *48*, 309–319. doi:10.1016/j.egypro.2014.02.036

Donkers, P. A. J., Sögütoglu, L. C., Huinink, H. P., Fischer, H. R., & Adan, O. C. G. (2017). A review of salt hydrates for seasonal heat storage in domestic applications. *Applied Energy*, *199*, 45–68. doi:10.1016/j.apenergy.2017.04.080

Donkers, P. A. J., Sögütoglu, L. C., Huinink, H. P., Fischer, H. R., & Adan, O. C. G. (2017). A review of salt hydrates for seasonal heat storage in domestic applications. *Applied Energy*, *199*, 45–68. Advance online publication. doi:10.1016/j.apenergy.2017.04.080

Freni, A., Maggio, G., Sapienza, A., Frazzica, A., Restuccia, G., & Vasta, S. (2016). Comparative analysis of promising adsorbent/adsorbate pairs for adsorptive heat pumping, air conditioning and refrigeration. *Applied Thermal Engineering*, *104*, 85–95. doi:10.1016/j.applthermaleng.2016.05.036

Gordeeva, L., & Aristov, Yu. (2010). Novel sorbents of ethanol—Salt confined to porous matrix for adsorptive cooling. *Energy*, *35*(6), 2703–2708. doi:10.1016/j.energy.2009.04.001

Gordeeva, L., Freni, A., Krieger, T., Restuccia, G., & Aristov, Y. I. (2008). Composites lithium halides in silica gel pores: Methanol sorption equilibrium. *Microporous and Mesoporous Materials*, *112*(1-3), 264–271. doi:10.1016/j.micromeso.2007.09.040

Gordeeva, L. G., Aristov, Yu. I. (2012). Composites "salt inside porous matrix" for adsorption heat transformation: a current state-of-the-art and new trends. *International Journal of Low-Carbon Technologies Advanced Access*, 1 – 15.

Gordeeva, L. G., & Aristov, Yu. I. (2018). Adsorptive heat storage and amplification: New cycles and adsorbents. *Energy*, *167*, 440–453. doi:10.1016/j.energy.2018.10.132

Grekova, A. D., Gordeeva, L. G., & Aristov, Yu. I. (2017). Composite "LiCl/vermiculite" as advanced water sorbent for thermal energy storage. *Applied Thermal Engineering*, *124*, 1401–1408. doi:10.1016/j.applthermaleng.2017.06.122

Hiremath, C. R., & Kadoli, R. (2013). Experimental studies on heat and mass transfer in a packed bed of burnt clay impregnated with $CaCl_2$ liquid desiccant and exploring the use of gas side resistance model. *Applied Thermal Engineering*, *50*(1), 1299–1310. doi:10.1016/j.applthermaleng.2012.08.002

Hongois, S., Kuznik, F., Stevens, P., & Roux, J.-J. (2011). Development and characterization of a new $MgSO_4$-zeolite composite for long term thermal energy storage. *Solar Energy Materials and Solar Cells*, *95*(7), 1831–1837. doi:10.1016/j.solmat.2011.01.050

Jabbari-Hichri, A., Bennici, S., & Auroux, A. (2015). Enhancing the heat storage density of silica–alumina by addition of hygroscopic salts ($CaCl_2$, $Ba(OH)_2$, and $LiNO_3$). *Solar Energy Materials and Solar Cells*, *140*, 351–360. doi:10.1016/j.solmat.2015.04.032

Jänchen, J., Ackermann, D., Stach, H., & Brösicke, W. (2004). Studies of the water adsorption on Zeolites and modified mesoporous materials for seasonal storage of solar heat. *Solar Energy*, *76*(1-3), 339–344. doi:10.1016/j.solener.2003.07.036

Jochen, J., & Helmut, S. (2012). Adsorption properties of porous materials for solar thermal energy storage and heat pump applications. *Energy Procedia*, *30*, 289–293. doi:10.1016/j.egypro.2012.11.034

Kakiuchi, H., Shimooka, S., Iwade, M., Oshima, K., Yamazaki, M., Terada, S., Watanabe, H., & Takewaki, T. (2005). Water vapor adsorbent FAM-Z02 and its applicability to adsorption heat pump. *Kagaku Kogaku Ronbunshu*, *31*(4), 273–277. doi:10.1252/kakoronbunshu.31.273

Kakiuchi, H., Shimooka, S., Iwade, M., Oshima, K., Yamazaki, M., Terada, S., Watanabe, H., & Takewaki, T. (2005). Novel water vapor adsorbent FAM-Z01 and its applicability to an adsorption heat. *Kagaku Kogaku Ronbunshu*, *31*(5), 361–364. doi:10.1252/kakoronbunshu.31.361

Karyakin, A. V., & Kriventsov, G. A. (1972). *Water state in organic and inorganic compounds (by IR-absorption spectrums)*. Science.

Kayal, S., Baichuan, S., & Saha, B. B. (2016). Adsorption characteristics of AQSOA zeolites and water for adsorption chillers. *International Journal of Heat and Mass Transfer, 92*, 1120–1127. doi:10.1016/j.ijheatmasstransfer.2015.09.060

Kleiner, F., Posern, K., & Osburg, A. (2017). Thermal conductivity of selected salt hydrates for thermochemical solar heat storage applications measured by the light flash method. *Applied Thermal Engineering, 113*, 1189–1193. doi:10.1016/j.applthermaleng.2016.11.125

Lu, Z. S., & Wang, R. Z. (2013). Performance improvement and comparison of mass recovery in $CaCl_2$/activated carbon adsorption refrigerator and silica gel/LiCl adsorption chiller driven by low grade waste heat. *International Journal of Refrigeration, 36*(5), 1501–1511. doi:10.1016/j.ijrefrig.2013.03.008

Mrowiec-Bialon, J., Jarzebski, A. B., & Pajak, L. (1999). Water vapor adsorption on the SiO_2-$CaCl_2$ sol-gel composites. *Langmuir, 15*(19), 6505–6509. doi:10.1021/la990231n

Neymark, I. Ye., & Shenfine, I. Yu. (1982). *Silica gel, its obtaining, properties and application*. Naukova dumka.

Palomba, V., & Frazzica, A. (2018). Recent advancements in sorption technology for solar thermal energy storage applications. *Solar Energy*. Advance online publication. doi:10.1016/j.solener.2018.06.102

Patent US6973963 B2. (2005). *Adsorber generator for use in sorption heat pump processes.*

Patent WO 02/066910 A1. Adsorption heat pump and use of adsorption material as adsorption material for adsorption heat pump, Kakiuchi, H., Takewaki, T., Takumi, H., Yamazaki, M., Watanabe H., 2003.

Rammelberg, H. U., Osterland, T., Priehs, B., Opel, O., & Ruck, W. K. L. (2016). Thermochemical heat storage materials – performance of mixed salt hydrates. *Solar Energy, 136*, 571–589. doi:10.1016/j.solener.2016.07.016

Restuccia, G., Freni, A., Russo, F., & Vasta, S. (2005). Experimental investigation of a solid adsorption chiller based on a heat exchanger coated with hydrophobic zeolite. *Applied Thermal Engineering, 25*(10), 1419–1428. doi:10.1016/j.applthermaleng.2004.09.012

Sapienza, A., Freni, A., Vasta, S., Restuccia, G., & Cacciola, G. (2007). Performance of a novel Functional Adsorbent Material for automotive adsorption air conditioning. *IMPRES – International Symposium on Innovative Materials for Processes in Energy Systems.*

Sapienza, A., Glaznev, I. S., Santamaria, S., Freni, A., & Aristov, Yu. I. (2012). Adsorption chilling driven by low temperature heat: New adsorbent and cycle optimization. *Applied Thermal Engineering*, *32*, 141–146. doi:10.1016/j.applthermaleng.2011.09.014

Scapino, L., Zondag, H. A., Van Bael, J., Diriken, J., & Rindt, C. C. M. (2017). Sorption heat storage for long-term low-temperature applications: A review on the advancements at material and prototype scale. *Applied Energy*, *190*, 920–948. doi:10.1016/j.apenergy.2016.12.148

Sharonov, V., Veselovskaya, J. V., & Aristov, Yu. I. (2006). Ammonia sorption on composites '$CaCl_2$ in inorganic host matrix': Isosteric chart and its performance. *The International Journal of Low Carbon Technologies*, *1*(3), 191–200. doi:10.1093/ijlct/1.3.191

Simonova, I. A., Freni, A., Restuccia, G., & Aristov, Y. I. (2009). Water sortion on composite "silica modified by calcium nitrate". *Microporous and Mesoporous Materials*, *122*(1-3), 223–228. doi:10.1016/j.micromeso.2009.02.034

Slisarenko, A. (Ed.). (1968). *Physicochemical investigations of natural sorbents*. Saratov State Teacher's Training College.

Sukhyy, K., Belyanovskaya, E., Kovalenko, V., Kotok, V., Sukhyy, M., Kolomiyets, E., Gubynskyi, M., Yeromin, O., & Prokopenko, O. (2018). The study of properties of composite adsorptive materials "silica gel – crystalline hydrate" for heat storage devices. *Eastern-European Journal of Enterprise Technologies*, *91*(1), 52–58. doi:10.15587/1729-4061.2018.123896

Sukhyy, K. M., Belyanovskaya, E. A., Kozlov, Ya. N., Kolomiyets, E. V., & Sukhyy, M. P. (2014). Structure and Adsorption Properties of the Composites 'Silica Gel – Sodium Sulphate', obtained by Sol – Gel Method. *Applied Thermal Engineering*, *64*(1-2), 408–412. doi:10.1016/j.applthermaleng.2013.12.013

Tso, C. Y., Chan, K. C., Chao, C. Y. H., & Wu, C. L. (2015). Experimental performance analysis on an adsorption cooling system using zeolite 13X/$CaCl_2$ adsorbent with various operation sequences. *International Journal of Heat and Mass Transfer*, *85*, 343–355. doi:10.1016/j.ijheatmasstransfer.2015.02.005

Veselovskaya, J. V., & Tokarev, M. M. (2011). Novel ammonia sorbents "porous matrix modified by active salt" for adsorptive heat transformation: 4. Dynamics of quasi-isobaric ammonia sorption and desorption on $BaCl_2$/vermiculite. *Applied Thermal Engineering*, *31*(4), 566–572. doi:10.1016/j.applthermaleng.2010.10.018

Wang, L. W., Wang, R. Z., & Oliveira, R. G. (2009). A review on adsorption working pairs for refrigeration. *Renewable & Sustainable Energy Reviews*, *13*(3), 518–534. doi:10.1016/j.rser.2007.12.002

Whiting, G., Grondin, D., Bennici, S., & Auroux, A. (2013). Heats of water sorption studies on zeolite-MgSO4 composites as potential thermochemical heat storage materials. *Solar Energy Materials and Solar Cells*, *112*, 112–119. doi:10.1016/j.solmat.2013.01.020

Xu, J., Wang, R. Z., & Li, Y. (2014). A review of available technologies for seasonal thermal energy storage. *Solar Energy*, *103*, 610–638. doi:10.1016/j.solener.2013.06.006

Xue, B., Ye, S., Zhang, L., Wei, X., Nakaso, K., & Fukai, J. (2019). High-temperature steam generation from low-grade waste heat from an adsorptive heat transformer with composite zeolite-13X/$CaCl_2$. *Energy Conversion and Management*, *186*, 93–102. doi:10.1016/j.enconman.2019.02.040

Yu, N., Wang, R. Z., & Wang, L. W. (2013). Sorption thermal storage for solar energy. *Progress in Energy and Combustion Science*, *39*(5), 489–514. doi:10.1016/j.pecs.2013.05.004

Zamengo, M., & Kato, Yu. (2017). Comparison of magnesium hydroxide/expanded graphite composites for thermal energy storage in cogeneration nuclear power plants. *Energy Procedia*, *131*, 119–126. doi:10.1016/j.egypro.2017.09.463

Zettl, B., Englmair, G., & Somitsch, W. (2015). An open sorption heat storage concept and materials for building heat supply. *Energy Procedia*, *73*, 297–304. doi:10.1016/j.egypro.2015.07.692

Zhang, Y. N., Wang, R. Z., Zhao, Y. J., Li, T. X., Riffat, S. B., & Wajid, N. M. (2016). Development and thermochemical characterizations of vermiculite/$SrBr_2$ composite sorbents for low-temperature heat storage. *Energy*, *115*, 120–128. doi:10.1016/j.energy.2016.08.108

Zhao, Y. J., Wang, R. Z., Zhang, Y. N., & Yu, N. (2016). Development of $SrBr_2$ composite sorbents for a sorption thermal energy storage system to store low-temperature heat. *Energy*, *115*, 129–139. doi:10.1016/j.energy.2016.09.013

Section 2

Basic Principles of Design and Performance of Adsorptive Heat Converters

Chapter 4
Design of Adsorptive Heat Storage Devices

ABSTRACT

The chapter is devoted to prospects of application of adsorptive heat storage devices, principles of operating the adsorptive heat storage systems, design of adsorptive storage devices and main factors determining the design of adsorptive heat storage unit. Perspectives of application of adsorptive heat storage devices in heat supply systems were analyzed. Basic principles of operating of heat storage devices were considered. Adsorptive heat storage units operating in close and open modes were compared. Constructions of adsorptive heat storage units operating in open and close mode were described. An efficient algorithm for calculating the volume of the adsorptive thermal energy storage device for a decentralised heat supply system of a private house is suggested by authors. The following procedure of computation is proposed to involve: calculation of thermal load for heating including the determination of thermal losses through external fences, thermal losses due to infiltration and internal heat dissipation, the evaluation of maximal adsorption, adsorption heat, and determination of adsorbent mass and adsorbent volume. The maximal adsorption value is suggested to be calculated by the characteristics of the adsorbent, that is, its maximal adsorption or in the absence of data for a composite adsorbent, as a linear superposition for a mechanical mixture. The adsorbent mass is suggested to calculate as a ratio of a thermal load for

DOI: 10.4018/978-1-7998-4432-7.ch004

heating and heat of adsorption. The adsorbent volume is calculated as ratio of mass and density of adsorbent. An evaluative calculation of the heat load for a private house was carried out with the proposed algorithm. Mass and volumes of conventional silica gels were compared with composite adsorbents 'silica gel – sodium sulphate' and 'silica gel – sodium acetate' obtained by sol gel method developed by authors. Mass and volume of silica gels occur to surpass the suggested composite at least by 1.5 – 5 times. This is shown to result from higher maximal adsorption and heat of adsorption of suggested composite adsorbents. The optimal composition of the composite adsorbents 'silica gel – sodium sulphate' and 'silica gel – sodium acetate' was determined according to the minimal volume of the layer of heat storage material. Both the lowest volume values and the highest efficiency of a composite adsorbents with a mass ratio of silica gel and Na_2SO_4 *or* CH_3COONa *20: 80 are explained by the maximum value of adsorption heat. Suggested composite adsorbent 'silica gel – sodium sulphate' and 'silica gel – sodium acetate' are shown to be promising for heat supply systems.*

NOMENCLATURE

A_{max} – maximal adsorption, kg/kg;
ΔH_{ads} is adsorption heat, kJ/kg;
K_{vol} – volumetric coefficients of the building, m^3/m^2;
k_{win}, k_c, k_f and k_{wall} W /(m^2 • þC) – heat transfer coefficients of walls, windows, ceilings and floors, W/ (m^2 • þC);
M_{ads} – adsorbent mass, kg;
q_{ihg} – specific heat (per 1 m^2 of living space), W/m^2;
Q'_o – load upon space heating, W;
Q'_h – heat loss through outer shell, W;
Q'_{inf} – heat loss resulted from infiltration, W;
Q_{ihg} – internal heat generation, W;
$t_{in.c}$ – temperature of the internal air, °C;
t_o – temperature of the external air, °C;
φ – glazing factor (the ratio of the surface of the windows to the total surface of vertical outer shell)

Ψ_{wall}, Ψ_{win}, Ψ_c, Ψ_f – coefficients of reducing the design temperature split for walls, windows, ceiling and floor.

PROSPECTS FOR APPLICATION OF ADSORPTIVE HEAT STORAGE DEVICES

Unrestrained growth of world prices for organic fuel stipulates the development of methods to limit rates for electric energy in power systems. The constant trends of technical solutions are generally focused on energy saving or equalization the schedules of electrical loads of the grid.

A typical daily schedule of electric load of many power systems resulted from the daily rhythms of social life is characterized by three time zones (Figure 1): a zone of minimum load (night hours or night-time valley) with a required power not more than the minimum P_{min}, the zone of medium or semicircular load with the capacity from minimum to maximum $P_{min} < P < P_{max}$ and zone of maximum or peak load with power not more than P_{max} (Malyarenko & Nechmoglod, 2011).

Figure 1. Typical daily plot of load of electrical power system (Malyarenko & Nechmoglod, 2011).

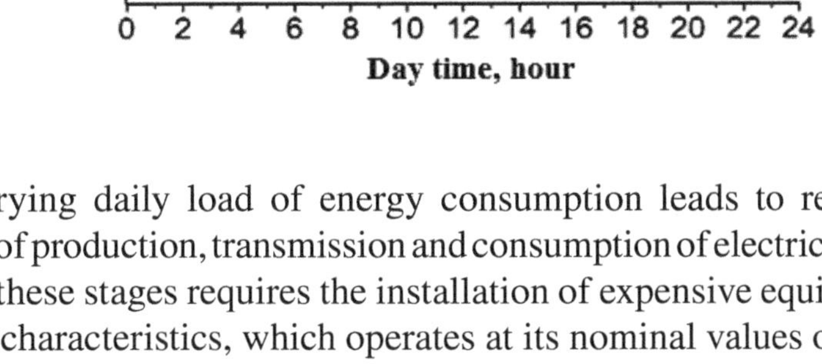

The varying daily load of energy consumption leads to reducing the efficiency of production, transmission and consumption of electricity, because of each of these stages requires the installation of expensive equipment with over-rated characteristics, which operates at its nominal values only a small amount of time. The consolidation of the load graphs of the varying load is proved to reduce the need for generating capacities, as well as the current costs of equipment production due to the better use of equipment. The ways of regulation of the electric load graphs are mainly focused on the adjusting the load of the organization (Hronusov, 1998). by shifting production processes according to constraints at low cost intervals due to the use of energy storage devices – rechargeable batteries and large capacity capacitors (Bragin, 2013), which occur to be rather expensive and difficult in operation units. An effective way to fill a period of a night-time valley may be the use of excess electricity to organize decentralized heat supply for residential buildings. For the power system consumers used electric installations for water heating can become controlled customers-regulators, which store energy in the off-peak hours of electricity consumption and use it at hours of maximum load, thereby decreasing the daily load of electrical power system.

One of the promising technical solutions for those tasks can become adsorptive thermal accumulators which heat the water for hot water or heat supply during peak hours of the network and consume electricity for the regeneration of the adsorbent at night. Moreover, the electric energy demand of these devices can be substantially reduced by using the solar energy to regenerate the adsorbent. As another heat energy source for adsorbent regeneration, combustion gases can be considered.

Figure 2. Principle of operation of adsorptive system of closed type
(Kerskes, 2006).

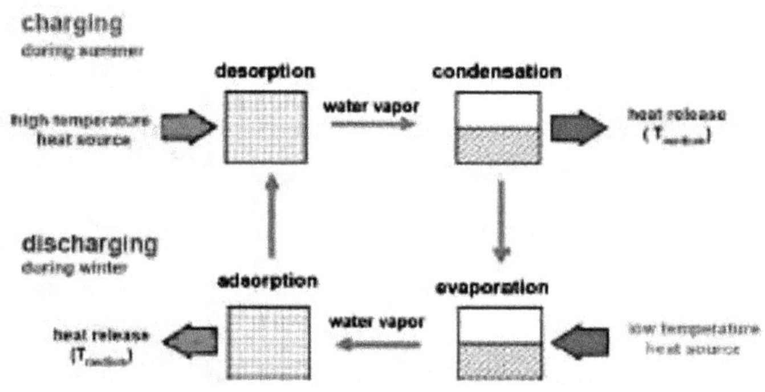

PRINCIPLES OF OPERATING THE ADSORPTIVE SYSTEMS

The systems of closed and open adsorptive heat storage were developed systems of decentralized heating and hot water supply. Specific examples of apparatus design and the results achieved in practice will be discussed below.

The first prototype of the adsorptive system for decentralized heating and hot water supply in conjunction with solar collectors of 20.4 m^2 was built in Austria (Hauer, 2007). This was a closed adsorption system based on the working pair 'silica gel – H_2O' (Figure 2). Charging the system results from silica gel heating by a heat exchanger with hot water from solar collectors up to 88 °C, the water desorbs and condenses in the condenser of the external cooler. Then the condensate and dry adsorbent are stored separately. During the period of discharge the water evaporates in the evaporator connected to the adsorber with silica gel, which adsorbs the vapor with heat release. The results that were experimentally achieved by 20% less than the theoretical predictions from modeling, the energy storage density being 150 kWh/m^3 silica gel.

The system for the sorption energy storage for centralized heating was developed in Munich (Hauer, 2002): a system designed by ZAE Bayern (Center for Applied Energy Research in Germany, specially for short-term accumulation of solar energy. This is an open adsorption system used as a buffer in a centralized heat supply network for storing 1300 kWh for heating during 14 hours. from a school building with a maximum capacity of 135 kW (Figure 3).

Figure 3. Open adsorption thermal storage system on district heating net in Munich: (a) charge mode for heating; (b) discharge mode for heating
(Hauer, 2002).

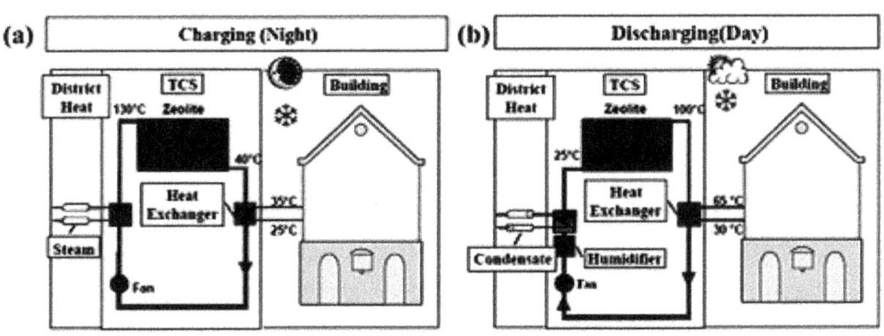

This system stores heat without loss and is used in peak hours and, consequently, makes the building independent of the network during periods of peak demand (Hauer, 2002). The system, developed in 1997 – 1998, uses zeolite 13X as an adsorbent. The heat energy storage density is 124 kWh/ m^3 for heating and 100 kWh/m^3 for cooling with efficiency of 0.9 and 0.86, respectively.

Recently a new design, namely CWT-NT (Chemische Wärmespeicherung – Niedertemperatur: chemical heat storage – low temperature) concept (Kerskes, 2006), for a long-term thermochemical energy storage integrated into a solar thermal combisystem for the composite was proposed for composite of zeolite and salt (Figure 4).

Figure 4. Schematic of the CWS-NT-concept
(Mette, Kerskes, Drück, & Müller-Steinhagen, 2013).

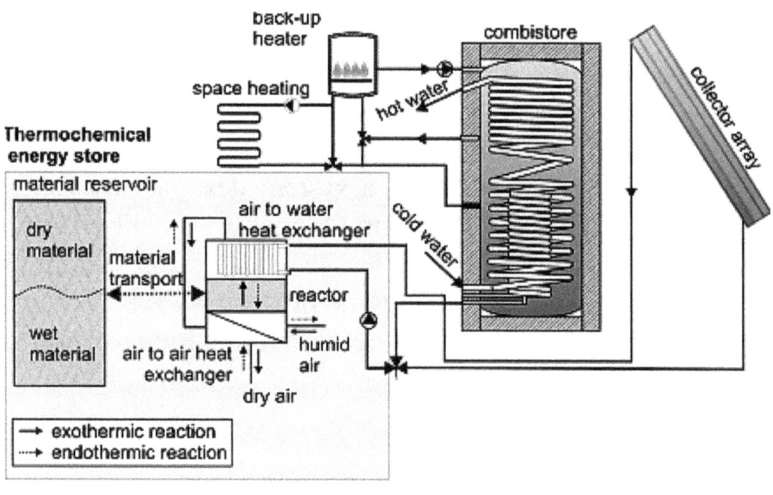

The heat energy storage process is designed as an open sorption/hydration system using ambient or exhaust room air for the charging and discharging process. No further energy for the evaporation of water is needed. Furthermore, good heat and mass transfer between air and storage material is enhanced through forced convection. In the CWS-NT- concept, an external reactor concept has been used. The storage material reservoir is separated from the reactor. So, the heat and mass transfer during the charging and discharging process is reduced to only a small part of the total amount of storage material. The thermal heat capacities and heat losses especially during the regeneration process are reduced. The storage material transport between the material reservoir and reactor is realized by a vacuum conveying system which allows transporting a very gentle material with low parasitic energetic requirements. Developed regeneration strategy allows reducing regeneration temperature from 180 to 130°C.

Closed sorption system based on sodium aluminosilicate $Na_2O \cdot Al_2O_3 \cdot 2SiO_2$ in the form of granule were developed by Stritih and Bombač (2014) (Figure 5). As adsorptive, water is used. System is operated in two stages. During the process of adsorption, adsorption heat is exhale from the adsorber and evaporation heat supplied into the evaporator. If this is not possible, thermodinamical equilibrium of process is achieved and adsorption is stopped. As the adsorbent saturated with water vapour, the process finished and the tank is discharged. During the regeneration process, adsorbent is heated by the heat exchanger. Surface bounded adsorbate is desorbed, adsorbent is dehumidified, heat being transferred from the condenser. When the adsorbent is dried out, desorption mode stops. At that, adsorbate is condensed in the condenser, or the adsorbent and adsorbate are separated. Then, discharge begins again (adsorption).

Figure 5. Working principle of a closed-cycle desorption/adsorption heat storage (Stritih & Bombač, 2014).

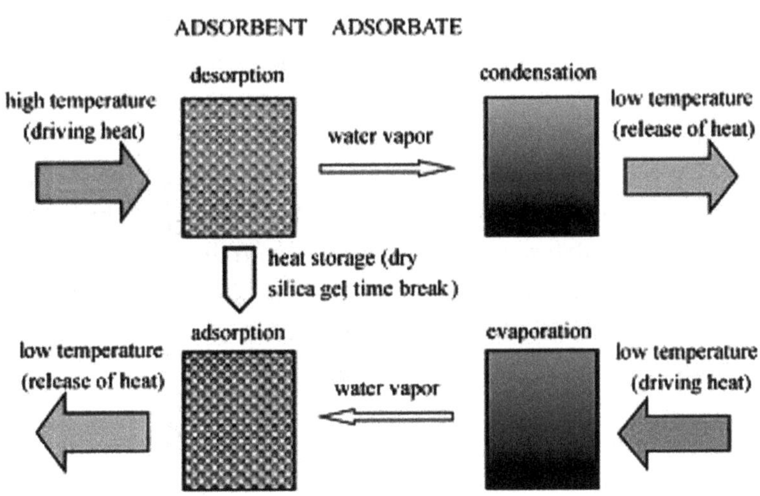

Cycle of adsorption storage process shown in Figure 6 coincides with a cycle of a heat pump, with exception of a 'time break' during the adsorption or desorption processes. This break is corresponded to the time when the storage heat is not needed.

Figure 6. p-T-x diagram of thermochemical storage cycle
(Stritih & Bombač, 2014)

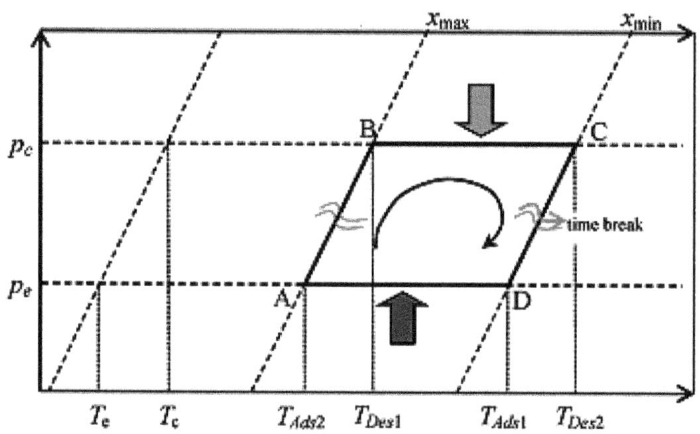

The mass of the adsorbed fluid (an adsorbate) varies between a minimum (line C-D) and the maximum value (line A-B). Adsorption is plotted as line D-A at the evaporation pressure p_e. Desorption is presented by line B-C at a pressure of condensation p_c (Wang, Li, Xia, & Zhang, 2010). This cycle includes the next stages:

1. The first stage is isosteric heating of the adsorbent through the heat exchanger. The valve separating the tank from the evaporator/condenser is closed. The temperature and pressure in the adsorber are increased along the line A – B up to the temperature T_{des1} is reached. At this temperature, pressure becomes equal to the pressure of condensation of the adsorbed adsorbate (point B).
2. At point B the valve between the condenser and the adsorber is opened. Adsorbent is heated (the line between the points B-C) until the adsorbate is removed and the adsorbent is dried. The pressure is equal to the pressure of condensation p_c. Maximum available temperature T_{des2} is achieved. Heat generated by the condensation Q_c is throwed into the environment as a waste heat. This influence on the temperature of condensation T_c

have to be kept as low as possible in order to make optimal desorption. At maximum achieved temperature T_{des2} the value of adsorption is the lowest (x_{min}), so cycle is stopped as the valve between condenser and adsorber closed. When the valve is closed, the adsorbate and adsorbent are separated, and the tank is filled. So, the stored energy can be used without loss, even after a longer period.

3. Depending on energy storage duration and ambient temperature, the temperature of the heat storage falls. Temperature drop and, as a consequence, the pressure drop are displayed by isosteres plotted between points C and D. The temperature may decrease to T_{ads1} or even lower.
4. Discharge process begins from the opening of the valve separating the evaporator and the adsorber. A higher pressure level in the evaporator as compared with the adsorber is established due to the added heat. Heat can be supplied from a variety of sources (solar, earth, etc.). Liquid is evaporated at a pressure p_e and temperature T_e. When the valve opened; adsorbate is re-adsorbed by the adsorbent, adsorption heat being released (line D-A). Heat generated is considered as an useful heat. It can be transferred or used in the heating system through a heat exchanger in the adsorber (Ziegler, 2002).

The system is developed for seasonal heat storage. Operation of adsorption system is cyclical. So, in the summer time tank is being charged, it being discharged in the winter. It can be connected to a heating or a solar cycle. The heat supplied by solar collector is stored in adsorption heat storage unit, and it being used later for house heating. Scheme of closed system for residential

Figure 7. System concept of the sorption reservoir for solar heating
(Stritih & Bombač, 2014)

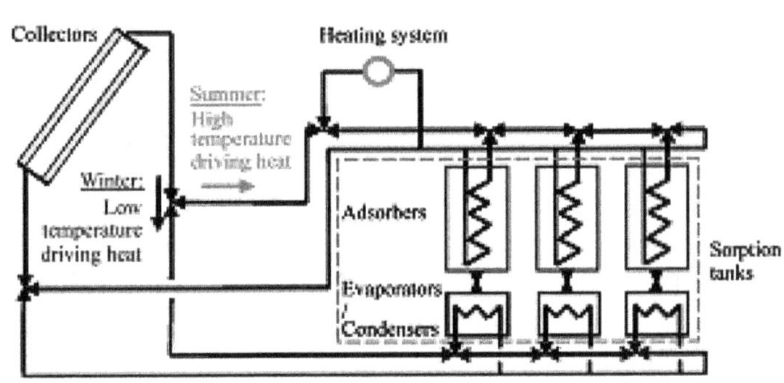

house heating is shown in the Figure 7. It includes adsorbers, sorption tanks, condensers and evaporators. Heat is supplied by solar collector.

Some elements of this system operate only in the summer time mode and others in the winter. When it is sunny and warm and the level of solar radiation is highest (up to 900 W/m^2), the water passed through the solar collector and heats up Solar collector used for heating buildings and sanitary hot water is medium temperature system converting solar radiation into heat. This system works in the range from 45 to 120 °C. Heated water goes through the valve for winter/summer arrangement into sorption tank. Then it is fed from here to the adsorber, where the adsorption proceeds.

The water transfers the heat through the heat exchanger and the dry adsorbent. Water returning from the heat exchanger is cooled and is led into the collector. Described circle is called the primary circuit. In the primary circuit water in the evaporator/condenser is evaporated with the heat from solar collector. In the secondary circuit, due to the heating of adsorbent with hot water, adsorbate separates and is supplied to the evaporator /compensator where it condensates. The heat released during condensation at a sufficiently high potential can be used for preheating domestic hot water. Condensate collected in the evaporator/ condenser is separated from the adsorbent with valve (depending on the design). As the adsorbent is dried, the phase of charging the storage ends. Usually, the discharge phase begins when the temperature of the surroundings is sufficiently low and building heating is required. In the secondary circuit, due to the heating of adsorbent with hot water, adsorbate separates and is led to the evaporator /compensator where it condensates. If the released heat during condensation is at a sufficiently high temperature level it can be used for example to preheat domestic hot water. Condensate, which is collected in the evaporator/ condenser, is separated from the adsorbent with valve (depending on the design). Once the adsorbent is dried, the phase of charging the storage ends. Usually, the discharge phase begins when the temperature of the surroundings is sufficiently low and building heating is required. Then the valve for winter/summer regime is moved to the winter circuit.

During the cold season, the collector does not get sufficient energy and additional energy source, which acts as a reserve, is needed. These can be electric heaters or heat pumps. Heated water flows into an evaporator/ condenser and condensate is evaporated through the heat exchanger. Cooled water from the evaporator/condenser returns to the solar collector or to the additional source. In the secondary circuit, the valve between the adsorber and the evaporator/condenser is open and water vapor is taken up with the adsorbent, adsorption begins, heat being released. This generated heat of adsorption warms up the water in heat exchanger. The heated water is supplied the water heating system (e.g. radiators, pipes for floor heating). Until the maximal value of adsorption is achieved, the adsorption heat can be used. When adsorbent hydrated, the tank is empty and the next container has to be switch (depending on model).

Posern and Kaps (2008) presented the $MgSO_4/H_2O$ system which can employ either a separate reactor or an integrated reactor (Figurer 8). The hydration and dehydration of the heat storage material proceed in the reactor. A separate reactor needs a much smaller heat exchange area. It requires to heat up only the necessitive quantity of storage material. So, the charging and discharging processes become much more efficient.

So, designs of adsorptive heat energy storage systems can be classified as the next types: one-bed and multiple-bed systems; systems with separate and integrated reactors. Obviously, multibed systems involved at least two adsorptive-desorptive units will be most promising when adsorbent with low maximal adsorbate uptake used. They can strongly expand operating period.

Figure 8. Chemical sorption heat storage system using $MgSO_4/H_2O$ [137]. (a) Separate reactor. (b) Integrated reactor
(Posern & Kaps, 2008).

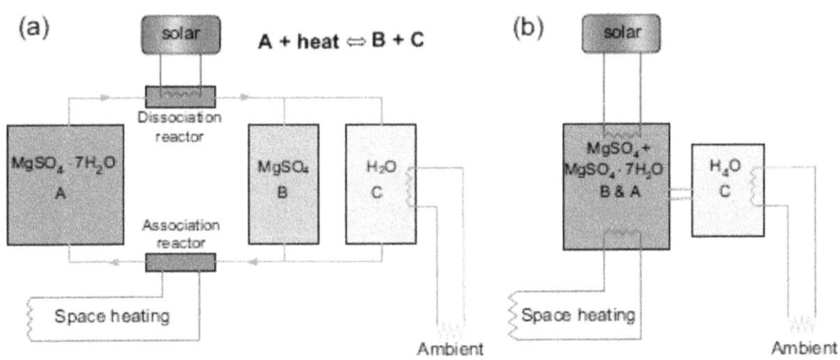

However, their performance is rather complicated. Systems with separated adsorptive-desorptive reactor exhibit the same disadvantage. The most promising type of adsorptive heat storage system is one-bed system with integrated reactor. However, its commercialization is strongly inhibited by properties of conventional adsorbents.

CONSTRUCTIONS OF ADSORPTIVE HEAT STORAGE UNITS

Generally, adsorptive heat storage devices work according to: closed or open-cycle. The working phases of a closed-type adsorptive unit are given on Figure 9 (Vasta et al., 2018). When charged, the adsorber, in which the adsorbtive material is saturated with adsorbate, is regenerated exploiting heat, Q_{des} which supplying from the heat source (Figure 9a). The desorbed vapor condenses in the condenser, and the heat of condensation, Q_{cond}, is dissipated into the ambient medium or delivered to the load, if the temperature level is adequate. When the charging process completed and the adsorbent dried, the connection between condenser and adsorber is closed. In this condition, the system can stored the heat energy for an indefinite time, because of the thermal energy is conserved as the adsorption potential between adsorbate and adsorbent material. To recover the stored thermal energy, as shown in Figure 9b, the connection between the liquid adsorbate reservoir, which at this step acts as an evaporator, and adsorber is opened once again. When discharged, the adsorbate is evaporated, adsorbing heat from the ambient, Q_{evap} and the vapor flows into the adsorber, where it is adsorbed. As the adsorption process is exothermic, the heat of adsorption, Q_{ads}, is released to the load.

Obviously, this process is defined as closed because the adsorbate is continuously condensed or evaporated in a closed system without any mass exchange with the ambient.

Vice versa, the open-type adsorptive heat device continuously exchanges mass (adsorbate) with the ambient medium (Figure 10) (Vasta et al., 2018).

Figure 9. A closed adsorption heat storage cycle: (a) charging phase; (b) discharging phase
(Vasta et al., 2018).

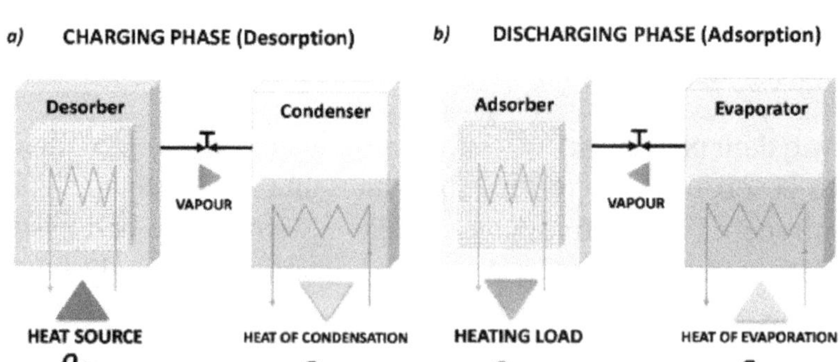

In fact, the two charge/discharge stages are similar to those already described for the closed cycle. In this case, heat is supplied and extracted by fluxing air through the adsorbent bed. Specifically, during the charging (desorption) stage, a hot and dry air flow enters the storage device, causes desorption of the adsorbed water and exits at a lower temperature and higher humidity content.

When discharged (adsorption stage), a humid and cooled air flux is led to the dry adsorbent. So, the adsorption and consequent release of the stored heat are triggered. The stored heat is released by hot and dry air flux exiting the system.

Some clear differences of closed and open adsorptive devices ought to highlight.

- Open adsorptive heat storage unit can use only water as working fluid. Per contra, closed-type adsorptive device can use a different adsorbate, even if water is the most widely employed due to its high latent heat as well as its environmental safety.
- Open adsorptive heat storage devices are strongly affected by the external ambient conditions. Consequently, the higher the absolute humidity of the external air, the higher the heat storage density can be achieved.
- Closed adsorptive heat storage devices when employed under short-term heat storage conditions, can also exploit the heat-pumping effect, related to the energy recovered from the condensation of water vapor during the charging phase. So, the energy storage density is enhanced. Per contra, for the open adsorptive heat storage system, the heat of condensation is dumped to the ambient medium and not recovered.
- Closed adsorptive heat storage devices are usually more complex systems, because they employ different heat exchangers for providing/extracting heat to the adsorber and the evaporator/condenser. Moreover, working in a closed cycle, they need to keep a saturated adsorbate atmosphere, and any air leakage must be prevented, making systems more complex and expensive. Contrastingly, open adsorptive heat storage devices are less complex and expensive systems more suitable for long-term heat storage.
- When operated for short-term heat storage, closed-type adsorptive heat storage units can also exploit the heat pumping effect resulted from the energy recovered the condensation of water vapor during the charging phase. So, the energy storage density is enhanced. Per contra,

for the open-type adsortive heat energy storage devices, the heat of condensation is dumped to the ambient medium and not recovered.

A few examples of adsorptive heat storage devices are considered further.

In adsorptive heat storage devices heat is stored by separation of substances (Li, Qian, Lee, Hwang, & Radermacher, 2014, Hadorrn, 2005). To obtain heat, the substances are poured together.

Figure 10. An open adsorption heat storage cycle: (a) charging phase; (b) discharging phase
(Posern & Kaps, 2008).

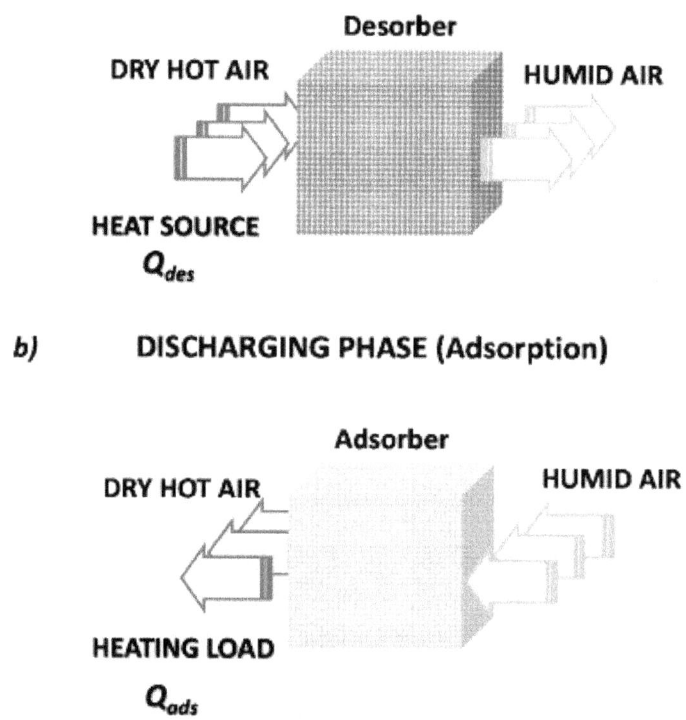

Heat storage devices can work both in closed and open circuits. Figure 11 schematically presents the principle of operation of a adsorption heat storage device of a closed type given in (Fumey, Weber, Gantenbein, Daguenet-Frick, Williamson, & Dorer, 2014).

Figure 11. Principle of operation of close-type heat storage device (Fumey et al., 2014).

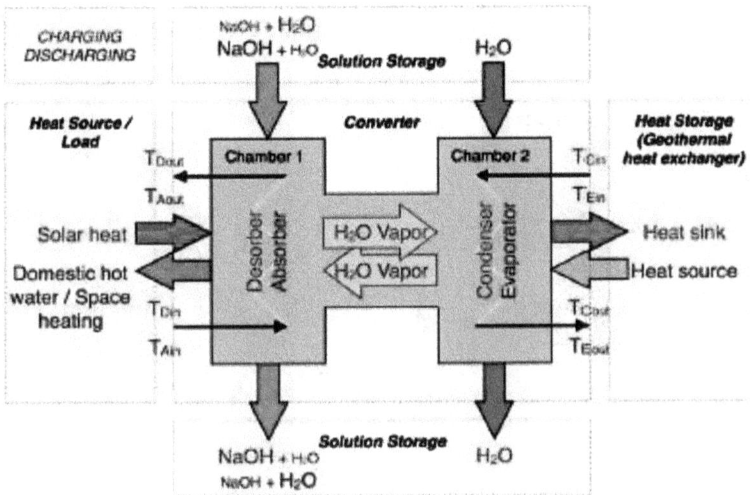

The device includes two tanks: the tank 1 acts as a desorber and a shock-absorber and contains an adsorbent, an aqueous solution of sodium hydroxide, and reservoir 2 being used as a condenser and evaporator, it contains water.

The operating process is based on a reversible chemical reaction:

$$NaOH \cdot (m + n)H_2O + \text{heat«}\ NaOH \cdot mH_2O + nH_2O \quad (1)$$

The working pair of $MgCl_2 - H_2O$ in an open-type storage device is tested (Zondag, Kikkert, Smeding, de Boer, & Bakker, 2013). However, HCl traces were revealed as a consequence of the irreversible decomposition of $MgCl_2 - H_2O$ during the dehydration reaction of the system at elevated temperatures. Therefore, $MgCl_2$ was replaced by zeolite 13X (Santori et al., 2013).

The technological scheme and hardware design of the adsorptive heat storage unit, which operates under an open circuit, is presented in Figure 12. The device consists of two vessels filled with zeolite 13X which contains thermocouples (de Boer, Smeding, Zondag, & Krol, 2009). Through the vessels, bottom-up, air is flowed for charging and discharging. Each vessel has an inlet electric valve to control the airflow through it. To prevent natural convection, a non-return valve is installed at the outlet. The air temperature and humidity at the entrance of the vessel are partly controlled by the location of heat exchangers on the left side of the unit.

Figure 12. Process flowsheet and device design of the adsorptive thermal battery, which operates under an open circuit
((de Boer et al., 2009))

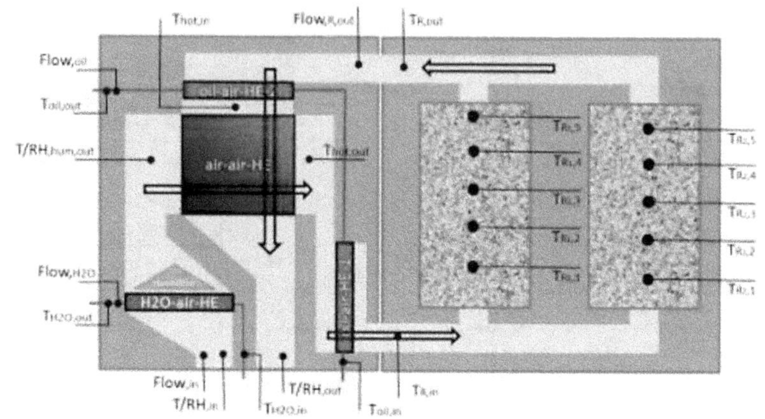

Atmospheric air flows to the device from below, through the lower H_2O-air heat exchanger. The air humidity entering the device is controlled by the first heat exchanger and the system of humidification. Further, the air flow is heated in an air-heat exchanger and then in an oil-air heat exchanger. Here the air temperature of air can be increased by the solar collector. During the adsorption stage when air passed through a vessel with zeolite, water from the air is adsorbed by zeolite and its temperature changes. On the way out of the vessel, the heated air passes through an oil air heat exchanger that removes heat from the air. Heat from the air is exchanged with the inlet air in the air-air heat exchanger.

A fan installed at the inlet at the bottom of the unit controls the airflow. Two thermostats are connected to the device. The first transfers heat into an oil-air heat exchanger at high temperature and the second maintains a constant temperature of the water-air heat exchanger.

Nevertheless, the use of adsorptive heaters in heating and hot water supply systems is an alternative to the use of a centralized heat supply system, but the high temperature of the regeneration of used adsorbents strongly inhibits the using low-potential and alternative energy sources for the desorption stage. Synthesis of composite adsorbents with lower regeneration temperature and higher adsorption capacity is a promising way to solve this problem.

In the framework of the MODESTORE project a concept is developed for a closed system with a spiral heat exchanger layout (Figure 13) loaded with approximately 200 kg of silica gel (Scapino et al., 2017).

Figure 13. MODESTORE project reactor layout
(Scapino, Zondag, Van Bael, Diriken & Rindt, 2017)

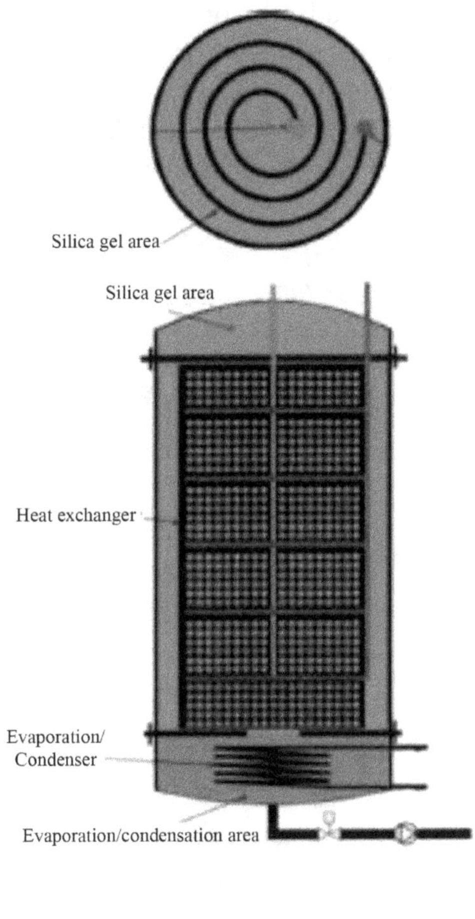

The reactor consisted of a perforated copper sheet with copper pipes soldered to it. A vertical channel in the center was used for vapor diffusion. The heat exchanger used as evaporator and condenser was placed at the bottom of the reactor to have a compact design. Experimental results during the adsorption phase showed a temperature increase of only 5 K, and a maximum transferred power of 400 W, highlighting the bad thermal conductivity of silica gel and the limited temperature increase achievable by the material. Nevertheless, silica gel was shown to be not suitable for heat storage purposes.

The design of heat storage tank developed by Stritih and Bombac (2014) was inspired by solar cooling tubes described in (Wang et al., 2010). The modernization consists in mainly in the fact that the adsorber and the evaporator/condenser are mounted in a single container. It is shown in Figure 14. According to Figure 14, when the valve for supplying the water opened, water in tank evaporates and adsorbs on the surface of the adsorbent. As consequence, the temperature of the adsorbent is rapidly (after 3 min and 15 s) risen up to 50 °C and soon reached maximum at 54.9 °C. Then, the temperature of the adsorbent occurs to be gradually decreased, corresponding to the degree of saturation of the adsorbent. The temperature of the adsorbent

Figure 14. Adsorptive heat storage tank
(Stritih & Bombac, 2014)

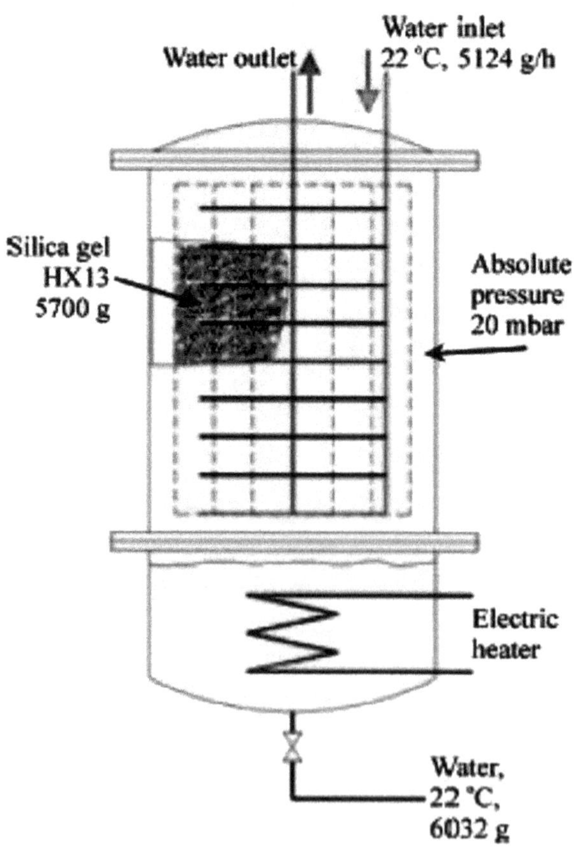

reaches a common point with the temperature of the water flowing from the heat exchanger after about three hours.

Later, the temperature of the adsorbent is lowered to 23.1 °C, which means that the water vapor remained adsorbed, but in very small quantities. The difference between the water supplied into the exchanger, and the adsorbent after 24 h was 1.3 °C.

The temperature of the water on the inlet of the heat exchanger remains practically constant throughout the measurement. The temperature of the water on the outlet of the heat exchanger gradually increases due to the adsorption heat, which is transferred through a pipe of heat exchanger. Maximal temperature 43.1 °C is achieved after 22 min and 45 s. Because of a temperature drop of

Figure 15. Adsorptive heat storage unit. 1 – adsorptive-desorptive reactor; 2 – hydraulic circuit; 3 – sinusoid profile; 4 – adsorbent; 5 – mesh; 6 – condenser; 7 – cooling jacket; 8 – evaporator; 9 – pipe
(Pat. 18578 UA, 1997).

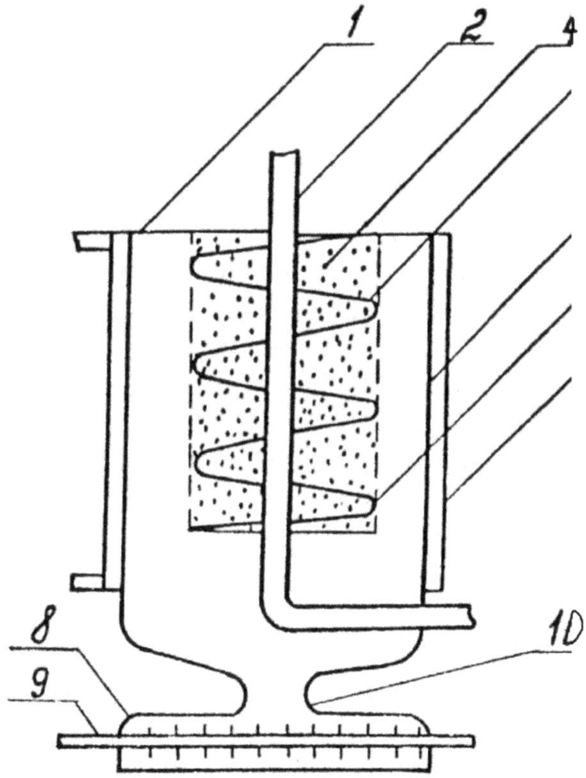

the adsorbent, the amount of heat transferred is less than at the beginning. So, the water temperature at the outlet of the exchanger is lowered.

Similar design of adsorptive heat storage device was developed in Institute of Technical Thermophysics of Academy of Sciences of Ukraine (Pat. 18578 UA, 1997). The adsorption heat storage device (Figure 15) contains a cylindrical reactor of an adsorptive-desorptive type in the center of which the hydraulic circuit 2 is installed. The hydraulic circuit 2 is the axis of a single edge of the profile of the sinusoid 3, which provides a sufficient rate of heat transfer from the granules of the adsorbent 4, located in the cavities of a single edge of the profile of the sinusoid 3 on the outer contour which fastens the bounding net 5, as well as the condenser 6, formed by a cooling jacket 7 disposed on the outer surface of the reactor 1 and an inner surface of the reactor 1 and an evaporator 8 in the shape of a cylindrical tank, where a finned hydraulic circuit 9 is situated.

Evaporator 8 and reactor are connected through the narrowing neck 10 (Pat. 18578 UA, 1997). The hydraulic circuit 2 is welded to the reactor body 1. The adsorbent 4 may be NaX or CaX zeolite, or calcium chloride salt. Water vapor is used as a working fluid in evaporator 8. The hydraulic circuit 2 and the only edge of the profile of the sinusoid 3 are made of Steel 3 and welded to each other, which provides the necessary intensification of the heat transfer from granules of the adsorbent 4 to the working fluid in the hydraulic circuit 2, as well as the required heating time of the adsorbent 4. On the outer contour of the single edge the profile of the sinusoid 3 is fixed restricting the mesh 5, after which the adsorbent 4 is filled into the cavity of the sole rib 3 bounded by the net 5. The condenser 6 has a cylindrical shape consisting of a cooling jacket 7, is welded to the outside the second wall of the reactor 1. The evaporator 8 is made in the form of a cylindrical tank with ribs with a hydraulic circuit made of steel of Steel 3, welded to the reactor 1 through the narrowing neck 10. The adsorptive battery of heat is carried out in two stages. The first stage is adsorption. The adsorbate vapor is fed from the evaporator 8 to the adsorptive-desorptive reactor 1, it being adsorbed by the granules of the adsorbent 4. It results in releasing the heat of adsorption which is transferred from the granule to the granule in the adsorbent layer to the sole profile of the sinusoid 3 and then to heat carrier in the hydraulic circuit 2, used for the needs of the consumer. The second stage is regeneration of the adsorbent. The heat from the heat carrier of the external heat source is transmitted through the hydraulic circuit 2 to the single edge of the profile of the sinusoid 3 to the granules of the adsorbent 4. At that, desorbed water vapor condenses on the inner surface of the reactor 1, in interaction with the heat carrier in the cooling jacket 7. Heat of condensation can be used for needs of consumer. The condensate flows down the inner surface of the reactor 1 into the evaporator 8 through the narrowing neck 10.

This design of a heat storage device has several drawbacks: the fins in the form of trays in an adsorptive-desorptive reactor do not provide sufficiently efficient heat exchange with the adsorbent granules because of hydraulic resistance when vapor penetrated into the adsorbent layer, as well as the lack of reliable thermal contact with the bottom of the mesh resulted from the sinking of granules, which increases the desorption time of the adsorbent layer, the thermal inertia of the reactor. Moreover, the condenser and the evaporator have separate hydraulic circuits for heat transfer. It results in increase of steel intensity of the device and reduces its effectiveness.

Another modified construction of heat storage device involves a condenser made in the form of a cylindrical spiral (Figure 16), which is located inside the reactor and is separated from the hydraulic circuit with ribs by an additionally installed cylindrical partition that does not reach the top cover of the reactor and forms the cavity of the receiver between the reactor wall and the partition wall in the form of flat cylindrical perforated chambers with the adsorbent, between which there is a gap for the passage of refrigerant vapor (Pat. 9922 UA, 1996).

Operation of the heat storage unit proceeds in two modes. The first mode is discharge, i.e. adsorption. Adsorbate vapors are fed from receiver 8 to reactor

Figure 16. Adsorptive heat energy storage unit: 1 – adsorptive-desorptive reactor; 2 – hydraulic circuit; 3 – cylindric perforated camera; 4 – adsorbent; 5 – gap; 6 – condenser; 7 – cylindrical screen; 8 – receiver.
(Pat. 9922 UA, 1996).

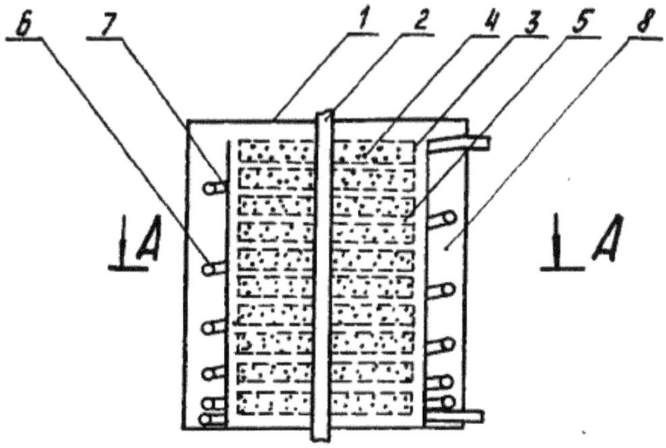

1 due to energy supplied to cylindrical spiral 6, which is an evaporator. Then, they are adsorbed by adsorbent granules 4. At the same time, adsorbent 4 layer releases heat of adsorption transferred from granule to granule to ribs in the form of flat cylindrical perforated chambers 3 to the heat carrier in the hydraulic circuit 2, and used for the needs of the consumer.

The second mode is discharge, i.e. desorption. The heat from external energy source is transmitted by the heat carrier through the hydraulic circuit 2 with fins in the form of flat cylindrical perforated chambers 3 and granules of the adsorbent 4. In this case, water vapor desorbs and condenses on the surface of the cylindrical helix 6, which is a condenser when interacted with a cooling heat carrier circulating inside with the release of heat of condensation, used for the needs of the consumer. Condensate fills the cavity of the receiver 8, flooding as the vapor desorbs by the cylindrical helix 6 and reducing its useful surface as desorption rate decreased.

Another design of open-type adsorptive heat storage device is given in Figure 17 (Pat. 83438 UA, 2013).

Operation is carried out in two stages. At the first stage, the air from the environment by means of the fan 5 is fed to the humidifier 7, where its relative humidity increases to 70 – 80%. Further, the humidified air is heated in an electrical resistive heating element of 8 °C to 30 °C, because at lower

Figure 17. Design of an adsorptive heat storage device operating under an open scheme: 1 – insulated case; 2 – mesh partitions; 3 - insulated cover; 4 - mesh cover; 5 - fan; 6 - guide cover; 7 - humidifier; 8 - resistive heating element; 9 - hydraulic circuit; 10 – heat storage material; 11 – branch pipe for pouring water into a humidifier; 12 - pipe fitting cover.
(Pat. 83438 UA, 2013).

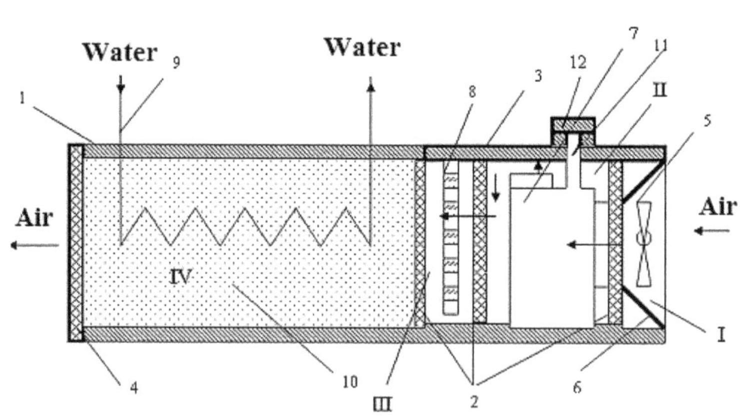

temperatures the adsorption process does not occur. Then the moist air is fed into the heat storage material. After a brief heating of the heat-accumulating material (15 min.), the water from the air is adsorbed with the heat-storage material. This leads to an increase in the temperature of both heat storage material and air. In this case, the air temperature at the outlet from section IV can reach 90 – 115 °C.

On the second stage, regeneration of the heat storage materials proceeds. The humidifier 7 and the electrical resistive heating element 8 are disconnected. Then, when the fan 5 is turned on, the water in the hydraulic circuit (coil) 9 is supplied with a temperature of 40 to 80°C, increasing the temperature of the heat storage material, which promotes its rapid dehydration.

Figure 18. Open-type heat storage device
(Pat. 119167 UA, 2017)

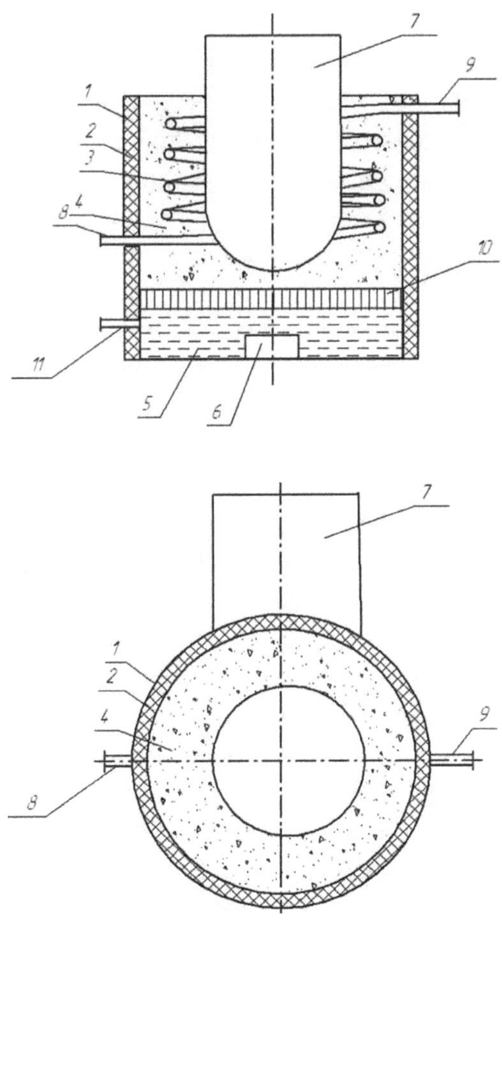

Another design of an open-type adsorptive heat storage device is shown in Figure 18 (Pat. 119167 UA, 2017). This device is developed for autonomous systems of heating, hot water supply and ventilation and utilization of heat of combustion gases.

The device includes a case 1 covered with thermal insulation 2. A hydraulic circuit 3 is installed in the middle of the case on the perimeter, it being filled with liquid heat carrier. The case is filled with adsorptive heat storage material (for example, silica gel, zeolites or composite adsorbents of the type 'silica gel – crystalline hydrate'). 4. The bottom of the case is a reservoir filled with water 5, in which the evaporator is installed. 6. The chimney pipe is installed in the center of the case. 7. Drainage and feeding the liquid heat carrier occur through the pipes 8 and 9. The case is divided into two parts by the mesh partition 10. The water supply to the tank 5 is carried out with the help of the pipe 11.

The device operated in the following way. The heat storage device works according to a two-stage cycle, charging and discharging. In the first stage, the gaseous heat carrier (e.g., combustion gases or heated air) passes through the chimney flue 7, as a result of which the heating and regeneration of the heat-accumulating material is carried out. Next, the liquid heat carrier is supplied through the pipe 8 and discharged through the pipe 9, and the heat storage material is cooled to a temperature beginning of adsorption. The second stage is an adsorption of water with heat storage material. The heat released as a result of this process is discharged by means of a liquid heat carrier, which is supplied and discharged through the pipes 8 and 9, respectively.

In pipe 7, not only combustion gases can be supplied, but also other gaseous heat-carrier, for example, from a solar collector. Water in the hydraulic circuit can be supplied either from a centralized heating or hot water supply system, but also from a solar collector.

The application of this device can almost eliminate the electric energy consumption when the heat storage material regenerated.

Design of closed type heat storage device is given on Figure 19 (Pat. 83436 A UA, 2013).

The heat storage device works in two stages. At the first stage, water in the tank 9 evaporates, the water vapor rises upwards and is absorbed by the heat-accumulating material; this process is exothermic. Water from the water supply system is pumped through the hydraulic contour 6, mounted in the case 1, which heats up to a temperature of 40 – 60°C and is fed to the consumer. The water vapor, passes through the heat storage material, enters the condenser 3, and condenses. The condensate is collected in a plate for collecting condensate 4, then enters the pipe 13 and returns to the tank 9, where it again evaporates. Water losses in the system are filled through the branch pipe 14 on the pipe 13. The process is controlled by measuring the temperature of the heat storage material and water in the hydraulic circuit 6 at the outlet of the heat accumulator. With its decrease begin the second stage.

At the second stage, blinds 15 are closed. Then, on coil 7, water is supplied with a temperature of 40 – 70 ° C, which warms the heat storage material, which leads to desorption of moisture. Desorbed water in the form of a water vapor rises upwards, enters the condenser 3 and condenses. The condensate

Figure 19. Constructive performance of an adsorptive heat storage device operating under a closed scheme: 1 – a heat-insulated case; 2 – mesh partitions; 3 - condenser; 4 – a plate for collecting condensate; 5 – adsorptive-desorptive reactor filled with adsorbent; 6, 7 – hydraulic contours; 8, 14 – branch pipes; 9 – a tank with water; 10 – evaporator; 11, 12 – covers; 13 – pipe; 15 – steel blinds.
(Pat. 83436 A UA, 2013).

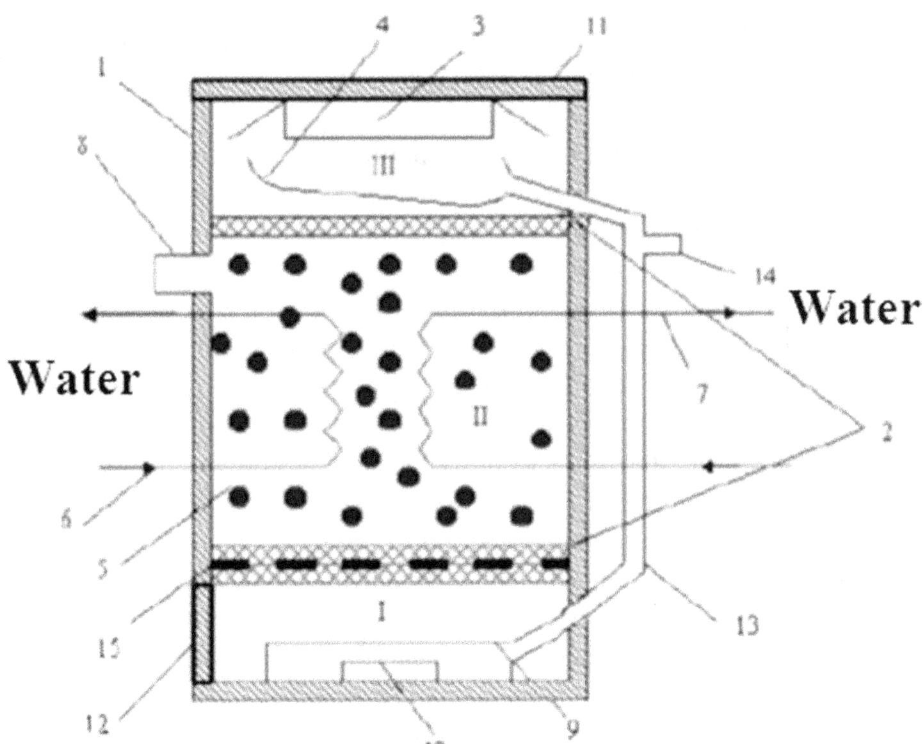

flows into a plate for collecting condensate 4 and into the pipe 13 and returns to the tank 9. Then the device is ready for use again.

In spite of a variety of designs and assignments, all of considered adsorptive heat storage devices are operated in two modes, i.e. charge and discharge. Most of constructions are developed when water used as working fluid. Generally, small-scale adsorptive heat storage devices are under development for domestic application, such as space heating. When such adsorptive devices properly connected with a heat distribution system, they can also provide a thermal energy storage, but also allow to obtain a heat-pumping effect during the winter season and cooling energy during the summer season. So, such a component of heat supply system is promising for fully-integrated heat and cold storage, throughout the year. Nevertheless, efficiency of adsorptive heat storage is strictly limited by operating characteristics of traditional adsorbents. Among the other factors, the main challenge is to enhance properties of available materials, in order to make adsorptive heat conversion systems more competitive.

DETERMINING OF STRUCTURAL CHARACTERISTICS OF ADSORPTIVE HEAT STORAGE UNIT

The effectiveness of heat storage devices is mainly affected by the performance characteristics of the heat storage material, such as: adsorptive capacity, i.e. maximal adsorption or water uptake, heat of adsorption and temperature of regeneration.

During more than decades adsorbents were considered as promising for the absorption, accumulation and conversion of heat energy (Gordeeva, Grekova, Krieger, & Aristov, 2013, Zamengo & Kato, 2017). Heat energy storage in modern heat supply systems is mainly based on sensible heat storage materials. At the same time, devices based on such media are strictly limited in terms of temperature range and energy storage density. More promising are phase-shift heat storage devices, the obstacles for their wide industrial use are the high corrosion activity of the materials and heavy restrictions of the source of thermal energy during regeneration. The most perspective and environmentally propitious are adsorptive heat storage devices which combine the storage of thermal energy on the basis of reverse adsorption – desorption processes with sensible heat and phase-transition mechanism. Nevertheless, the algorithms of calculations of heat supply systems with heat accumulators are

Figure 20. Control-flow chart for calculation of adsorbent mass and volume of adsorptive heat storage device
(Sukhyy, Kolomiyets, Prokopenko, Gavrilko, & Belyanovskaya, 2017)

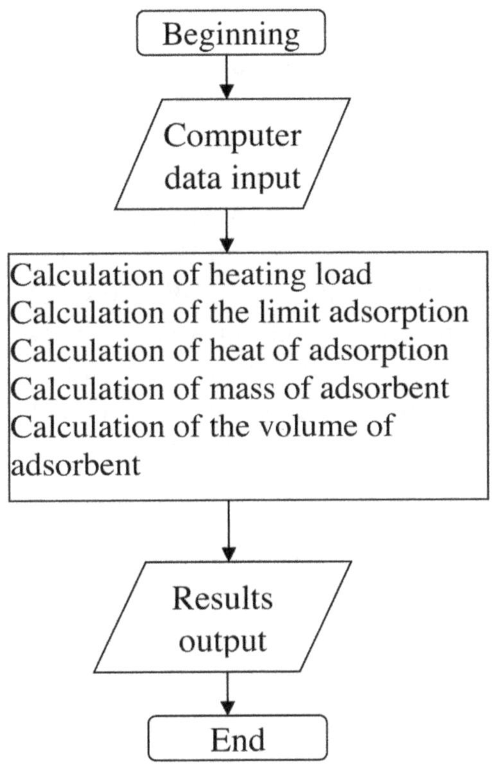

mainly developed sensible heat storage devices (Ganga & Khiminko, 2012). Therefore, the task of developing an algorithm for calculating adsorption heat-storage devices is of crucial importance.

The control-flow chart of the suggested algorithm is given in Figure 20.

Obviously, the first step is focused on calculation of the mass and volume of adsorbent. So, the proposed algorithm ought to involve determining the heat load for warming. The initial data includes the outside space of the building, the area of the heated premises and the material of the external fences, as well as the scheme of decentralized heating system. Then, the load upon space heating Q'_o, W is obviously calculated according to heat balance of the premise (Sokolov, 2001):

$$Q'_o = Q'_h + Q'_{inf} - Q_{ihg}, \tag{2}$$

where Q'_h is heat loss through outer shell, W, Q'_{inf} is heat loss resulted from infiltration, W, Q_{ihg} is internal heat generation, W.

In these conditions a heating system with mechanical circulation is used. The main key element of such systems is a circulation pump. Under its influence the heat carrier is delivered to the heating units, and then to the heat storage device. When such a system installed, there is no need for constant slope of pipes. The external volume of the house is proposed to accept 289.8 m^3.

The glazing factor (the ratio of the surface of the windows to the total surface of vertical outer shell) φ is 0.25. Heat transfer coefficients of walls, windows, ceilings and floors: k_{wall} = 0.49 W/ (m^2 • °C); k_{win} = 1.48 W / (m^2 • °C); k_c = = 0.31 W / (m^2 • °C); k_f= 0.42 W /(m^2 • °C). The coefficients of reducing the design temperature split for walls, windows, ceiling and floor: $\Psi_{wall} = \Psi_{win} = 1$; $\Psi_c = 0.8$; $\Psi_f = 0.6$. The temperature of the internal air $t_{in.c}$ = 18°C, and the temperature of the external air is calculated for heating, t_o = -23°C. The volumetric coefficients of the building are assumed K_{vol} = 5 m^3/m^2, and the specific heat (per 1 m^2 of living space) – q_{ihg} = 20 W/m^2.

The heat losses through outer shells and load upon space heating are calculated according to traditional technique (Safonov, 1985). The load upon space heating is 3929.3 W, which corresponds to about 339.5 MJ per a day.

Mass of the adsorbent corresponded to the load upon space heating is calculated as (Sukhyy et al., 2017)

$$M_{ads} = \frac{Q_0^{'}}{\Delta H_{ads}} \tag{3}$$

where ΔH_{ads} is adsorption heat, kJ/kg.

Adsorption heat is calculated as follows:

$$\Delta H_{ads} = \Delta h \cdot A_{\max} \cdot \frac{1000}{\mu}, \tag{4}$$

where Δh is adsorption heat of substance (water), kJ/mole, A_{max} is maximal adsorption, g/g, μ is molar mass of adsorbed substance, g/mole.

Maximal adsorption can be estimated according to adsorbent properties or as a linear superposition for mechanical mixture:

Figure 21. The scheme of a decentralized heating system of a private house: 1 – heat accumulator; 2 – pump; 3 – expansion tank; 4 – polypropylene tubes; 5 – an aluminum battery
(Sukhyy et al., 2017)

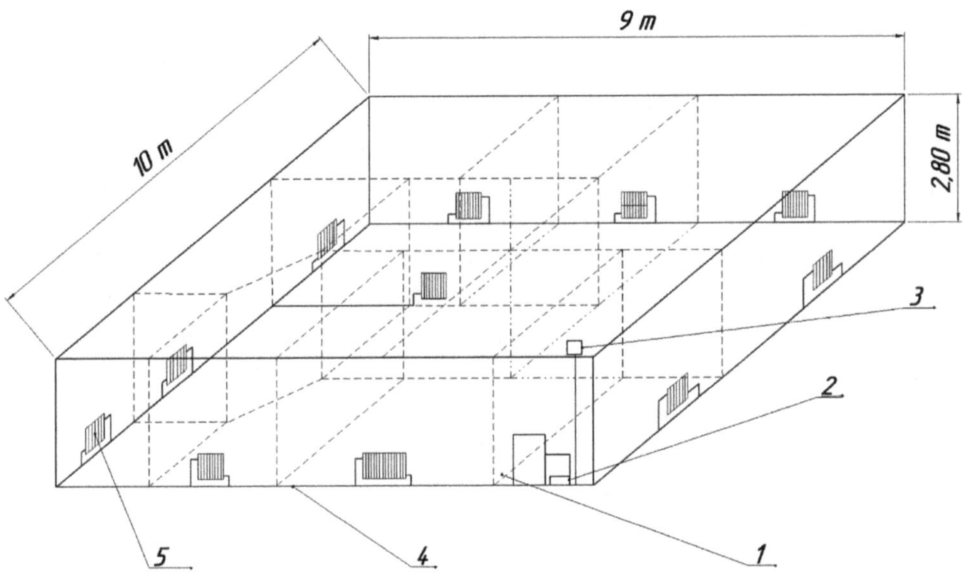

$$A_{max} = \sum_{i=1}^{N} A_i \cdot x_i \tag{5}$$

where A_i is maximal adsorption of i-th component, x_i is its mass content in an adsorbent.

Adsorbent volume V, m^3 can be calculated as:

$$V = \frac{M}{\rho} \tag{6}$$

where ρ is adsorbent density, kg/m^3.

So, proposed algorithm is suggested to calculate the load upon space heating which involves estimation of heat loss through outer shell and resulted from infiltration and internal heat generation, maximal adsorption, heat of adsorption, mass and volume of adsorbent.

The suggested algorithm can be used to determine the optimal composition of the adsorbent, which corresponds to the load upon space heating with minimum total dimensions of the heat storage device or unit. The scheme of a decentralized heating system is shown in Figure 21.

Results of calculation of mass of traditional silica gels are given in Table 1. According to results of calculation, silica gels of domestic and foreign grades are shown to be required at least of 282,9 kg to suit a daily load for space heating. Obviously, this results from low values of adsorption and as a consequence, of adsorption heat. More promising materials are composite adsorbents 'salt inside porous matrix'. The composites 'silica gel – sodium sulphate' or 'silica gel – sodium acetate' synthesized by sol gel technique (Sukhyy et al., 2018, Sukhyy, Belyanovskaya, Kozlov, Kolomiyets, Sukhyy, 2014) are suggested to use as heat storage materials. The results of calculating the heat of adsorption and the mass of the adsorbent required for a load upon space heating of 339.5 MJ per a day and a mass ratio of silica gel and Na_2SO_4 or CH_3COONa 20:80, 40:60, 60:40, and 80:20, are presented in Tables 2 and 3.

Apparently, the mass and volume of the adsorbent layer are determined by the heat of adsorption. As content of sodium sulphate or sodium acetate increased, the volume of the adsorbent layer occurs to be decreased. Consequently, the

Table 1. Adsorption heat and adsorbent mass and volume for load upon space heating of 339.5 MJ per a day

Adsorbent	Maximal Adsorption A_{max}, kg/kg	Heat of Adsorption, kJ/kg	Adsorbent Mass, kg	Adsorbent Volume, m³	Regeneration Temperature, °C
Silica gel KCM (State Standard 3956-76)	0.20	666.67	509.3	0.71	150
Silica gel Type 3A (Ng et al., 2001)	0.32	1066.67	318.3	0.44	100
Silica gel Type RD (Ng et al., 2001)	0.36	1200.00	282.9	0.39	100

Table 2. Results of the calculation of the adsorption heat and the mass of the adsorbent 'silica gel – Na_2SO_4' with a load upon a space heating 339.5 MJ/day

Silica Gel Content, %	Content of Na_2SO_4, %	Maximal Adsorption, A_{max}, kg/kg	ΔH_{ads}, kJ/kg	M_{ads}, kg	V, m³
20	80	1.046	3486.67	97.368	0.135
40	60	0.822	2740.00	123.9	0.172
60	40	0.598	1993.33	170.31	0.237
80	20	0.374	1246.70	272.31	0.378

(Sukhyy et al., 2017)

Table 3. Results of the calculation of the adsorption heat and the mass of the adsorbent 'silica gel – CH_3COONa' with a load upon a space heating 339.5 MJ/day

Silica Gel Content, %	Content of CH_3COONa, %	Maximal Adsorption, A_{max}, kg/kg	ΔH_{ads}, kJ/kg	M_{ads}, kg	V, m³
20	80	0.557	1856.10	182.91	0.277
40	60	0.455	1517.07	223.79	0.339
60	40	0.353	1178.05	288.19	0.437
80	20	0.252	839.02	404.64	0.613

(Sukhyy et al., 2017)

most efficient adsorbents are composites with a mass ratio of silica gel and Na_2SO_4 or CH_3COONa 20:80.

Higher values of the heat of adsorption and, consequently, smaller masses and volumes of adsorbents are stated for the composite 'silica gel – sodium sulphate', which is more effective compared to the composite 'silica gel – sodium acetate'. At the same time, the regeneration temperatures of the composite 'silica gel – sodium sulphate' and 'silica gel – sodium acetate' occur to be 90 ° C and 60°C.

The maximal adsorption and the heat of adsorption of the composite adsorbents 'silica gel – sodium sulphate' and 'silica gel – sodium acetate' occur to suppass the traditional silica gels by a factor of 1.5 – 20. So, to support the same load upon space heating the less mass of suggested composites 'silica gel – sodium sulphate' and 'silica gel – sodium acetate' is required, it being evidently resulted in less size of adsorptive-desorptive unit.

REFERENCES

Bragin, A.A. (2013). Algoritm formirovaniya grafikov elektricheskih nagruzok predpriyatiya s primeneniem akkumulyatornyh batarej v kachestve potrebitelej-regulyatorov moshchnosti. Elektrotekhnicheskie kompleksy i sistemy. Dis. na soisk. kand. tekhn. Nauk, SPb, 130.

De Boer, R., Smeding, S., Zondag, H., & Krol, G. (2009). Development of a prototype system for seasonal solar heat storage using an open sorption process. *Eurotherm Seminar. Advances in Thermal Energy Storage*, *6*, 321 – 327.

Fumey, B., Weber, R., Gantenbein, P., Daguenet-Frick, X., Williamson, T., & Dorer, V. (2014). Development of a closed sorption heat storage prototype. *Energy Procedia*, *46*, 134–141. doi:10.1016/j.egypro.2014.01.166

Ganga, I.G., & Khiminko, A.G. (2012). Teplovoe akkumulirovanie kak sposob povysheniya energeticheskoy effectivnosty system teplosnabgeniya [Heat storage as a method of enhancing of energy efficiency of heat supply system]. *Energosberegenie – Energetika – Energoaudit, 3*(97), 16- 21.

Gordeeva, L., Grekova, A., Krieger, T., & Aristov, Yu. (2013). Composites “binary salts in porous matrix” for adsorption heat transformation. *Applied Thermal Engineering*, *50*(2), 1633–1638. doi:10.1016/j.applthermaleng.2011.07.040

Hadorn, J.-C. (2005). Thermal energy storage for solar and low energy buildings / State of the art, *IEA SHC. Servei de Publicacions Universidad Lleida*, *32*, 123–136.

Hauer, A. (2002). Thermal energy storage with zeolite for heating and cooling applications, *Proceedings of 3rd workshop of annex 17 ECES IA/IEA*, 258 – 267.

Hauer, A. (2007). Adsorption systems for TES/design and demonstration projects, *Thermal Energy Storage for Sustainable Energy Consumption*, 409 – 27.

Hronusov, G. S. (1998). Formirovanie effektivnyh rezhimov elektropotrebleniya promyshlennyh predpriyatij: Nauch.-prakt. izd. *M-vo obshch. i prof. obrazovaniya RF. Ural. gos. gor.-geol. akad.*, Ekaterinburg, CH. 1, 340.

Kerskes, H. (2006). Institut fËur Thermodynamik und WËarmetechnik (ITW). Seasonal sorption heat storage, *DANVAK seminar (solar heating systems – Combisystems – heat storage), DTU Lyngby. Danemark*, *3*, 123–128.

Li, G., Qian, S., Lee, H., Hwang, Y., & Radermacher, R. (2014). Experimental investigation of energy and exergy performance of short term adsorption heat storage for residential application. *Energy*, *65*, 675–691. doi:10.1016/j.energy.2013.12.017

Malyarenko, V.A., Nechmoglod, I. E. (2011). Neravnomernost’ grafika zagruzki energosistemy i sposoby ee vyravnivaniya. *Svitlotekhnika ta elektroenergetika*, *4*, 61 – 66.

Ng, K. C., Chua, N. T., Chung, C. Y., Loke, C. H., Kasiwagi, T., Akisawa, A., & Saha, B. B. (2001). Experimental investigation of silica gel water adsorption isotherm characteristics. *Applied Thermal Engineering*, *21*(16), 1631–1642. doi:10.1016/S1359-4311(01)00039-4

Pat. 119167 UA, IPC (2017.09) F24H 7/02. Heat energy storage device / Belyanovskaya E.A., Lytovchenko R.D., Sukhyy K. M., Sukha I.V., Sukhyy M. P. (Ukraine); State Higher Education Institution 'Ukrainian State University of Chemical Engineering' – Appl. N° u 2017 03884; applied 19.04.17; published 11.09.17, Bul. N° 27, 5 pp.

Pat. 18578 UA IPC F25B 17/00, F25B 27/00 ADSORPTION ACCUMULATOR OF HEAT // Institute of Technical Thermophysics of Academy of Sciences of Ukraine (Ua). Zhuravlenko Viktor Yakovych (UA); Rakitin Oleh Ivanovych (UA); Chalaiev Dzhamalutdin Murshydovych (UA), appl. 94321616, 11.05.1993. Publ. 25.12.1997, bull. N° 6. 2 pp. (in Ukrainian)

Pat. 83436 A UA, IPC F 24 H 7/00. Heat storage device / Sukhyy K. M., Sukhyy M. P., Belyanovskaya E.A., Kozlov Ya.M., Kolomiyets E.V.; (UA); State Higher Education Institution 'Ukrainian State University of Chemical Engineering' – Appl. N° u 2013 03474; appl. 21.03.13; published 10.09.2013, bul. N°17, 5 pp.

Pat. 83438 UA, IPC F 24 H 7/00. Heat energy storage device / Sukhyy K.M., Sukhyy M.P., Belyanovskaya E.A., Kozlov Ya.M., Kolomiyets E.V.; State Higher Education Institution 'Ukrainian State University of Chemical Engineering'.– Appl. u 2013 03488; applied 21.03.2013; published 10.09.2013, bull. N° 17, 4 pp.

Pat. 9922 UA IPC F 25 B 17/08: F 25 B 27/00. Adsorptive accumulator of heat energy // Institute of Technical Thermophysics of Academy of Sciences of Ukraine (Ua). Zhuravlenko Viktor Yakovych (UA); Naumov Sergey Yevgenjevich (UA); Rakitin Oleh Ivanovych (UA); Chalaiev Dzhamalutdin Murshydovych (UA) 94321615,11.05.93, publ. 30.09.98.Bull. N° 3, 3 pp. (in Ukrainian)

Posern, K., & Kaps, C. (2008). Humidity controlled calorimetric investigation of the hydration of $MgSO_4$ hydrates. *Journal of Thermal Analysis and Calorimetry*, *92*(3), 905–919. doi:10.100710973-007-8640-4

Safonov, A. P. (1985). Sbornik zadach po teploficatsii I teplovim setyam [Book of tasks for district heating and heat networks]. Moskow: Energoatomizdat.

Santori, G., Frazzica, A., Freni, A., Galieni, M., Bonaccorsi, L., Polonara, F., & Restuccia, G. (2013). Optimization and testing on an adsorption dishwasher. *Energy*, *50*, 170–176. doi:10.1016/j.energy.2012.11.031

Scapino, L., Zondag, H. A., Van Bael, J., Diriken, J., & Rindt, C. C. M. (2017). Sorption heat storage for long-term low-temperature applications: A review on the advancements at material and prototype scale. *Applied Energy*, *190*, 920–948. doi:10.1016/j.apenergy.2016.12.148

Sokolov, E.Ya. (2001). *Tepoficatsiya i teplovie seti* [District heating and heat networks]. Moskow: MEI.

State Standard 3956-76. (n.d.). *Silica gel for industrial use. Specifications.* Publishing Office of State Standards, 13.

Stritih, U., & Bombač, A. (2014). Description and Analysis of Adsorption Heat Storage Device, Strojniški vestnik –. *Jixie Gongcheng Xuebao*, *60*, 619–628. doi:10.5545v-jme.2014.1814

Sukhyy, K., Belyanovskaya, E., Kovalenko, V., Kotok, V., Sukhyy, M., Kolomiyets, E., Gubynskyi, M., Yeromin, O., & Prokopenko, O. (2018). The study of properties of composite adsorptive materials "silica gel – crystalline hydrate" for heat storage devices. *Eastern-European Journal of Enterprise Technologies*, *91*(1), 52–58. doi:10.15587/1729-4061.2018.123896

Sukhyy, K. M., Belyanovskaya, E. A., Kozlov, Ya. N., Kolomiyets, E. V., & Sukhyy, M. P. (2014). Structure and Adsorption Properties of the Composites 'Silica Gel – Sodium Sulphate', obtained by Sol – Gel Method. *Applied Thermal Engineering*, *64*(1-2), 408–412. doi:10.1016/j.applthermaleng.2013.12.013

Sukhyy, K.M., Kolomiyets, O.V., Prokopenko, O.M., Gavrilko, A.V., & Belyanovskaya, E.A. (2017). Algoritm rozrahunku ob'єmu adsorbcijnogo teplovogo akumulyatora dlya sistemi decentralizovanogo opalyuvannya, *Komp'yuterne modelyuvannya, analiz, upravlinnya, optimizaciya. Zbirnik naukovih prac'*, *2*, 72 – 76.

Vasta, S., Brancato, V., La Rosa, D., Palomba, V., Restuccia, G., Sapienza, A., & Frazzica, A. (2018). Adsorption Heat Storage: State-of-the-Art and Future Perspectives. *Nanomaterials (Basel, Switzerland)*, *8*(7), 522–535. doi:10.3390/nano8070522 PMID:30002331

Wang, D. C., Li, Y. H., Xia, Y. Z., & Zhang, J. P. (2010). A review on adsorption and refrigeration technology and adsorption deterioration in physical adsorption systems. *Renewable & Sustainable Energy Reviews*, *14*(1), 344–353. doi:10.1016/j.rser.2009.08.001

Zamengo, M., & Kato, Yu. (2017). Comparison of magnesium hydroxide/expanded graphite composites for thermal energy storage in cogeneration nuclear power plants. *Energy Procedia*, *131*, 119–126. doi:10.1016/j.egypro.2017.09.463

Ziegler, F. (2002). State of the art in sorption heat pumping and cooling Technologies. *International Journal of Refrigeration*, *25*(4), 450–459. doi:10.1016/S0140-7007(01)00036-6

Zondag, H. A., Kikkert, B., Smeding, S. F., de Boer, R., & Bakker, M. (2013). Prototype the thermochemical heat storage with open reactor system. *Applied Energy*, *109*, 360–365. doi:10.1016/j.apenergy.2013.01.082

Chapter 5
Performance of Adsorptive Heat Storage Devices for Heat Supply

ABSTRACT

The chapter is focused on modelling of performance of adsorptive heat storage devices and estimation of performance of heat storage devices. Two groups of models of adsorptive heat storage units suggested previous researchers are analyzed. The first one is focused on predicting the heat energy storage density, it being based on Dubinin-Polanyi theory. The second one is devoted to analyzing the kinetic of adsorption processes and performance of the adsorber or adsorptive-desorptive reactor filled with traditional adsorbent or salt which forms crystalline hydrates. The key drawback of both groups of models concerns with considering only one stage of exploitation of adsorptive heat storage device in spite of its operating in two-stage mode, that is, alternating discharge (adsorption) and charge (regeneration). It inhibits estimation of efficiency of adsorptive heat storage device basing on full operating cycle and its involving in heat supply system. Two algorithms for estimation of operating parameters are proposed by authors for closed-type and open-type

DOI: 10.4018/978-1-7998-4432-7.ch005

heat storage devices. The algorithm for calculation of operating parameters of closed type adsorptive heat storage device is proposed: calculation of the mass transfer coefficient, adsorption, useful heat, that is, heat of adsorption, determination of the heat input, it being calculated as heat inputs for heating the adsorbent, device housing, water in the tank, evaporation of water in the tank, heating of the adsorbed water and desorption. Then efficiency factor is calculated. The operating characteristics of a closed-type heat energy storage device were studied when the composite adsorbent 'silica gel – sodium sulphate' used. The effect of the humid airflow velocity on the efficiency factor is taken into account by introducing a coefficient equal to the value of the adsorption. An increase in the efficiency coefficient was stated when the velocity and relative humidity of the airflow. It is shown that the humid air flow temperature practically does not affect its value. Having been used the suggested algorithm, the optimal operating characteristics of an adsorptive heat storage device of a closed type based on a composite adsorbent 'silica gel – sodium sulphate' for a private house heating system are revealed to be humid air velocities of 0.6 – 0.8 m/s and relative humidity 40 – 60%. When these operational data applied, the efficiency coefficient is shown to reach the maximum values (about 55%). Algorithm of calculation of operating parameter of open-type heat storage device includes computation of mass transfer coefficient, adsorption, useful heat (heat of adsorption), heat input for heating the adsorbent, device casing, water in the humidifier, evaporation of water, heating the adsorbed water, desorption, and calculating efficiency coefficient. Performance of open-type heat storage device based on the composite adsorbent 'silica gel – sodium sulphate' is estimated. The optimal operating conditions of the heat accumulating device which allow operating with maximal magnitudes of efficiency coefficients 53 – 57% are stated to be humid airflow speed of 0.6 – 0.8 m/s and relative humidity of 40 – 60%. Correlation between efficiency factors obtained by experiments and calculated with suggested algorithm is confirmed. The possibility of reducing the power consumption when heat storage devices applied in 2,4 – 90 times versus decentralized heating systems on basis of solid fuel boiler, gas boiler and electric boiler is stated when open-type sorptive heat storage device used. Results of the study can be used to develop adsorptive storage devices in decentralized heat supply and ventilation systems and adsorption units for utilization of low-temperature waste heat.

NOMENCLATURE

a and c – concentrations of adsorbate in the mobile and stationary phases;
A – adsorption, kg/kg;
A_{max} – adsorption capacity of adsorbent (maximal adsorption), kg/kg;
A_w is the cross-sectional area of the body of the reactor wall;
c – water vapor concentration, mol/m^3;
C_0 and C – absolute humidity of the air flow on the inlet and outlet of the adsorbent layer, kg/m^3;
c_c – molar heat capacity of heat collector;
C_{Pg} – specific heat capacity of air, J/kg·K;
C_{pv} – specific heat capacity of water vapor, J/kg·K;
C_{pw} – specific heat capacity of the rector wall, J/kg·K;
c_s – molar heat capacity of salt;
C_w – specific heat of water, kJ/kg·K,
D – effective diffusion coefficient;
d_i and d_o – inner and outer diameters of the reactor;
D_P, D_M and D_K – the macropore, molecular and Knudsen diffusivity, respectively.
D_z – the axial dispersion coefficient, m^2/s;
F_{hs} – cross-section area of heat storage device, m^2;
H – height of heat storage material (adsorbent), m;
G_0 – maximal loading (adsorbate mass per adsorbent mass), g/g;
k_{LDF} – linear-driving-force kinetics coefficient, s^{-1};
k_{cin} – kinetic constant;
M_{ads} – mass of adsorbent, kg,
n – molar density;
n_h – molar density of the humid air;
n_a – molar density of the dry air;
p – partial pressure of the vapor phase, Pa;
p_S – saturation partial pressure, Pa;
p_t – total pressure, Pa;
p_{eqSG} – equilibrium pressure corresponded to equilibrium solid – gas;
p_v – partial pressure of water;
Q_u – useful heat, kJ;
Q_{in} – heat inputs, kJ;

$Q_{des}, Q^h_{ads}, Q^h_{cas}, Q^h_{cp}, Q^h_{w}, Q^h_{s.w}, Q^h_{s.w}, Q^{ev}_{w}$ – heat inputs for desorption, heating of adsorbent, device casing, coil pipe, water in the evaporator, adsorbed water, water evaporation, kJ.
R – universal gas constant, J/mol · K.
Re_{PM} – Reynolds number in the porous media;
R_i and R_o – heat resistances at the inner and outer sides of the wall,
$R_{cond, tef}$ – conductive heat resistance in the Teflon layer;
R_{conv} – convective heat resistance at the inside wall;
q – averaged amount of adsorbed water per kg of zeolite, mol/kg;
q_{eq} – equilibrium loading of water in solid phase per kilograms of dry zeolite, mol/kg;
t – time, sec;
t_{reg} and $t_{env.}$ – temperatures of regeneration and environment, °C;
T – temperature, K;
T_w – temperature of the reactor wall, K;
T_{amb} – temperature of the ambient air, K;
u is the velocity, m/s;
V_{air} – air volume, m^3;
W – adsorbed specific volume, cm^3/g;
W_0 – maximal value of the adsorbed specific volume, cm^3/g;
w – humid air flow speed, m/s;
X – reaction advancement;
X_{salt} – conversion of the salt;
y_v, – molar water content in the humid air flow;
z – axial coordinate, m;
α_{ads} is the thermal expansion coefficient of the adsorbate, K^{-1};
β is mass transfer coefficient, s^{-1};
β_y, β_p and $\beta_{c.s.}$ – mass transfer coefficients for gas phase, pores and coplanar stirring, s^{-1}
ΔH_{ad} – molar adsorption enthalpy, J/mol;
ΔH_{ads} – specific heat of adsorption determined according, kJ/kg;
ΔH_{des} – specific heat of desorption, kJ/kg;
ΔH_n^v – molar enthalpy of evaporation, J/mol;
Δh_T^0 is the standard enthalpy of the reaction.
ΔS_n – molar adsorption entropy, J/mol·K;
η – efficiency factor;
Φ_v refers to the Vapour sink term.
φ_{ad} – adsorption potential, J/mol;

ϕ_{ad}^{sp} – specific adsorption potential, J/g;
ζ – tortuosity;
λ_c is the thermal conductivity of the heat collector;
λ_{eff} is the effective thermal conductivity;
λ_s is the thermal conductivity of the reactive solid;
λ_{SS} – thermal conductivity of stainless steel layer, W/(m·K);
λ_z –effective thermal conductivity in axial direction, W/(m·K);
μ_{ads} – molar mass of an adsorbate;
μ_{H2O} – molecular mass of water vapor, kg/mol;
ρ_{ads} – adsorbent density, g/cm^3;
ρ_g – air density, kg/m^3;
ρ_w – density of the reactor wall,
ρ_p – particle density, kg/m^3;
τ – adsorption time, s;
ω – gas source or sink due to the reaction (kg/m^3·s);
ω_c – sink or a source of heat collector;
ω_s – sink or a source of salt;
ω_v – sink or a source of vapor,
$\in_b$ – bed porosity;
$\in_{dif}$ – porosity of the mass diffusers;
$\in_p$ – the porosity of particle;
$\in_s$ is the porosity of salt;
$\in_{s0}$ – the porosity of dehydrated salt.
$\in_{s1}$ – porosity of hydrated salt and
$\overline{\rho C_p}$ – overall volumetric heat capacity, J/m^3·K.

MODELLING OF PERFORMANCE OF ADSORPTIVE HEAT STORAGE DEVICES

The most important part of adsorptive heat storage system is obviously an adsorptive heat storage device, i.e. an adsorber or a reactor where adsorbent reacts with adsorbate such as water vapor. In order to achieve a high efficiency of the adsorptive heat storage system, the reactor ought to be optimally designed for the intended reaction and operational condition. This requires an indepth understanding of the physical processes inside the adsorber, which can be done by developing a validated numerical model.

The suggested models can be divided into two groups. The first one is focused on predicting the heat energy storage density. The second one is devoted to analyzing the kinetic of adsorption processes and performance of the adsorber or adsorptive-desorptive reactor filled with traditional adsorbent or salt which forms crystalline hydrates.

According to analyze of the impact of density models on the correlation of computed and experimental values of adsorption enthalpy because calculation of this value is extremely sensitive to the choice of the density model (Nagel, Beckert, Böttcher, Gläser, & Kolditz, 2015).

Adsorption equilibrium can be described by a plane as a functional dependence of adsorption or so called the loading of the adsorbent sample vs. vapor pressure and temperature. By evaluating the change in the molar free enthalpy (i.e. the chemical potential) induced by an isothermal transition of a fluid from the free liquid state to the adsorbed state, the adsorption potential A, J/mol was found as follows:

$$\phi_{ad} = RT\ln\frac{p_s}{p} \tag{1}$$

where φ_{ad} is the adsorption potential, J/mol; p is partial pressure of the vapor phase, Pa; p_s is saturation pressure, Pa; T is the temperature, K; R is the universal gas constant, J/mol · K.

According to the Dubinin-Polanyi theory the two-dimensional description in Eq. (1) can be reduced to a so-called characteristic curve adsorption potential A vs. adsorbed volume per unit mass of adsorbent W (specific volume), i.e. A(W). This characteristic curve is temperature-independent for a given adsorption working pair and expresses the dependence between the differential molar work and the adsorbed specific volume W. The maximal value of the adsorbed specific volume is noted as W_0 and represents the maximum uptake defined by the pore space available for the adsorptive. It can be calculated for zeolites basing on their crystal structure. The temperature dependent adsorbate density is now determined such that it maps the temperature-dependent maximum loading into this temperature independent available volume:

$$\rho_{Ads}\left(T\right) = \frac{G_0\left(T\right)}{W_0} \tag{2}$$

where ρ_{ads} corresponds to the adsorbent density, g/cm^3; G_0 is the maximal loading (adsorbate mass per adsorbent mass), g/g; W_0 is the maximal value of the adsorbed specific volume, cm^3/g.

The exact functional dependence of $\rho_{ads}(T)$ is unknown. Its determination is complicated because of the fact that C_0 is difficult to determine, e.g. due to the influence of capillary condensation at higher loadings. Nagel et al. (2015) recommended for energy storage applications to use (i) a sufficiently general mathematical expression for an accurate fit of the experimentally determined characteristic curve instead of a generic (albeit physically motivated) model and (ii) a (quasi) linear density model as applied in (Hauer, 2002, Mugele, 2005):

$$\rho_{Ads}\left(T\right) = \rho_{10^o} \cdot \left[1 - 3.781 \cdot 10^{-4} \cdot \left(T - 283.15\right)\right] \tag{3}$$

$$\rho_{Ads}\left(T\right) = \frac{\rho_{20^o}}{1 + \alpha_{ads20^o}\left(T - 293.15\right)} \tag{4}$$

The molar enthalpy of adsorption Nagel et al. (2015) suggested to calculate as the sum of the enthalpy of evaporation, the adsorption potential and an entropic term:

$$\Delta H_{ad} = \Delta H_n^v + A - T\Delta S_n \tag{5}$$

where ΔH_{ad} is the molar adsorption enthalpy, J/mol; ΔH_n^v is the molar enthalpy of evaporation, J/mol; A is the adsorption potential, J/mol; T is the temperature, K; ΔS_n is the molar adsorption entropy, J/mol·K.

The molar adsorption entropy is determined by the slope of the characteristic curve:

$$\Delta S_n = \alpha_{ads} \cdot W \cdot \left(\frac{\partial A}{\partial W}\right)_T \tag{6}$$

where α_{ads} is the thermal expansion coefficient of the adsorbate, K^{-1}; W is the adsorbed specific volume, cm^3/g; A is the adsorption potential, J/mol.

The thermal expansion coefficient of the adsorbate directly is determined by the temperature dependence of the density, namely the slope in the ρ-T-diagram, as

$$\alpha_{ads} = -\frac{1}{\rho_{ads}} \cdot \frac{\partial \rho_{ads}}{\partial T} \tag{7}$$

The suggested model had been validated by the experimental data of water adsorption isotherms of binderless granulated 13X zeolite spheres (KÖSTROLITH® 13XBF by CWK Bad Köstritz, Germany) with a mean diameter of 2.0 mm (Nagel, 2015).

Lehmann et al. (2017) modified the model described in work by Nagel et al. (2015). They compared experimentally determined adsorption enthalpies obtained from heat storage density measurements and those computed from adsorption equilibriums using the Dubinin-Polanyi theory for two granulated microporous adsorbents zeolite Na-X and zeolite Ca-X. The main aim was to determine whether the values predicted by the theory match the experimental data and thus are useful for predictive simulations based on the measurement of equilibrium isotherms alone. Since the heat storage density is strongly affected by the process conditions, measurements and simulations were performed when partial vapour pressure varied.

Lehmann et al. (2017) used the fundamentals of the Dubinin-Polanyi theory similarly to (Nagel et al., 2015). Adsorption materials exhibit a bivariate adsorption equilibrium, i. e., for every partial pressure of the adsorptive and every temperature value there exists an equilibrium value of the loading $C_{eq} = C_{eq}(p, T)$. The Dubinin-Polanyi theory splits this functional relation into two simpler ones, $C_{eq} = \rho_{ads}\left(T\right) \cdot W\left(A_m\right)$. The first is corresponded to the adsorbate density ρ_{ads} and the second refers to the adsorbed specific volume W, i. e., the adsorbent pore volume occupied by adsorbate per unit mass of adsorbent, and A_m is the specific adsorption potential, i. e., the change in specific free enthalpy during the isothermal transition of a fluid from the free liquid to the adsorbed state J/g, . The specific adsorption potential can be calculated by

$$\phi_{ad}^{sp} = \frac{R}{\mu_{ads}} T \ln \frac{p_s}{p} \tag{8}$$

where R denotes the universal gas constant, J/mol·K; μ_{ad} is the molar mass of the adsorbate, p_S the saturation partial pressure, Pa and p the partial pressure of the adsorbate atmosphere, Pa. The central assumption of the Dubinin-Polanyi theory is, that the relation adsorbed specific volume vs. specific adsorption potential, i.e. $W(\phi_{ad}^{sp})$ also denoted as characteristic curve – is independent of temperature, which has been empirically confirmed for many adsorption working pairs.

The adsorbate density ρ_{ads} is different from the density of the corresponding free liquid. Specifically, it is defined as the ratio of the temperature-dependent maximum loading, C_0, and the maximal value of the adsorbed specific volume, i.e. the temperature-independent specific pore volume available, W_0 according to Eq. (2). It is not easily accessible by direct measurements, and several different adsorbate density models have been proposed. Lehmann et al. (2017) supposed that the most adequate one is the density model introduced by Hauer (2002).

Besides providing a simple description of the adsorption equilibrium, the Dubinin-Polanyi theory can be used to compute adsorption heats solely from adsorption equilibrium data, i. e., without the need for additional calorimetric experiments. Within the theory, the adsorption enthalpy comprises three parts: the enthalpy of evaporation, the adsorption potential, and an entropic term according to Eq. (5).

The change in entropy is related to the slope of the characteristic curve as determined by Eq. (6). The thermal expansion coefficient of the adsorbate directly follows from the temperature dependence of the density, namely the slope in the ρ-T-diagram, according to Eq. (7).

To optimize and up-scale from the lab-scale reactor size, an adsorber model is needed to predict the thermal performance. Because of long term calculations of the energy provided by the thermochemical system, a simple and fast model requiring small computational power is required. Generally, on reactor scale, adsorption models are developed of water into zeolite 13X in closed sorption heat storage systems for various type of applications. Examples are a waste heat adsorption cooling system associated with the coupled heat and mass transfer mechanisms in the adsorbent (Zhang & Wang, 1999), and the coupled heat and mass transfers occurring in closed storage systems for entire cycles under extreme conditions (Duquesne, Toutain, Sempey, Ginestet, & del Barrio, 2014). These models are based on a one dimensional model first introduced in (Sun, Amar, & Meunier, 1995) for the heat pump process, and further developed into a two dimensional model in (Amar, Sun,

& Meunier, 1996). Extensive experimental and numerical investigations on the water vapor adsorption into zeolite 13X in an open system are reported in (Mette, Kerskes, Drück, &Müller-Steinhagen, 2014). All of these reactor models provide local thermodynamic equilibrium which is maintained for heat and mass transfer between the solid and gas phases.

The characteristics of the sorption process are described in a more detailed reactor model by a material model. Several material models are developed for the water vapor adsorption into zeolite. An elaborate model for the equilibrium amount of adsorbed water was developed by Ahn and Lee (2004) when the effect of capillary condensation on adsorption and thermal desorption of water in zeolite 13X studied. Dawoud, Vedder, Amer and Dunne (2007) suggested a nonisothermal model for adsorption of water vapor into a consolidated zeolite layer and studied the temperature dependence of the micro-pore diffusion. In all above-mentioned works, a Linear Driving Force (LDF) model is used as the kinetics model. The LDF-model is usually used for adsorption modeling because of its simplicity and physical consistency (Sircar, 1983). More omnifaceted kinetic models have been developed to consider fundamental peculiarities of solid-state reactions in a grain scale, but they are quite cumbrous to handle at a continuum-scale calculation (Lan, Zondag, van Steenhoven, & Rindt, 2015).

To study the thermal dynamics of the heat storage device zeolite 13X based on a non-isothermal and non-adiabatic model including mass and energy transfer equations has been developed by Gaieni, Zondag and Rindt (2016). At first, the heat and mass balances used in the model are set first. Then the Linear Driving Force (LDF) kinetics model and its coefficient values are explained.

The equations in the model describing the dynamics of the a fixed-bed reactor filled with zeolite 13X are based on the next assumptions: (a) the flow in the reactor can be described by an axially dispersed plug flow model (Delgado, 2006); (b) heat transfer in radial direction through the bed and the reactor walls can be modeled as a one-dimensional resistance model; (c) the gas phase behaves as an ideal gas and is in thermal equilibrium with the solid phase; (d) the adsorbent beads have identical characterizations and the bed properties are uniform.

The temperature distribution problem is defined by two partial differential equations, one for the bed temperature (T) and the other one for the wall temperature (T_w), connected by heat resistances. A schematic view of the model is shown in Figure 1. The bed and wall temperatures are calculated in axial direction in the nodal points and are in contact through the resistances in radial direction.

Figure 1. The temperature distribution problem is determined by two PDEs for the bed temperature T and the wall temperature T_w connected by heat resistances (Gaeini et al., 2016).

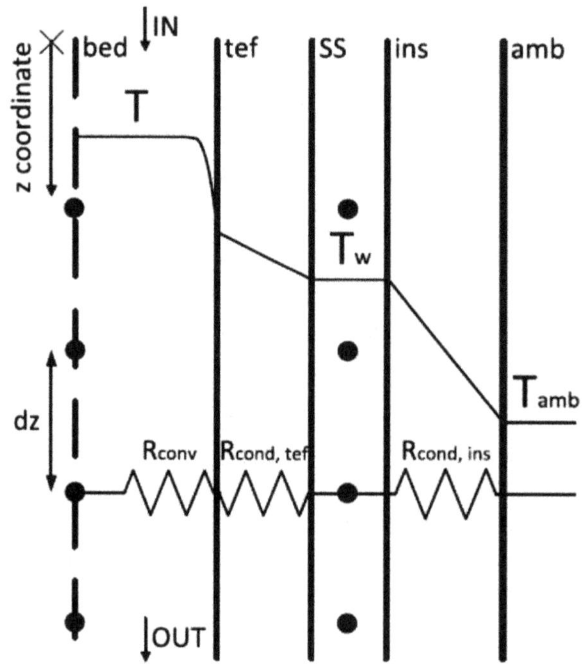

According with these assumptions, the governing heat and mass balances can be written by a set of partial differential equations along the vertical axial coordinate (z) and the time (t). The overall equation of mass balance is based on the air density, ρ_g:

$$\frac{\partial \rho_g}{\partial t} + u \cdot \frac{\partial \rho_g}{\partial z} = 0 \tag{9}$$

where u is the velocity, m/s; t is time, s; ρ_g is the air density, kg/m^3; z is the axial coordinate, m.

The water mass balance is expressed on basis of the water vapor concentration, c:

$$\frac{\partial c}{\partial t} + u \cdot \frac{\partial c}{\partial z} - D_z \cdot \frac{\partial^2 c}{\partial z^2} + \frac{\left(1 - \epsilon_b\right)}{\epsilon_b} \rho_p \cdot \frac{dq}{dT} = 0 \tag{10}$$

where c is water vapor concentration, mol/m^3; t is time, s; ϵ_b is the bed porosity, ρ_p is the particle density, kg/m^3; q is the averaged amount of adsorbed water per kg of zeolite, mol/kg and D_z the axial dispersion coefficient, m^2/s. The axial dispersion coefficient Gaeini et al. (2016) derived from Gunn's correlation (Delgado, 2006).

The energy balance in the bed is given as follows:

$$\overline{\rho C_p} \frac{\partial T}{\partial t} + \epsilon_b \rho_g C_p u \frac{\partial T}{\partial z} - \lambda_z \frac{\partial^2 T}{\partial z^2} - \left(1 - \epsilon_b\right) \rho_p \cdot \frac{dq}{dT} \Delta H + \frac{4}{\pi d_i^2} \cdot \frac{\left(T - T_w\right)}{R_i} = 0 \tag{11}$$

where $\overline{\rho C_p}$ is the overall volumetric heat capacity, J/m^3·K; T is temperature, K; t is time, s; u is the velocity, m/s; ϵ_b is the bed porosity, ρ_p is the particle density, kg/m^3; q is the averaged amount of adsorbed water per kg of zeolite, mol/kg; C_{Pg} is specific heat capacity of air, J/kg·K; ρ_g is the air density, kg/m^3; z is the axial coordinate, m; λ_z is effective thermal conductivity in axial direction, W/(m·K); ΔH is the adsorption enthalpy, J/mol; T_w is the temperature of the wall, K; d_i is the inner diameter; R_i is the inner heat resistance, m·K/W.

According to Gaeini et al. (2016), the effective thermal conductivity in axial direction, λ_z, was estimated by the model of Zehner and Schlünder (Kandula, 2011) which is satisfactory over a broad range of solid-to-fluid thermal conductivities and solid fractions.

The overall volumetric heat capacity, $\overline{\rho C_p}$, involves several terms, which are the heat capacity of the air in the bed voids (first term), the heat capacity of the air in the beads pores (second term), the heat capacity of the solid (third term) and the heat capacity of the adsorbed water (fourth term), it being dominated by the third and forth terms because of the higher density of the solid phase compared to the gas phase as shown by the next equation:

$$\overline{\rho C_p} = \epsilon_b \rho_g C_{pg} + \left(1 - \epsilon_b\right) \epsilon_p \rho_g C_{pg} + \left(1 - \epsilon_b\right) \rho_p C_{pv} + \left(1 - \epsilon_b\right) \rho_p C_{pv} \mu_{H2O} \tag{12}$$

where μ_{H2O} is the molecular mass of water vapor, kg/mol; C_{Pg} is specific heat capacity of air, J/kg·K; ρ_g is the density of air, kg/m^3; ρ_p is the density of particle, kg/m^3; q is the averaged amount of adsorbed water per kg of zeolite, mol/kg; C_{pv} is the specific heat capacity of water vapor, J/kg·K; ϵ_p is the porosity of particle. In Eq. (12), the last term represents the heat loss from the inside reactor wall, at inner diameter of the reactor d_i, per unit of reactor volume.

The heat loss through the reactor wall was simulated when wall temperature (T_w) and the thermal mass of the reactor wall in another energy balance for the reactor wall considered Gaeini et al. (2016) as:

$$\rho_w C_{p,w} \frac{\partial T_w}{\partial t} - \lambda_{SS} \frac{\partial^2 T_w}{\partial z^2} = \frac{T - T_w}{R_i A_w} - \frac{T_w - T_{amb}}{R_0 A_w} \tag{13}$$

where T is the temperature, K; T_w is the temperature of the reactor wall, K; T_{amb} is the temperature of the ambient air, K; λ_{SS} is the thermal conductivity of stainless steel layer, W/(m·K); ρ_w is the density of the reactor wall, C_{pw} is the specific heat capacity of the rector wall, J/kg·K; d_i and d_o correspond to the inner and outer diameters of the reactor, R_i and R_o are the heat resistances at the inner and outer sides of the wall, and A_w is the cross-sectional area of the body of the reactor wall.

The heat resistance at the inner side of the wall (R_i) is calculated as a sum of the conductive heat resistance in the Teflon layer and convective heat resistance at the inside wall ($R_i = R_{cond, tef} + R_{conv}$), see Figure 1. According to data given in (Ahn, Kim, & Lee, 2004) on heat transfer in a packed bed, the relation between the Nusselt and Reynolds numbers is assumed to be similar as the case of heat transfer of flow in a circular tube. So, in the model of Gaeini et al. (2016), the convective heat resistance at the inside wall is calculated using the Nusselt number relation presented in Eq. (5.6):

$$Nu_{PM} = 0.309 \, \mathrm{Re}_{PM}^{0.8} \, \mathrm{Pr} \tag{14}$$

where Re_{PM} is the Reynolds number in the porous media which is defined as ($Re_{PM} = d_p \cdot u \cdot \rho_g / \mu_g \cdot (1 - \epsilon_b)$) and Pr is the Prandtl number.

The heat resistance at the outer side of the wall (R_o) is mainly determined with the conductive heat resistance of the insulation layer, the convective heat resistance at the outside surface of the insulation layer being neglected ($R_o = R_{cond;ins}$), see Figure 1.

Similarly to many adsorption systems, Gaeini et al. (2016) used to describe the diffusion-controlled kinetics the linear driving force-approximation, first introduced by Glueckauf (1955):

$$\frac{dq}{dt} = k_{LDF}\left(q_{eq} - q\right) \tag{15}$$

where q and q_{eq} are the adsorbed and equilibrium loading of water in solid phase per kilograms of dry zeolite, and k_{LDF} is the LDF-kinetics coefficient.

In real adsorption systems, the overall kinetics is affected by several mass transfer resistances, the resulting kinetics coefficient k_{LDF} [s^{-1}] being a cumulative parameter that considers the external film resistance, the macropore resistance and the micropore resistance (Ruthven, 1984):

$$k_{LDF} = \left(\frac{d_p}{6k_f} + \frac{d_p^2}{60 \in_b D_p}\right)^{-1} \tag{16}$$

The first addend on the right hand side represents the mass transfer resistance in the external film of fluid around the particle. The value of k_f, can be calculated from the Sherwood number expressed by Wakao and Funazkri in the range of Reynolds numbers from about 3 to 10,000 (Wakao, & Funazkri, 1978). Nevertheless, the dominant mass transfer resistance for the water vapor adsorption of zeolite 13X is the macropore diffusion resistance (Ruthven, 1984) presented by the second term on the right hand side of the equation above, in which D_P is given by Thomas and Crittenden (1998):

$$D_p = \frac{1}{\zeta}\cdot\left(\frac{1}{D_M} + \frac{1}{D_K}\right)^{-1} \tag{17}$$

where ζ is tortuosity, and D_P;D_M and D_K are the macropore, molecular and Knudsen diffusivity, respectively. The molecular diffusivity for the binary gas mixture of air and water can be evaluated by Chapman–Enskog theory (Chapman & Cowling, 1970), and the Knudsen diffusion by the kinetic theory (Thomas & Crittenden, 1998).

The above mentioned kinetics coefficient k_{LDF} is usually only temperature dependent, however in some types of zeolite it is also concentration dependent

(Thomas & Crittenden, 1998). As with the zeolite systems the concentration dependency of the kinetics arises mainly from the thermodynamic correction factor; calculations by the corrected coefficient lead to better results (Ruthven, 1984):

$$k^{*}_{LDF} = k_{LDF}\left(\frac{\rho_p RT}{\epsilon_p}\cdot\frac{\partial q_{eq}}{\partial p_v}\right) \tag{18}$$

where p_v is the pressure of the water vapor.

So, according to Gaeini et al. (2016), the final set of equations includes four partial differential equations:

- Eq. (10) for modeling the water concentration in the gas phase (*c*);
- Eq. (11) for modeling the bed temperature (T);
- Eq. (13) for modeling the wall temperature (T_w);
- Eq. (15) for modeling the water loading in the solid phase (q).

Efficiency of the reactor was determined as a ratio of heat which leaved the reactor by means of air flow through the reactor ($Q_{convection}$) and heat energy produced during process of adsorption (Q_{ads}). The numerically calculated and experimentally measured temperatures are compared and are in a good agreement.

Similarly, the results of simulation of an adsorbtive heat storage unit based on zeolite 13XBF for refrigerating purposes are given in (Semprini, Asenbeck, Kerskes, Drück, 2017). The operation of the adsorber is modelled under the following assumptions:

- Equilibrium condition is described by the Dubinin-Astakhov model (Do, 1998)
- Adsorption kinetics follows the Linear Driving Force model (Kandula, 2011, Sircar, 1983)
- Adsorption enthalpy is corresponded the curve reported by Mette (2014)
- Heat capacity of the zeolite conforms the description reported by Vučelic and Vučelic (1985)
- Flow in porous media is computed according to Darcy's law
- Air flow is laminar

Michel, Neveu and Mazet (2014) developed a general and original two-dimension (2D) model for adsorber used hydrated salt and water as working pair. It considers heat and mass transfer within the reactive bed and with the moist air flow. Due to additional assumptions, this model was modified to represent the same porous bed reacting with pure vapor. Both dynamic models include heat balances, mass balances and a kinetic equation.

The model is based on the following assumptions:

(a) The reactive bed is pseudo-homogeneous, i.e. the reactive solid and gas phases (vapor or moist air) are assumed at the same temperature. This supposition is justified by the high solid/gas heat transfer rate which usually observed in such porous media, with small grain sizes (<0.5 mm).

(b) mass transfer in the porous bed is described the Darcy law. The inertial effects can be neglected because the vapor velocity is usually low in such porous bed of thermochemical systems,

Figure 2. Schematic representation of a global thermochemical reactor: a) open system (operating with moist air), b) closed system (operating with pure vapor) *By (Michel, Neveu, & Mazet, 2014).*

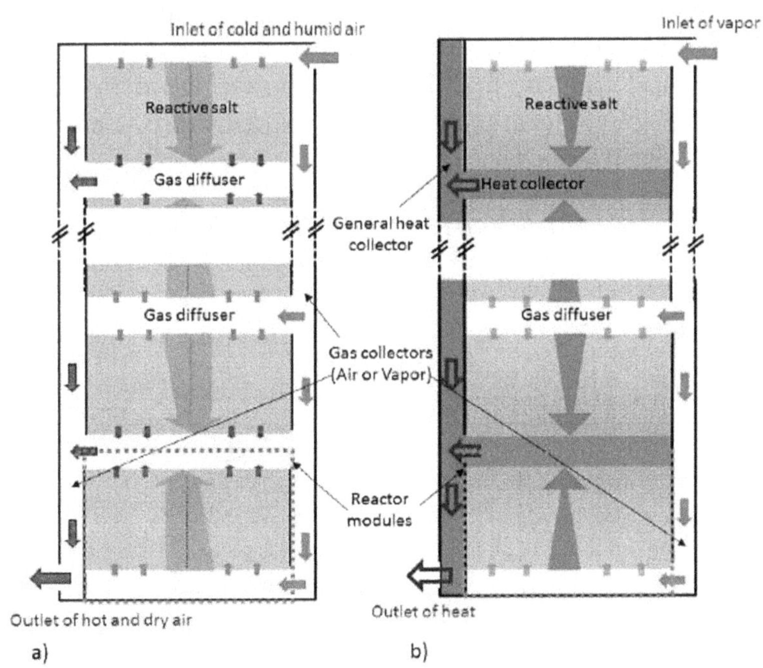

(c) heat transfer processes in the porous media are conductive, according to Fourier law, and convective.
(d) the kinetics of the reversible solid gas reaction involved in such a process is affected by the reaction advancement and on the equilibrium drop. The adsorber was designed according to a modular concept o by stacking several rectangular modules. Each module includes:
   - a reactive layer inserted between two gas diffusers, one connected to the air input and the other to the air output, for the open reactor (see Figure 2a),
   - a reactive bed inserted between a plate heat exchanger and a gas diffuser on the other side for the closed reactor (see Figure 2b).

All the modules are connected in parallel to the heat and mass input and output. Therefore, the study can be restricted to only one rectangular module

For an open thermochemical reactor, gas diffusers supply or collect the humid air flow at the inlet/outlet of the reactor. Humid air supplied to the reactor from the inlet diffuser side, and then it flows through the reactive layer from the top to the bottom side (Figure 3). Depending on the thermodynamic operating conditions, it results in either an exothermal hydration, or an endothermal dehydration of the salt. When the hydration step proceeded, a part of the reaction heat is transferred across the bed by conduction, most of it being transferred to the humid air flow. Thus, the temperature of the humid air flowing out by the bottom side of the bed is higher and air humidity is lower as compared with the inlet humid air flow (Figure 3a). On the other hand, during the dehydration step, the humid air flow provides heat required

Figure 3. Schematic representation of an open thermochemical reactor (operating with moist air) a) Hydration step, b) Dehydration step. (Numbers refer to boundary conditions described below)
(Michel, Neveu, & Mazet, 2014)

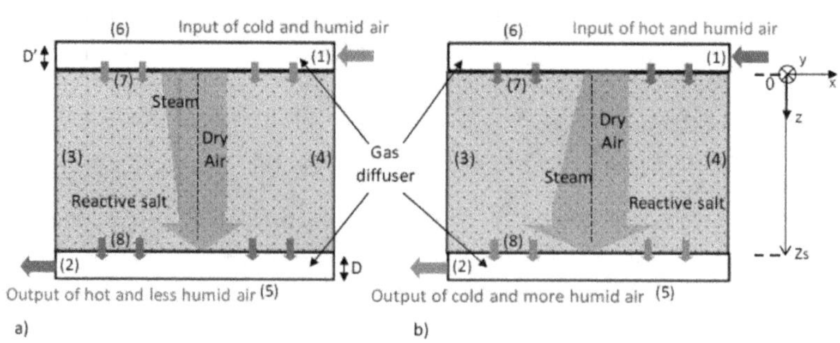

by the reaction in the bed. So, the outlet flow of humid air is colder and wetter than the inlet one (Figure 3b). Lastly, the humid air flowing out of the reactive bed is collected by the outlet diffuser and evacuated.

For the dynamic modeling of this open reactor an additional assumption have to be postulated:

(e) Humid air is a mixture of two ideal gases: dry air and water vapor. This assumption is usually used for this range of pressure (atmospheric pressure), temperature (<100°C) and partial pressure of water vapor (<10 kPa).

The set of equations required by this modeling is described below.

The molar conservation is applied to the moist air flow:

$$\frac{\partial\left(\in n_h\right)}{\partial t} = n - \nabla \cdot n_h u \tag{19}$$

where $\in$ is the porosity; t is time; n_h is molar density of the humid air; n is molar density; u is the velocity of humid air which evaluated through Darcy law:

$$u = -\frac{k_s}{\mu}\nabla p \tag{20}$$

where k_s is permeability of the salt; μ is air viscosity; p is pressure.

According to assumption (e), Eq. (19) can be split on two molar balances, the first for dry air and the second for vapor:

$$\begin{cases} dry\,air \dfrac{\partial\left(n_h\left(1-y_v\right)\in\right)}{\partial t} = \dfrac{\partial\left(\in n_a\right)}{\partial t} = -\nabla n_h u 1 - y_v \\ Vapor \dfrac{\partial\left(n_h y_v \in\right)}{\partial t} = \dfrac{\partial\left(\in n_v\right)}{\partial t} = \omega - \nabla n_h y_v u \end{cases} \tag{21-22}$$

where n_a is a molar density of the dry air; ω is the gas source or sink due to the reaction (kg/m³·s); y_v, is the molar water content in the humid air flow, which depends on the water partial pressure p_v as follows:

$$p_v = y_v p_t \tag{23}$$

where p_t refers to the total pressure.

When the molar balance for vapor developed, molar balance of the water contained in the moist air flow is obtained:

$$\in n_h \frac{\partial y_v}{\partial t} + n_h u \nabla y_v = 1 - y_v n \tag{24}$$

So, the molar balance in the reactive bed is given by Eq. (19) and Eq. (24). Indeed, it is easier to work with variables and balances involving the total flow of moist air on one hand (Eq. (19)) and the molar concentration of vapor on the other hand (Eq. 24).

For the reactive bed, this system can be written as:

$$\begin{cases} \dfrac{\partial(\in_s n_h)}{\partial t} = n - \nabla n_h u \\ \in_s n_h \dfrac{\partial y_v}{\partial t} + n_h u \nabla y_v = \left(1 - y_v\right) n \end{cases} \quad (25\text{-}26)$$

with $\in_s$ is the porosity of the reactive bed.

Eq. (25) and (26) also can be applied for the gas diffusers as the source term cancelled:

$$\begin{cases} \dfrac{\partial(\in_{dif} n_h)}{\partial t} = -\nabla n_h u \\ \in_{dif} n_h \dfrac{\partial y_v}{\partial t} + n_h u \nabla y_v = 0 \end{cases} \quad (27\text{-}28)$$

with $\in_{dif}$ refers to the porosity of the mass diffusers.

According to assumptions (a) and (c), the energy balance of the solid porous layer writes as:

$$n_s c_X + \in_s n_h c_h \frac{\partial T}{\partial t} = \nabla \cdot \lambda_{eff} \nabla T - n_h c_h u \nabla T + q \tag{29}$$

where q is sink or source of heat

$$q = \Delta h_T^0 \cdot \frac{dX}{dt} \cdot n_s = -n \cdot \frac{\Delta h_T^0}{\nu}$$

The three right side terms correspond respectively to the conductive heat flow across the porous solid layer, the enthalpy variation of the flow of moist air, and the heat source or sink due to the reaction (W/m^3). For the mass diffusers, heat transfer processes are not modeled, flow being supposed as isothermal.

The kinetic of this type of solid/gas-reaction is determined by the difference between the operating conditions p_v, T and the thermodynamic equilibrium p_{eqSG} defined by Clausius – Clapeyron relation. For the open operation mode, p_v is the partial pressure of water. As the solid is consumed during the reaction, the kinetics is also affected by the reaction advancement. According to the previous study (Mazet, Amouroux, & Spinner, 1991) has shown that it can be expressed by a 1st order kinetic law, respectively Equation (30) for hydration reaction and Equation (31) for dehydration:

$$\frac{\partial X}{\partial t} = k_{cin} 1 - X \left(1 - \frac{p_{egSG}}{p_v} T \right) \tag{30}$$

$$\frac{\partial X}{\partial t} = k_{cin} X \left(1 - \frac{p_{egSG}}{p_v} T \right) \tag{31}$$

where X is the reaction advancement; k_{cin} refers to the kinetic constant; t corresponds to the time; p_v is the partial pressure of water; T is the temperature; p_{eqSG} is equilibrium pressure corresponded to equilibrium solid – gas.

To compare the two working modes, the model for closed system was derived from the previous one involving humid air. Alongside with previous assumptions from (a) to (d), additional assumptions, (f) to (h), are need to simplify the open system modeling accordingly.

The assumptions are listed as follows:

(f) no inert gas supplies the reactor

(g) because of the low vapor flow through the porous reactive bed, heat transfers are mainly conductive, and are represented with an effective conductivity. As a consequence, convective heat transfers due to the flow of pure vapor in the porous layer can be neglected

(h) as compared to the solid reactant sensible heat, the sensible heat of the vapor is neglected,

Furthermore, as distinct from the open reactor, for a closed reactor a heat exchanger is required to collect or supply the reaction heat, in addition to a gas diffuser. So, these two elements ought to be modeled and combined with the reactive bed.

The scheme of the closed thermochemical reactor is presented on Figure 4. It consists of three basic elements of the same length L: the gas diffuser, the reactive layer and the heat conductive fin.

Figure 4. Schematic representation of a closed thermochemical reactor (operating with pure vapor). a) Hydration step, b) Dehydration step. (Numbers refer to boundary conditions described below). The thicknesses of the conductive fin, reactive layer and mass diffuser are respectively D, Z_s and D'.

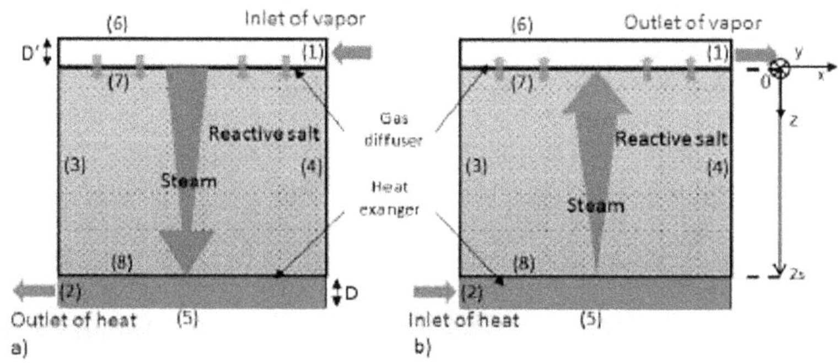

When discharged, vapor enters from the mass diffuser located at the top of the salt layer, then it flows through the reactive bed, i.e. layer of adsorbent, and an exothermal reaction proceeds. The releasing heat is transferred by conduction through the porous layer to the exchanger at the bottom of the layer (Figure 4a), and then it being supplied to users.

When charged (decomposition), this heat exchanger is connected with the external heat source and heat being transferred to the porous reactant. It leads to an endothermal decomposition that produces vapor which is diffused through the layer and then collected due to the mass collector (Figure 4b).

As the total gas flow passing through the porous layer considered, the mass balances are similar for both operating modes.

$$\begin{cases} \dfrac{\partial(\in_s \omega_v)}{\partial t} = n - \nabla \omega_v u \\ \dfrac{\partial(\in_{dif} \omega_v)}{\partial t} = -\nabla \omega_v u \end{cases} \quad (32\text{-}33)$$

where $\in_s$ is the porosity of salt; $\in_{dif}$ refers to the porosity of the diffuser; ω_v is the sink or a source of vapor, u is the velocity.

So, the vapor balance (Eq. 32)) for a closed system is the same as the humid air balance for the open-type reactor (Eq. (19)) with the corresponding variable, n_v. In addition, equation (33) describes the vapor mass balance in the diffuser at the top of the layer.

The velocity u is presented according Darcy law (Eq. (20)).

According to assumptions (g) and (h), the energy balance of the reactive layer was simplified to the Eq. (5.34), the energy balance of the heat collector has to be added (Eq. (35)):

$$\omega_s c_s \frac{\partial T}{\partial t} = \nabla \cdot \lambda_{eff} \nabla T + \Delta h_T^0 X \omega_s \tag{34}$$

$$\omega_c c_c \frac{\partial T}{\partial t} = \nabla \cdot \lambda_c \nabla T \tag{35}$$

where ω_s is a sink or a source of salt; ω_c is the sink or a source of heat collector; c_s is the molar heat capacity of salt; c_c is the molar heat capacity of heat collector; λ_{eff} is the effective thermal conductivity; λ_c is the thermal conductivity of the heat collector; Δh_T^0 is the standard enthalpy of the reaction.

The kinetic equations are unchanged compared to previous cases (Eq. (30) and (31)). They involve the thermodynamic equilibrium pressure $p_{eqSG}(T)$ while p_v is the total vapor pressure in this closed system case.

The described model was validated on storage system based on working pair '$SrBr_2$ – H_2O'. Simulations are given in Figure 5.

Results of the simulations fit adequately to experimental results for several successive hydrations. These experimental kinetics occur to decrease during the first cycles, then stabilize, this effect being well-known for such solid/

Figure 5. Reaction advancement versus time for several successive hydrations. Experimental and simulated results for hydrations 1, 2, 3 and 7
(Michel et al., 2014).

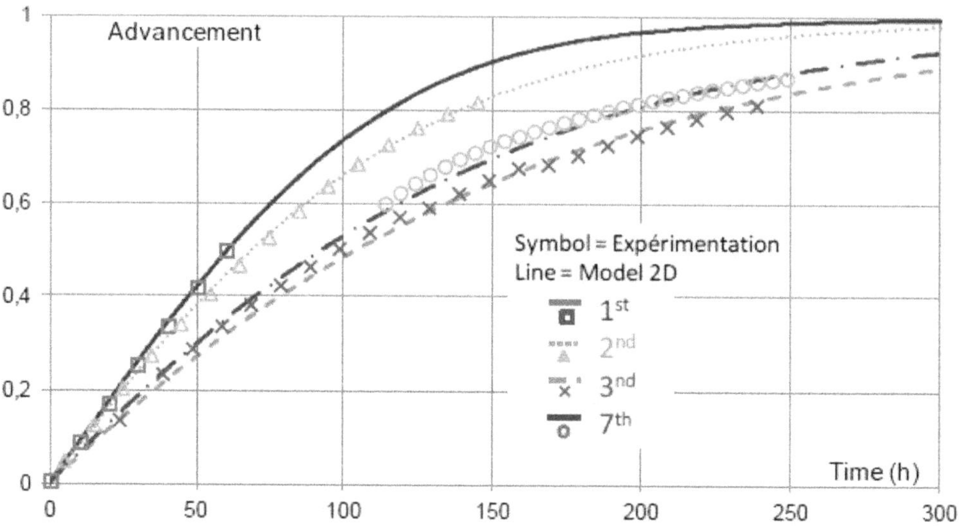

gas reactions. Such behavior can be explained by breakages of grains and re-organizations of the porous network in the salt bed during the first cycles could explain. These breakages are resulted from the strong mechanical stress that results from the change in grain volume during hydration. As a consequence, different values of the kinetic parameter have to be identified for these successive cycles. Due to these values, the difference between experimental and simulated rates of reaction is rather low: at X=0.5, it is 1%, 2% and 6%, respectively for hydrations from 1 to 3.

The model represented in (Michel et al., 2014) obtained a further development in (Farcot, Le Pierres, Michel, Fourmigue, Papillon, 2018) who designed a precise 2D dynamic model of the moving bed reactor using hydrated salt and water as working pair. This model considers interrelated heat and mass transfers within the reactive bed and the reaction kinetics.

To develop the reactor model, the following hypotheses are postulated as follows by Michel et al. (2014):

a The system is pseudo-homogeneous: the reactive solid and the gas phases are assumed to be at the same temperature, because the solid velocity is very low.

b The heat capacity of hydrated and dehydrated solids is constant.

c The heat capacity of dry air and water vapour is constant.
d The heat of the reaction, $\Delta h_r^{\ 0}$, is constant.
e The air velocity in the reactive bed follows Darcy's law.
f Solid flows homogeneously from the top to the bottom of the reactor (u_s = constant).
g Heat transfers in the porous media are conductive (according to the Fourier law) and convective.

Both governing equations are obtained from the mass balance of the water vapour and the humid air, and the hydrated and dehydrated solids. The evolution of the water vapour fraction in the air y_v, is described by (Eq. (36)):

$$\in n_h \frac{\partial y_v}{\partial t} = \$_v \left(1 - y_v\right) - n_h u_h \nabla y_v \tag{36}$$

where Φ_v refers to the Vapour sink term.

The evolution of the conversion of the salt is given by (Eq. (37)):

$$\left(1 - \in\right)\frac{\partial X}{\partial t} = -\frac{\$_v}{v} - u_s \nabla X_{salt} \tag{37}$$

where X_{salt} refers to the conversion of the salt which is calculated as a ratio of the molar densities of hydrated salt and solid $X_{salt} = n_{s1}/n_s$.

The model was implemented by the reactive bed properties on the fraction of the hydrated and dehydrated salt, i.e., conversion X.

So, λ_{eff}, the effective thermal conductivity of the solid bed, is determined by Archie's law (Mazet et al., 1991, Farcot et al., 2018) which was validated for reactive porous beds by Farcot et al. (2018):

$$\lambda_{eff} = \lambda_s \left(1 - \in \left(X\right)\right)^{1.5} \tag{38}$$

where λ_s is the thermal conductivity of the reactive solid.

As a consequence of the expansion of the solid during hydration, the porosity of the reactive bed, ε, depends on the conversion:

$$\in \left(X\right) = X \in_{s1} + \left(1 - X\right) \in_{s0} \tag{39}$$

where $\in_{s1}$ is the porosity of hydrated salt and $\in_{s0}$ is the porosity of dehydrated salt.

Similarly, other characteristics of the reactive bed are affected by the conversion:

$$C_{p_s}\left(X\right)=\frac{XM_{s1}C_{p_{s1}}+\left(1-X\right)M_{s0}C_{p_{s0}}}{XM_{s1}+\left(1-X\right)M_{s0}} \tag{40}$$

$$\rho_s\left(X\right)=\frac{D_{ec}M_s\left(X\right)}{\Delta h_r^0} \tag{41}$$

where

$$M_s\left(X\right)=XM_{s1}+\left(1-X\right)M_{s0} \tag{42}$$

$$k_s\left(X\right)=\frac{1}{\frac{1-X}{k_{s0}}+\frac{X}{k_{s1}}} \tag{43}$$

The reactor performance can be estimated by three parameters: the average outlet temperature of the air, $T_{h,out}$, the thermal power of the reactor, P_{th} (Eq. (44)), and the power needed for air ventilation, P_q (Eq. (45)), related to the power supplied by the reactor, P_{th}:

$$p_{th}=Q_h\rho_a\left[\left(Cp_a+\frac{y_{v,out}M_v}{\left(1-y_{v,out}\right)M_a}\right)T_{h,out}-\left(Cp_a+\frac{y_{v,in}M_v}{\left(1-y_{v,in}\right)M_a}\right)T_{h,in}\right] \tag{44}$$

where

$$p_q=\left(p_{in}-p_{out}\right)\cdot u_h\cdot S_h\cdot\eta_{fan} \tag{45}$$

with $S_h = 1\ m^2$ and the fan efficiency, $\eta_{fan} = 64\%$ (this value corresponds to the equipment that will be used for the experimental tests in our laboratory). The pressure drop created by the other process equipment (singularities, air pipes, diffuser, etc.) is negligible as compared to the pressure drop in the solid bed.

Thus, despite the fact of intense and deep work on simulation of operating the adsorptive heat storage devices, most of suggested models are commonly focused on only one phase of operating the adsorptive heat storage, the performance of heat storage device being estimated by only one characteristic. As a rule, such models are developed for systems based on traditional adsorptive materials such as zeolites or hydrated salt. All of these facts induce the developing of models which allows the integral estimation of the performance of adsorptive heat storage units based on the new composite materials 'salt inside porous matrix'.

ESTIMATION OF OPERATING PARAMETERS OF CLOSED-TYPE ADSORPTIVE HEAT STORAGE DEVICE

Algorithm to Determine the Operating Parameters of Closed-Type Adsorptive Heat Storage Device

The basic principles of modeling and optimization of cyclic adsorption processes are given in the works of Lukin and Novosel'skij (1989). However, the estimation of the operating characteristics of heat storage devices and their optimal design and operating parameters requires carrying out multifactor experiments of heat and mass transfer in these adsorptive heat storage media, which result in large material costs. Consequently, the development of an algorithm to determine the integral operating characteristics of adsorption heat storage devices becomes of primary importance. In pursuance of (Lukin & Novosel'skij, 1989), a system of differential equations which describes the internal diffusion kinetics of adsorption in the almost instantaneous establishment of equilibrium between the concentration of adsorbate in the gas that fills the pores and its concentration in the adsorbed state includes:

1. material balance equation: $\frac{\partial a}{\partial t} + \frac{\partial c}{\partial t} = \frac{1}{\theta^K} \cdot \frac{\partial}{\partial \theta}\left(D \cdot \theta^K \frac{\partial c}{\partial \theta}\right)$
2. adsorption isotherm equation: $a = f\left(c\right)$;
3. initial data: $a\left(\theta,0\right) = c\left(\theta,0\right) = 0$;

4. boundary data: $c\left(r,t\right)=c_{0},\left[\theta^{K}\cdot\frac{\partial c}{\partial\theta}\right]_{\eta\cdot 0}=0$

where a and c are the concentration of adsorbate in the mobile and stationary phases; θ equals x at the grain shape parameter K = 0 for a prismatic or cylindrical grain of length 2r, the lateral surfaces of which are impermeable (plate); $\theta=r$ corresponds to radial coordinate for cylindrical grain with impermeable ends (K = 1) or spherical grain (K = 2); t is the time; D is the effective diffusion coefficient. Saturation occurs from the outer surface of the granule, where a constant concentration is maintained.

There is no any amount of adsorbate at the initial instant of time in the grains, the diffusion coefficient D being assumed to be constant.

Obviously, it ought to choose operational parameters, at which the efficiency of the thermal battery reaches the maximal values.

The design of an adsorptive heat storage device of a closed type is given on Figure 6 (Belyanovskaya, Sukhyy, Kolomiyets, & Gubinskyi, 2019, Sukhyy, Kolomiyets, & Belyanovskaya, 2016).

The operation of the closed-type adsorptive heat storage device is realized in four-phase mode. The first stage is a heating the adsorbent (heat storage material) from the ambient temperature to the temperature of the beginning of adsorption.

On the second phase adsorption of water vapor with adsorbent occurs. The third phase is the regeneration, i.e. water desorption, the adsorbent is heated up to the regeneration temperature and water desorption proceeds. And the fourth phase is the cooling of the adsorbent down to the temperature of the beginning of adsorption.

The control-flow chart of the algorithm for calculating the efficiency factor of a heat storage device of closed-type is given in Figure 7 (Belyanovskaya et al, 2019).

According to these stages, the efficiency factor η was calculated as a ratio of the useful heat and heat input, %:

$$\eta=\frac{Q_{u}}{Q_{in}}\cdot 100 \tag{46}$$

where Q_u is useful heat (heat of adsorption), kJ:

$$Q_{u}=M_{ads}\cdot\Delta H_{ads}, \tag{47}$$

Figure 6. Design of the closed-type absorption heat storage device: 1 – heat-insulated case; 2 – mesh partitions; 3 – the condenser; 4 – a plate for collecting condensate; 5 – adsorption-desorption reactor; 6 – a coil pipe; 7 – a branch pipe; 8 – a tank of water; 9 – evaporator; 10, 11 – covers; 12 – a tube; 13 – a branch pipe; 14 – steel blinds; 15 – resistive heating element; 16 – steel blinds
(Belyanovskaya et al., 2019).

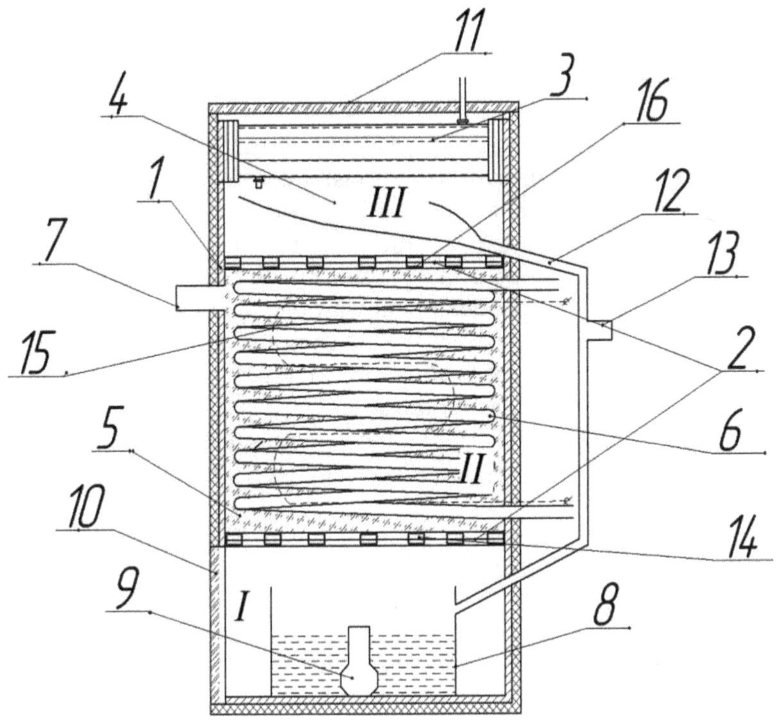

where M_{ads} refers to the adsorbent mass, kg; ΔH_{ads} is the specific heat of adsorption determined according to the Eq. (4.3), kJ/kg; Q_{in} is the heat input, kJ:

$$Q_{in} = Q_{des} + Q^h_{ads} + Q^h_{cas} + Q^h_{cp} + Q^h_{w} + Q^h_{s.w} + Q^h_{s.w} + Q^{ev}_{w}, \tag{48}$$

where

$$Q_{des}, Q^h_{ads}, Q^h_{cas}, Q^h_{cp}, Q^h_{w}, Q^h_{s.w}, Q^h_{s.w}, Q^{ev}_{w}$$

Figure 7. Control-flow chart for calculation of efficiency factor of closed-type heat storage device
(Belyanovskaya et al. 2019).

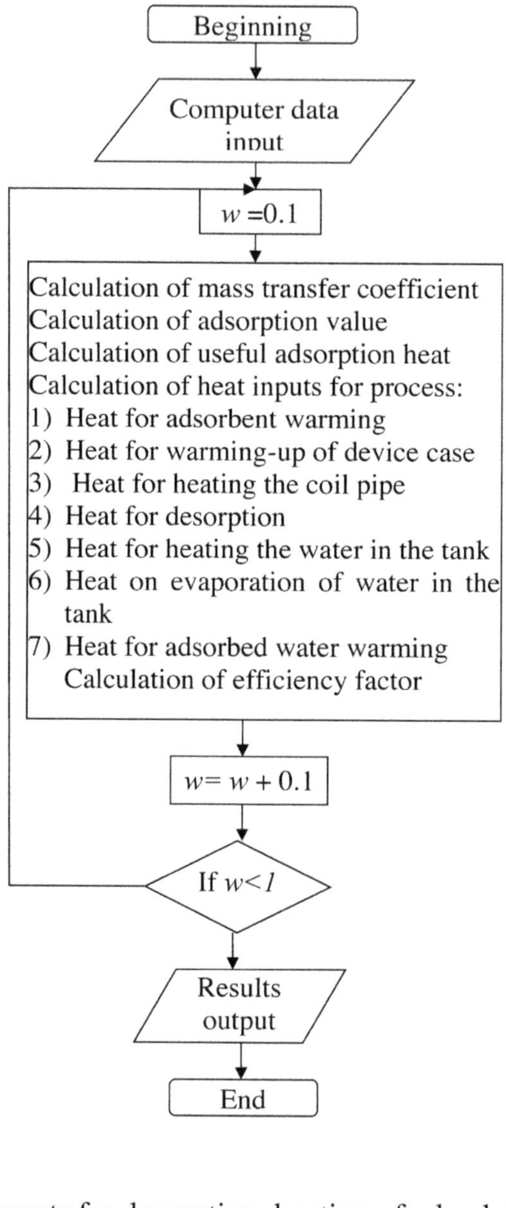

refers to the heat inputs for desorption, heating of adsorbent, device casing, coil pipe, water in the evaporator, adsorbed water, water evaporation, kJ.

Heat quantity for warming of materials and device components can be determined by conventional formula:

$$Q = M \cdot C \cdot \Delta t \tag{49}$$

The influence of humid air flow speed on efficiency coefficient is taken into account by adsorption value A, kg/kg. This coefficient is used for calculating of the useful heat Q_u according to Eq. (47) and heat input for desorption Q_{des} and adsorbed water $Q^h_{s.w}$ warming:

$$Q_{des} = M_{ads} \cdot \Delta H_{des} \cdot A, \tag{50}$$

$$Q^h_{s.w} = M_{ads} \cdot A \cdot C_w \cdot \left(t_{reg} - t_{env}\right), \tag{51}$$

where $\Delta H_{des} = 2850$ kJ/kg is heat of desorption, M_{ads} refers to the mass of adsorbent, kg, C_w is specific heat of water, kJ/kg·K, t_{reg} and $t_{env.}$ correspond to temperatures of regeneration and environment, °C.

Value of adsorption, i.e. water uptake, is calculated by formula (Belyanovskaya et al. 2019):

$$A = \frac{C_0 - C}{M_{ads}} \cdot V_{air} \tag{52}$$

where V_{air} refers to the air volume which is determined as a product of cross-section area of heat storage device, humid air flow speed and time of adsorption: $V_{air} = F_{hs} \cdot w \cdot \tau$, m^3; F_{hs} is cross-section area of heat storage device, m^2, w is humid air flow speed, m/s, τ is adsorption time, s, C_0 and C are absolute humidity of the air flow on the inlet and outlet of the adsorbent layer, kg/m^3 (Pavlov, Romankov, Noskov, 1987):

$$C = \frac{C_0}{1 + e^{\frac{\left(\beta \frac{-\tau w C_0}{A_{\max}} + H\right)}{w}}} \tag{53}$$

where H is height of heat storage material (adsorbent), m, A_{max} is adsorption capacity of adsorbent (maximal adsorption), kg/kg; β is mass transfer coefficient, s^{-1}, calculated according to (Matveykin, Pogonin, Putin, & Skvortsov, 2007):

$$\frac{1}{\beta} = \frac{1}{\beta_y} + \frac{1}{\beta_p} + \frac{1}{\beta_{c.s.}} \tag{54}$$

where β_y, β_p and $\beta_{c.s.}$ are mass transfer coefficients for gas phase, pores and coplanar stirring, s^{-1} [Matveykin et al., 2007].

Then performance of heat storage device was simulated. As an adsorbent, composite material 'silica gel sodium sulphate' was used. Results of calculations confirm that adsorption rate is limited with rate of diffusion in pores, it being confirmed by kinetic study of water vapor sorption (Sukhyy et al., 2018, Sukhyy, Belyanovskaya, Kozlov, Kolomiyets, Sukhyy, 2014).

Additionally, it is expediently to evaluate the influence of the constructional parameters of the heat storage device, particularly, the height of the layer of an adsorbent, on the value of adsorption, i.e. water uptake. The calculation

Figure 8. Adsorption vs. humid air flow speed. Height of composite layer: a) H=0.25 m; b) H=0.5 m; c) H=1.0 m
(Sukhyy et al., 2017)

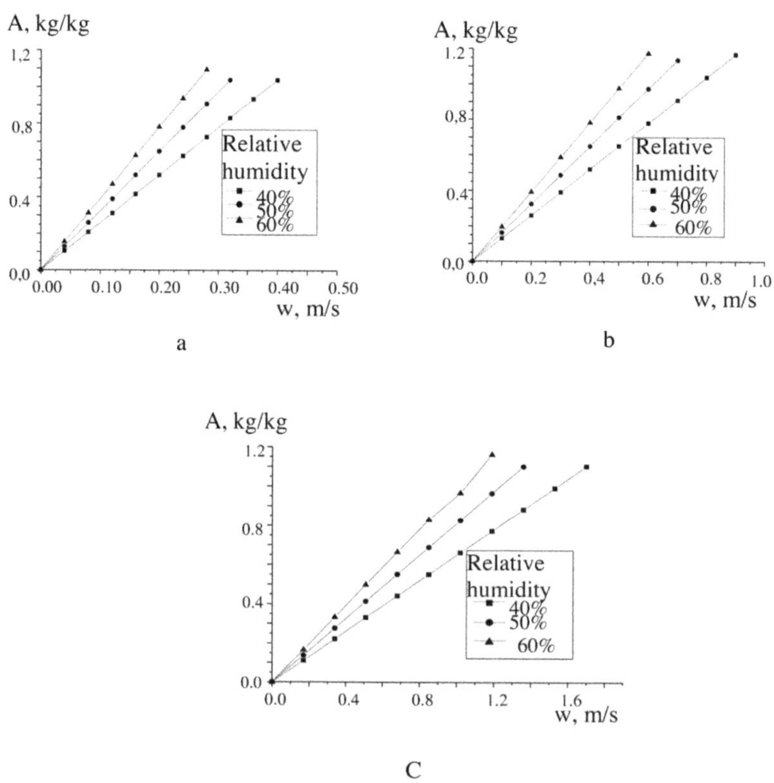

was carried out according to the suggested algorithm (Figure 8) (Sukhyy et al., 2016).

The dependence of the adsorption value vs. the velocity of the humid air flow with a relative humidity of 40, 50 and 60% at a temperature of 60þC was calculated (Figure 9).

The maximal values of adsorption for 8 hours was achieved at a relative humidity of 60% and vapor flow speeds w = 0.2 – 0.35 m/s (height of the composite layer H = 0.25 m), 0.6 – 0.85 m/s (H = 0.5 m) and 1.1 – 1.7 m / s (H = 1.0 m). Since it is maintenance-friendly to use w < 0.6 m/s, it is advisable to recommend the height of the composite layer of 0.25 – 0.5 m (Sukhyy, Kolomiyets, Prokopenko, Gavrilko, Belyanovskaya, 2017). In these conditions maximal values of the efficiency factors are achieved (Figure 10) (Sukhyy et al., 2016).

Figure 9. Efficiency factor vs. humid air flow speed (temperature 45°C). Rel0ative humidity $\varphi = 30 – 60\%$. Height of composite layer, m: a – 1; b – 0.5; c – 0.25 *(Belyanovskaya et al., 2019).*

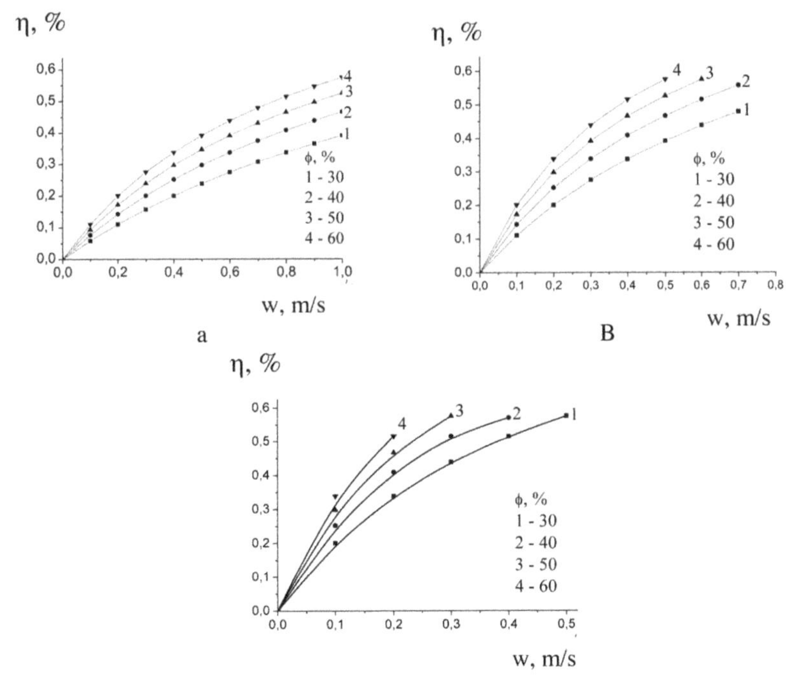

Efficiency factor vs. humid-air-flow speed relationships calculated according to Eq. (5.46) – (5.54) are given on Figure 10 and 11 when composite adsorbent 'silica gel – sodium sulphate' used (Sukhyy et al., 2016).

When the humid air flow speed and relative humidity increased, the efficiency factor of heat energy storage device of closed-type η raised. The value η is stated to be negligibly changed as humid air flow temperature increased. Maximal values of efficiency factor (about of 53 – 57%) are corresponded to the relative humidities of 40 – 60% at the humid air flow speeds of 0.6 – 0.8 m/s.

Closed-type heat storage unit was tested for heating the premise with a floor area 100 m^2 during period from 1 of November 2016 to 28 of February 2017 (Belyanovskaya et al., 2019). It was used instead of fan heater and oil heater with energy consumption of 1.5 – 2.5 kW. The initial temperature in the premise was 12°C. Vapour-air flow rate was of at 0.25 m/s. Mass of adsorbent 'silica gel – sodium sulphate' was 10 kg. During testing, temperature in the

Figure 10. Efficiency factor vs. humid air flow speed. Relative humidity: 1 – 40; 2 – 50; 3 – 60%. Regeneration temperature: a) 80 þC; b) 90 þC; c) 100 þC (Sukhyy et al., 2016).

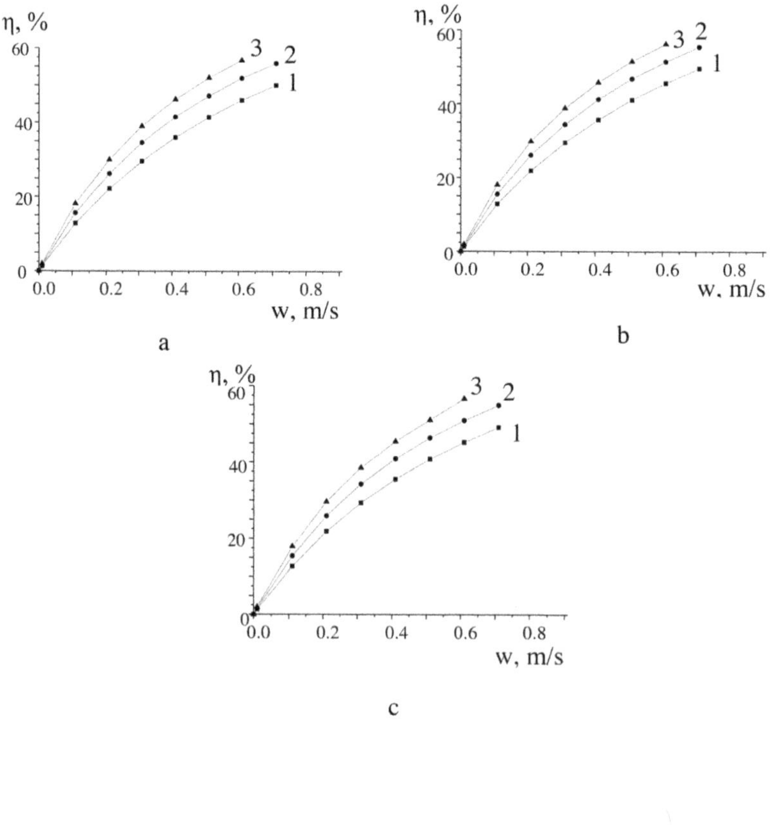

Figure 11. Efficiency factor vs. humid air flow speed at the temperatures: ■ – 40; ● – 50; ▲– 60 þC. Relative humidity: a – 40%; b – 50%; c – 60%
(Sukhyy et al., 2016).

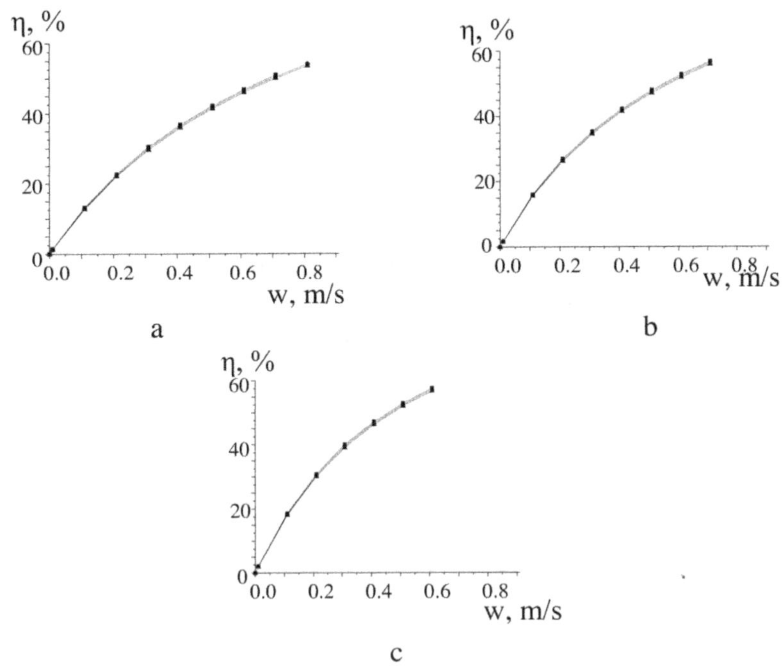

Figure 12. Structure of heat input for closed-type heat storage unit
(Belyanovskaya et al., 2019).

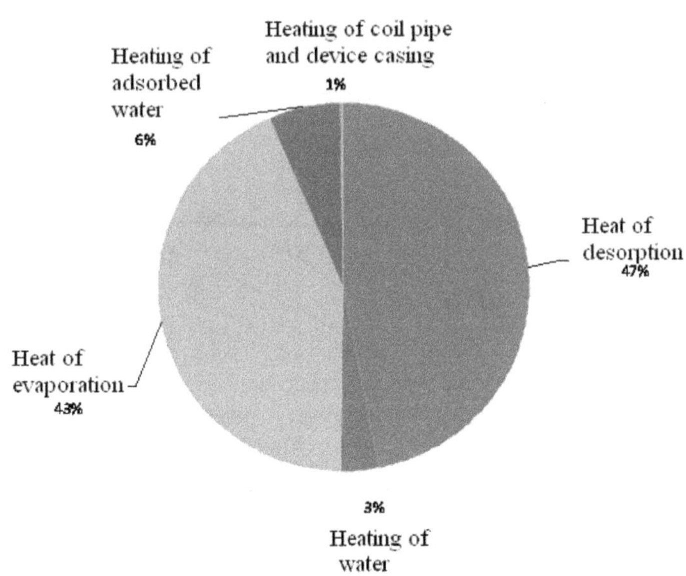

premise was changed from 20 °C to 22°C. Two peaks corresponded to discharge and charge of heat storage energy device are registered on temperature-time curves. Efficiency factor was calculated as the ratio of two peaks. They are almost 32%. These values are correlated with results obtained according to suggested algorithm.

Figure 13. Efficiency factor-vs.-humid air flow speed (absolute humidity: a – 0,04; b – 0,06; c – 0,08 g/m³) for close-type heat storage device: 1 –mass of water corresponds with limit adsorption, 2 – mass of water delivered with humid air to adsorbent layer; 3 – mass of water corresponds with 1.5A$_{max}$
(Belyanovskaya et al., 2019).

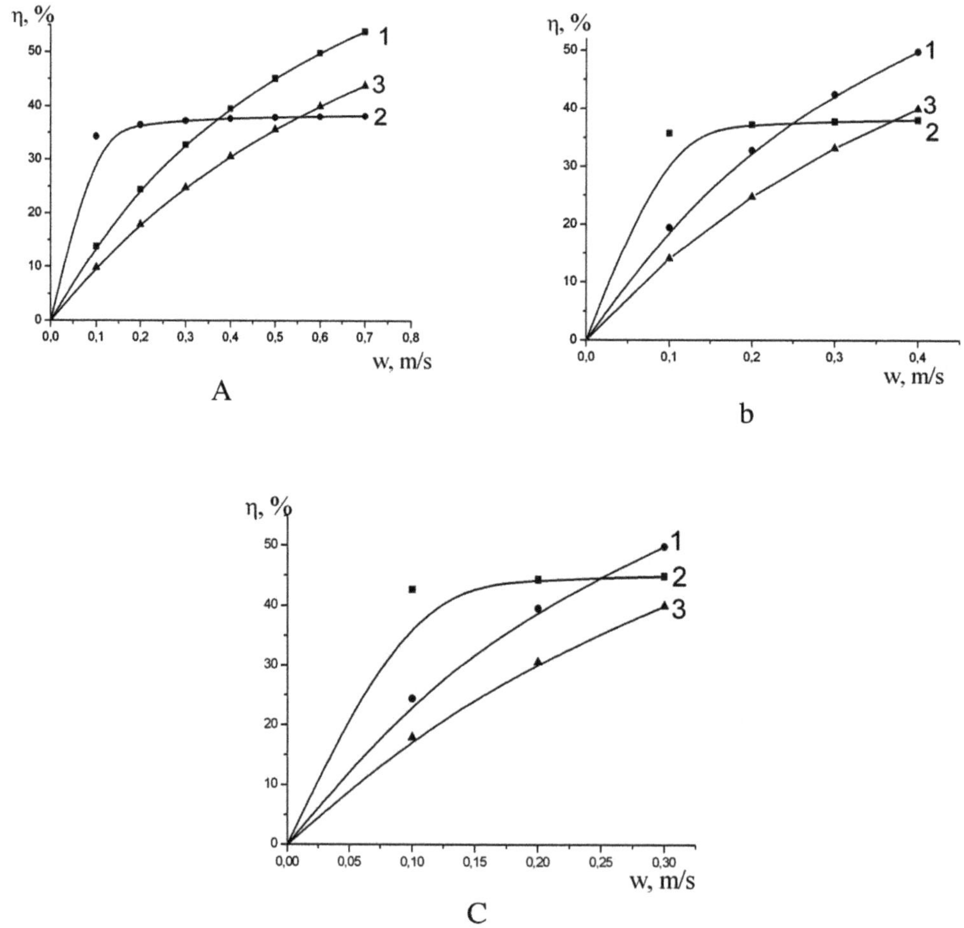

Improving the Efficiency of Close-Type Adsorptive Heat Energy Storage Device

Then, to develop methods of increasing of the efficiency coefficient structure of heat inputs was analyzed. According to Figure 12, the most significant input items correspond to the heat of evaporation and the heat of desorption.

This influence of water mass on efficiency factor confirms this conclusion (Figure 14). Maximal magnitudes of efficiency factors are stated when mass of water conformed to the maximal adsorption (curve 1). As water mass corresponds to sesquialteral value of maximal adsorption, efficiency factors are decreased (curve 3). When the water mass coincided to amount delivered with humid air to inlet of the composite adsorbent layer, efficiency factors are intermediate in between considered options (curve 2). At that, efficiency factor vs. humid air flow speed curves reach plateau. This form promotes more easy operation and maintaining of heat storage unit.

Figure 14. Test heat storage device: 1 – thermal insulated case; 2 – heat storage material; 3 – compressor; 4 – air humidifier; 5 – resistance element; 6 – weighing-machine
(Belyanovskaya, et al., 2017).

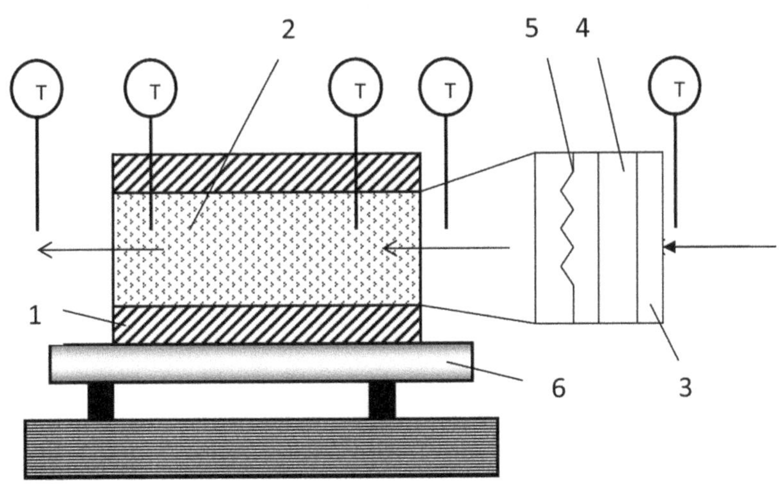

Figure 15. Control-flow chart for calculation of efficiency factor of adsorptive heat storage device
(Belyanovskaya, et al., 2017)

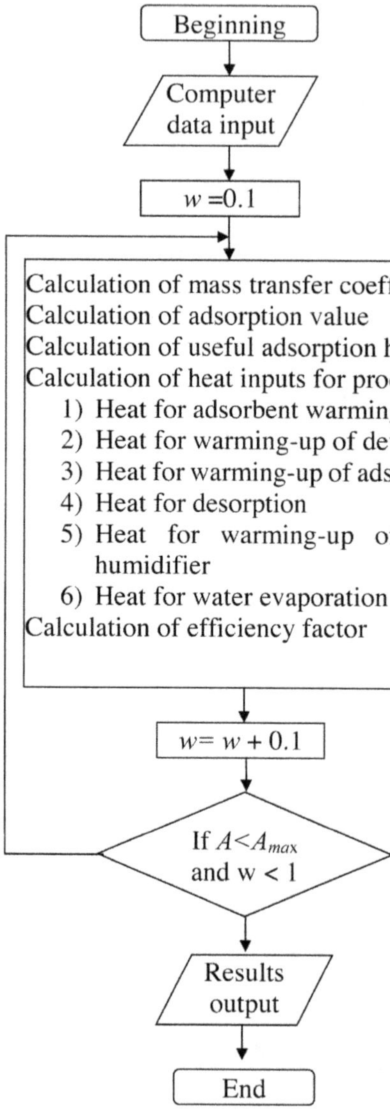

ESTIMATION OF THE OPERATING PARAMETERS OF OPEN-TYPE ADSORPTIVE HEAT STORAGE DEVICE

Algorithm to Evaluate the Operating Parameters of Open-Type Adsorptive Heat Storage Device

Similar algorithm was suggested to estimate the performance of open-type adsorptive heat energy storage device. Construction of open-type heat storage device is given in Figure 15 (Belyanovskaya, Sukhyy, Kolomiyets, & Sukhyy, 2017). As heat storage media the composite adsorbents 'silica gel – sodium sulphate' synthesized by sol – gel technique (Sukhyy et al., 2018).

The heat storage device is operated according to two-phase mode. On the first step external air fans by compressor 3 into humidifier 4 where its relative

Figure 16. Efficiency factor vs. humid air flow speed and relative humidity: 1 – 40%; 2 – 50%; 3 – 60%. Humid air flow temperature °C: 40 (a); 50 (b) and 60 (c) (Belyanovskaya, et al., 2017)

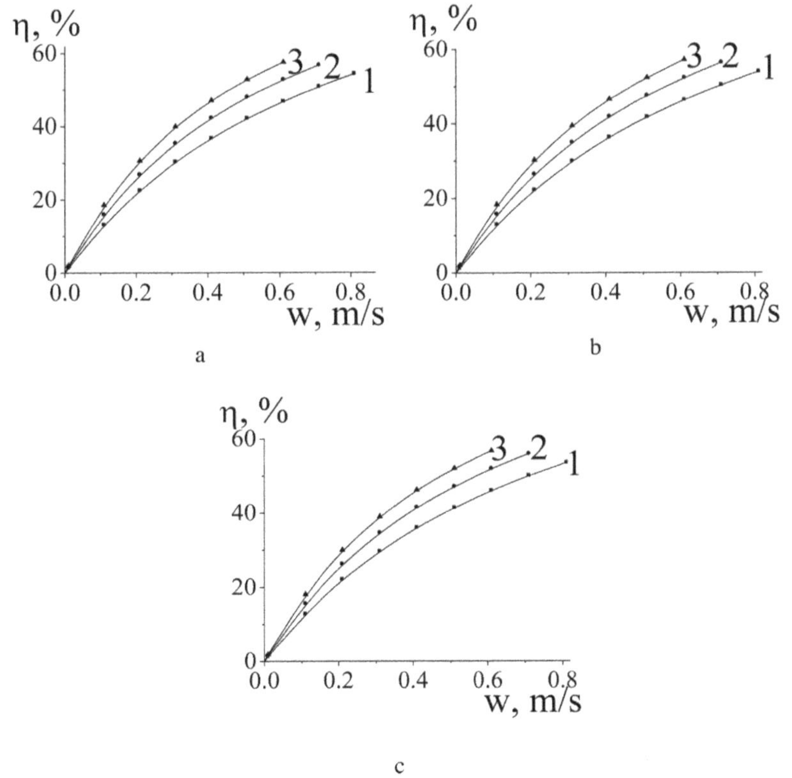

degree of humidity is increased up to 60 – 80%, heating of the humidified air by resistance element 5 to 30 – 40°C because of low rates of adsorption at lower temperatures. Then humidified air is supplied to the adsorptive heat storage material. After short-term warming of adsorbent during 15 min. adsorption of water from air passed through heat storage material occurs. Temperature of both heat storage material and air increases due to exothermic adsorption processes.

As consequence, temperature of the outlet air is raised up to 90 – 115 °C. In the second phase (regeneration) hot air is fanned through adsorbent layer for warming up adsorbent up to temperatures above 90°C which results in desorption.

According to these stages, efficiency factor of open-type heat storage device is suggested to calculate by the formula (46). But the heat input is calculated (Belyanovskaya et al., 2017):

Figure 17. Efficiency factor vs. air flow speed. Relative humidity degree: 40 (a), 50 (b) and 60% (c)
(Sukhyy, Belyanovskaya, & Kolomiyets, 2018).

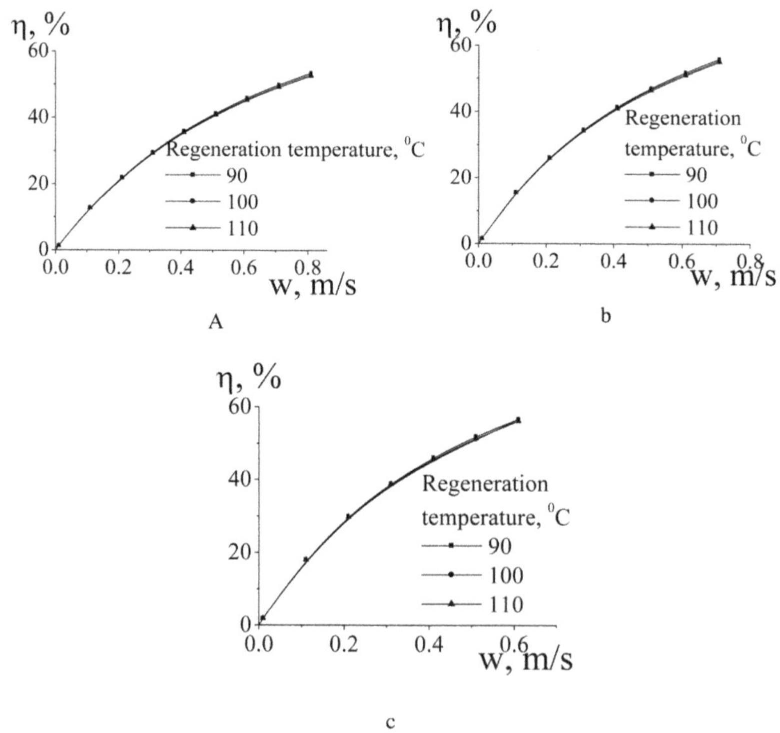

$$Q_{in} = Q_{des} + Q^h_{ads} + Q^h_{cas} + Q^h_w + Q^h_{s.w} + Q^{ev}_w \quad (55)$$

where

$$Q_{des}, Q^h_{ads}, Q^h_{cas}, Q^h_w, Q^h_{s.w}, Q^{ev}_w$$

are heat inputs for desorption, heating of adsorbent, device casing, water in the humidifier, adsorbed water, water evaporation, kJ.

Heat inputs are calculated by Eq. (49) – (51). Computing of adsorption processes is carried out according to Eq. (47), (52) – (54). According to results of calculations adsorption rate is limited with rate of diffusion in pores, that being conformed to kinetic study of water vapor sorption (Sukhyy et al., 2018). So, suggested algorithm for computation of efficiency factor of adsorptive heat storage device operating in open mode is given on Figure 15.

Relationships for efficiency factor of open-type heat storage device based on composite 'silica gel – sodium sulphate' vs. humid air flow speed calculated by suggested algorithm are given on Figure 16 and 17. The shape of these curves occurs to be almost identical to heat storage device operating by close mode (Figure 16 and 17).

Growth of efficiency factor of heat storage device η is determined by increase in humid air flow speed and its relative degree of humidity. Value

Figure 18. Averaged temperature sorbent curves when heat storage device started in the morning (1) and in the evening (2)
(Belyanovskaya et al., 2017).

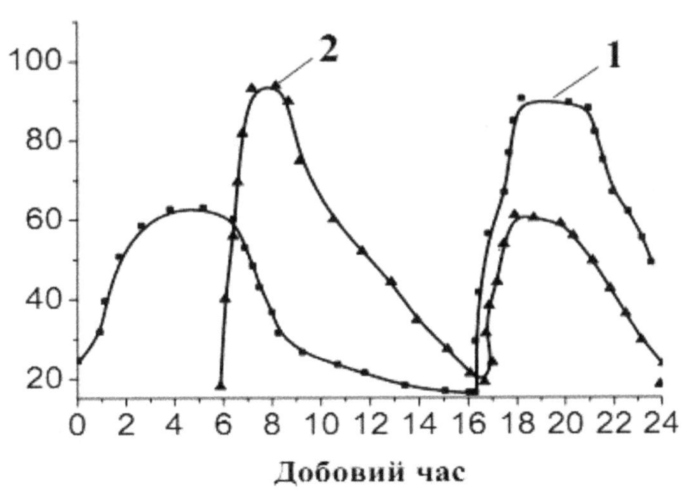

η appeared to be near constant when regeneration temperature increased from 90 to 110°C. Maximal values of the efficiency factors of 55 – 57% are estimated for humid air flow speeds of 0.6 – 0.8 m/s and relative humidity of 40 – 60%.

Heat storage device tests was tested for heating the space with a floor area 100 m^2 during period from 1 of November 2014 to 28 of February 2015 (Belyanovskaya et al., 2017). Tested device was used for load-factoring in power systems instead of fan heater and oil heater with energy consumption of 1.5 – 2.5 kW. The initial temperature in the premise was 12°C. Humid air flow speed was set at 0.25 m/s. Heat storage device was filled with 10 kg of adsorbent 'silica gel – sodium sulphate'. As thermal insulation material mineral wool (thickness 100 mm, thermal conductivity coefficient 0.035 W · m/K) was used.

Heat storage device was tested in two modes which matched to the load-factoring of electric energy in the morning or in the evening (first or second half-term). When both modes used, indoor temperature is maintained within limits from 20 °C to 22°C during day or night period. Average temperature graphs for both test modes are presented on Figure 18.

Two peaks corresponded to discharging and charging of heat energy device are observed on both curves, they being registered as the first and second after heat storage device start. When discharged, maximal temperature of

Table 1. Energy consumption when adsorptive open-type heat energy storage device used for space warming (floor area is 100 м^2)

Heat Source	**Solid Fuel Combustion Boiler PROTECHN TT**	**Gas-Fired Combustion Boiler PROSKURIV AOGV-16 B**	**Electric Boiler Dnipro-Mini 12kW, 380 V**	**Heat Energy Storage Device$_{min}$**	**Heat Energy Storage Device$_{max}$**
Capital input, Thous. UAH	7	7	5	6	6
Energy consumption, kg of standard fuel per day	22	26	17.69	0.29	7.5
Saving coal equivalent, kg coal equivalent per a day (min)	21.71	25.71	17.4		
Saving coal equivalent, kg coal equivalent per a day (max)	14.5	18.5	10.19		

(Belyanovskaya et al., 2017)

the adsorbent is almost 60°C, it being resulted from adsorbent cooling with air inflow.

Higher temperature when charged is a consequence of composite adsorbent regeneration temperature. During day hours the composite is regenerated by the heat provided by solar collector, in night hours it can be supplied by resistance element. Efficiency factors calculated with ratio of two peaks ratio are almost 32% for both operating modes. These values are correlated with computed values 33% for the same conditions.

Comparing of energy consumption with solid fuel combustion boiler, gas-fired combustion boiler and electric boiler gives evidence of advisability of adsorptive heat storage device for heating premises when heat demand equaled to 339.5 MJ per day for floor area 100 m^2. Indexes min and max refers to energy consumption at solar and cloudy weather respectively. Results are given in the Table 5.1 (Belyanovskaya et al., 2017).

Using the open-type adsorptive heat storage device allows to decrease energy consumption in heat supply system in minimum of 3.5 times versus gas-fired combustion boiler. As compared with solid fuel combustion boiler and electric boiler energy consumption can be decreased in no less than 2.9 and 2.4 times.

When heat for adsorbent regeneration supplied by solar collector, energy consumption was decreased about 90, 76 and 61 times versus gas-fired combustion boiler, solid fuel combustion boiler and electric boiler, respectively (Belyanovskaya et al., 2017).

So, suggested algorithms allow to estimate design characteristics and performance of open-type and close-type heat energy storage devices or units. Optimal operating parameters of performance for heat storage devices based on composite sorbent 'silica gel – sodium sulphate' are stated to be humid air flow speed of 0.6 – 0.8 m/s and relative humidity of 40 60%, efficiency coefficient being ranged from 53 to 57% (Belyanovskaya et al., 2017).

Power consumption is shown to be decreasing in 2,4 - 90 times versus decentralized heating systems based on solid fuel, gas and electric boilers when open-type adsorptive heat storage device used.

Figure 19. Influence of the speed of the humid air flow (absolute humidity: a - 0,04; b - 0,06; c - 0,08 g / m³) on the efficiency factor of the adsorptive heat storage device operating by the open mode: 1- mass of water corresponds to the maximal adsorption, 2 – mass of water which is fed with humid air to the adsorbent layer (Belyanovskaya, Lytovchenko, et. al., 2019).

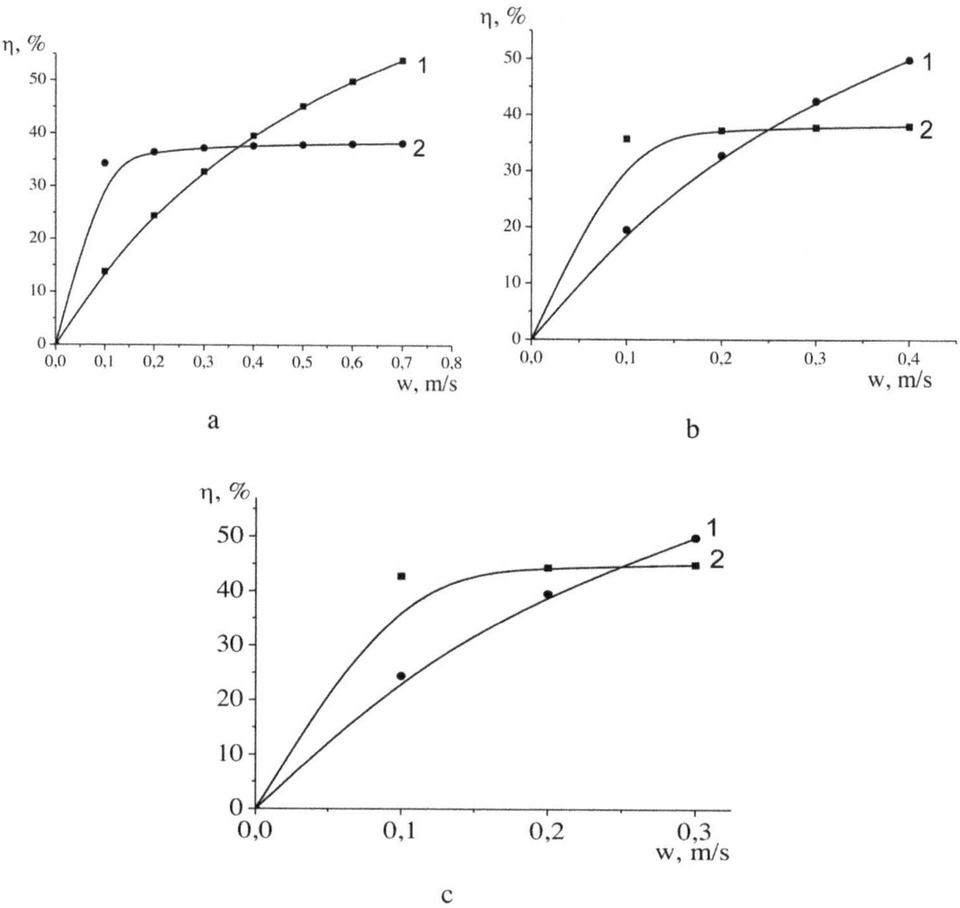

Improving the Performance of Open-Type Adsorptive Heat Energy Storage Device

Application perspectiveness of adsorptive heat storage devices is shown for heat supply systems. Nevertheless, range of airflow speeds corresponded to maximal values of the efficiency factor is rather narrow.

To develop methods for increasing the efficiency factor, an analysis of the structure of heat inputs for the operation of the open-type heat storage device

Figure 20. Influence of airflow speed and relative humidity on the efficiency of the open-type adsorptive heat storage device. The temperature of the air flow, ° C: a - 40, b - 50, c - 60. The thickness of the adsorbent layer is 0.25 m (Belyanovskaya, Lytovchenko, et al., 2019).

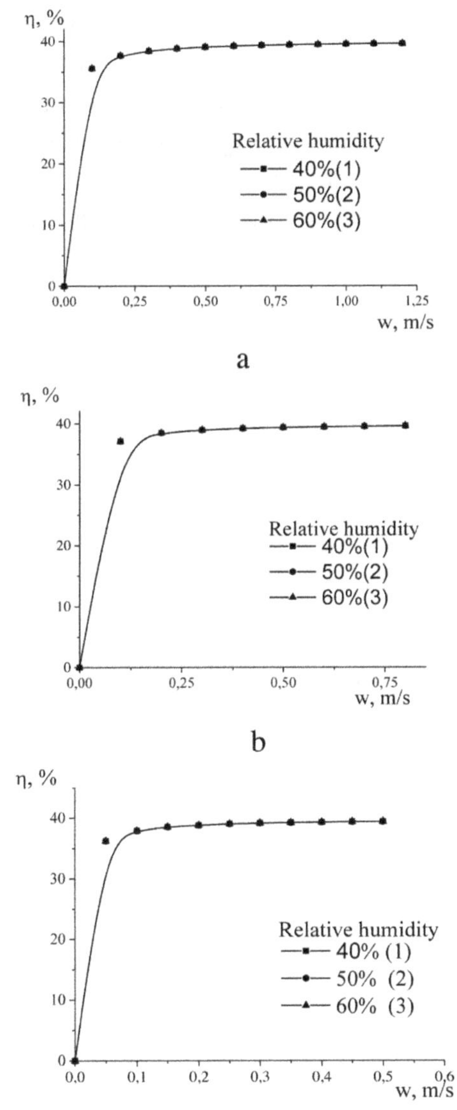

was performed. According to its results, the most significant are the inputs for evaporation of water (43%) and desorption (47%), followed by the inputs for heating the adsorbed water (6%), heating the water in the humidifier (3%) and heating the heat storage device case (1%). So, limitation of interval of maximal efficiency factors is suggested to mitigate by variation of the mass of water vapor which supplied to the adsorbent layer.

The dependence of the efficiency on the velocity of the steam flow supplied to the heat storage device is given on Figure 19. Line 1 shows that, if the mass of water is taken according to the maximal adsorption, the maximal value of the efficiency factor is observed in a rather narrow range of the airflow speeds.

Figure 21. The influence of design characteristics on the value of the efficiency factor of the open-type heat storage device. The temperature of the air flow is 50°C. The relative humidity of the air stream: a - 30%; b - 40%; c - 50% *(Belyanovskaya, Lytovchenko, et al., 2019).*

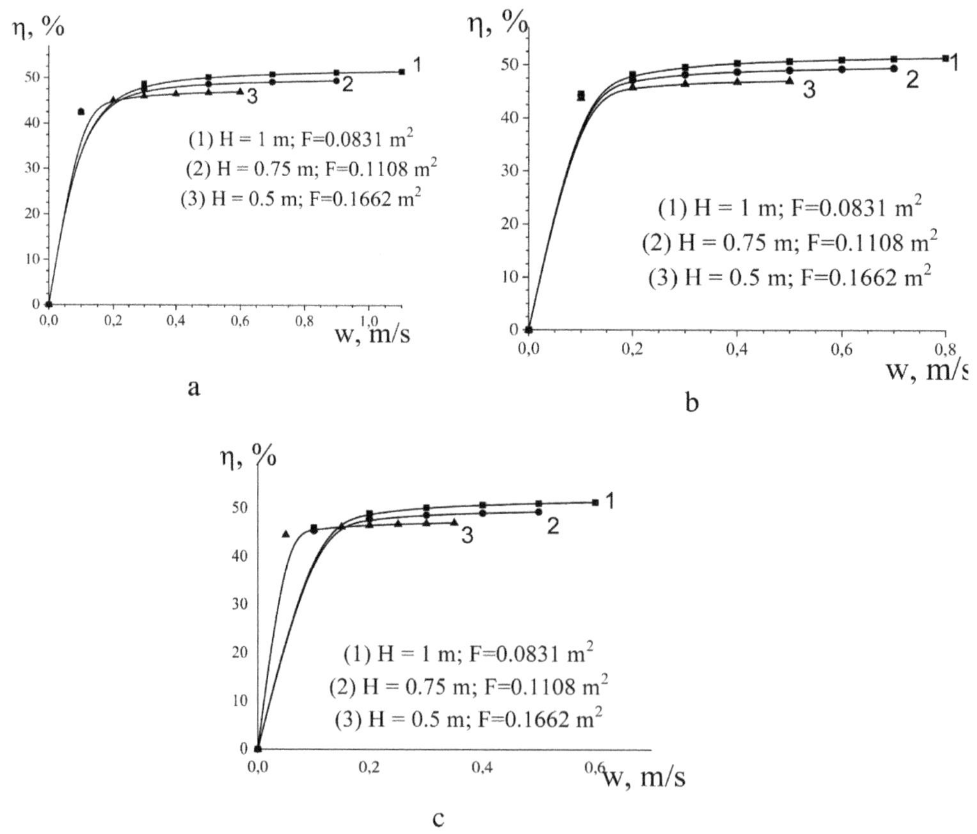

At the same time, when the water supply corresponds to the volume and absolute humidity of the air according to formula $M_{H2O} = V_{air} \cdot C_0$, on the plot of the dependence of the efficiency factor vs. the airflow speed, it reach the plateau at w = 0.1 m / s (Figure 19, curves 2). Under similar conditions (Figure 20), since the mass of water supplied to the layer of heat storage material not only depends on the flow velocity, but also exceeds the value corresponding to the adsorption of water vapor by a layer of the composite under these conditions, about three times, the efficiency factor does not exceed 40% (Belyanovskaya, Lytovchenko, Mikhaylov, Sukhyy, & Sukha, 2019).

Obviously, the amount of water supplied to the adsorbent layer, taking into account the effect of the absolute humidity on the adsorption, can be corrected by adjustment of the volume of air by changing the cross section of the heat storage device and, therefore, the thickness of the adsorbent layer. Thus, reducing the cross-sectional area from 0.1662 to 0.0831 m^2 results in an increase of the efficiency factor from 42 to 50% (Figure 21). In this case, the maximal values of the efficiency factors are observed in a wider range of velocities of the vapor flow as compared to the supply of moisture in accordance with the maximal adsorption.

However, the maximal adsorption values are obviously observed at values of w = 0.5 - 0.8 m / s (Sukhyy et al., 2017) at temperatures of 40 - 60 ° C and relative humidity of the air flow not lower than 40 - 50%. Therefore, as maximal heat load proveded, it is advisable to operate the heat storage device in accordance with the above-mentioned operating parameters.

REFERENCES

Ahn, H., Kim, M.-B., & Lee, C.-H. (2004). Effects of heat-transfer coefficients on thermal dynamics in a near-adiabatic fixed bed. *Separation Science and Technology*, *39*(11), 2627–2654. doi:10.1081/SS-200026757

Ahn, H., & Lee, C.-H. (2004). Effects of capillary condensation on adsorption and thermal desorption dynamics of water in zeolite 13X and layered beds. *Chemical Engineering Science*, *59*(13), 2727–2743. doi:10.1016/j.ces.2004.04.011

Amar, N. B., Sun, L., & Meunier, F. (1996). Numerical analysis of adsorptive temperature wave regenerative heat pump. *Applied Thermal Engineering*, *16*(5), 405–418. doi:10.1016/1359-4311(95)00045-3

Belyanovskaya, E., Sukhyy, K., Kolomiyets, E., & Gubinskyi, M. (2019). *Operating characteristics of heat conversion units in heat supply system. In The Actual Problems of the World Today, 2*. Sciemcee Publishing.

Belyanovskaya, E.A., Lytovchenko, R.D., Mikhaylov, A.G., Sukhyy, K.M., & Sukha, I.V. (2019). Algoritm rozrahunku konstruktivnih ta ekspluatacijnih harakteristik adsorbcijnogo akumulyatora teplovoï energiï vidkritogo tipu na osnovi kompozitnih adsorbentiv. *Komp'yuternoe modelirovanie i optimizaciya*, *1*(5), 3 – 9.

Belyanovskaya, E. A., Sukhyy, K.M., Kolomiyets, E.V., & Sukhyy, M.P. (2017). Operating processes parameters of open-type heat storage devices in heat supply systems. *Scientific Works*, *81*(1), 4 – 9.

Chapman, S., & Cowling, T. (1970). The Mathematical Theory of Non-uniform Gases: An Account of the Kinetic Theory of Viscosity, Thermal Conduction and Diffusion in Gases. Cambridge University Press.

Dawoud, B., Vedder, U., Amer, E.-H., & Dunne, S. (2007). Non-isothermal adsorption kinetics of water vapour into a consolidated zeolite layer. *International Journal of Heat and Mass Transfer*, *50*(11-12), 2190–2199. doi:10.1016/j.ijheatmasstransfer.2006.10.052

Delgado, A. (2006). A critical review of dispersion in packed beds. *Heat and Mass Transfer*, *42*(4), 279–310. doi:10.100700231-005-0019-0

Do, D. D. (1998). Adsorption Analysis: Equilibria and Kinetics. London: Imperial College Press.

Duquesne, M., Toutain, J., Sempey, A., Ginestet, S., & del Barrio, E. P. (2014). Modeling of a nonlinear thermochemical energy storage by adsorption on zeolites. *Applied Thermal Engineering*, *71*(1), 469–480. doi:10.1016/j.applthermaleng.2014.07.002

Farcot, L., Le Pierres, N., Michel, B., Fourmigue, J.-F., & Papillon, P. (2018). Numerical investigations of a continuous thermochemical heat storage reactor. *Journal of Energy Storage*, *20*, 109–119. doi:10.1016/j.est.2018.08.020

Gaeini, M., Zondag, H. A., & Rindt, C. C. M. (2016). Effect of kinetics on the thermal performance of a sorption heat storage reactor. *Applied Thermal Engineering*, *102*, 520–531. doi:10.1016/j.applthermaleng.2016.03.055

Glueckauf, E. (1955). Theory of chromatography. Part 10. Formula for diffusion into spheres and their application to chromatography. *Transactions of the Faraday Society*, *51*(0), 1540–1551. doi:10.1039/TF9555101540

Hauer, A. (2002). *Beurteilung fester Adsorbentien in offenen Sorptionssystemen für energetische Anwendungen* (PhD thesis). Fakultät III - Prozesswissenschaften, Technische Universität Berlin.

Kandula, M. (2011). On the effective thermal conductivity of porous packed beds with uniform spherical particles. *Journal of Porous Media*, *14*(10), 919–926. doi:10.1615/JPorMedia.v14.i10.70

Lan, S., Zondag, H., van Steenhoven, A., & Rindt, C. (2015). Kinetic study of the dehydration reaction of lithium sulfate monohydrate crystals using microscopy and modeling. *Thermochimica Acta*, *621*, 62144–62155. doi:10.1016/j.tca.2015.10.005

Lehmann, C., Beckert, S., Nonnen, T., Möllmer, J., Gläser, R., Kolditz, O., & Nagel, T. (2017). A comparison of heat storage densities of zeolite granulates predicted by the Dubinin-Polanyi theory to experimental measurements. *Energy Procedia*, *105*, 4334–4339. doi:10.1016/j.egypro.2017.03.909

Lukin, V.D., Novosel'skij A.V. (1989). *Tsiklicheskie adsorbcionnye processy: Teoriya i raschet*. Leningrad: Himiya.

Matveykin, V. G., Pogonin, V. A., Putin, S. B., & Skvortsov, S. A. (2007). Matematicheskoye modelirovaniye i upravleniye processom korotkotsyklovoy beznagrevnoy adsorbtsiyi. *Mashinostroyeniye*, *1*, 140.

Mazet, N., Amouroux, M., & Spinner, B. (1991). Analysis and experimental study of the transformation of a non-isothermal solid/gas reacting medium. *Chemical Engineering Communications*, *99*(1), 155–174. doi:10.1080/00986449108911585

Mette, B. (2014). *Experimentelle und numerische Unterschungen zur Reaktionsführung thermochemischer Energiespeicher.*

Mette, B., Kerskes, H., Drück, H., & Müller-Steinhagen, H. (2013). New highly efðcient regeneration process for thermochemical energy storage. *Applied Energy*, *109*, 352–359. doi:10.1016/j.apenergy.2013.01.087

Mette, B., Kerskes, H., Drück, H., & Müller-Steinhagen, H. (2014). Experimental and numerical investigations on the water vapor adsorption isotherms and kinetics of binderless zeolite 13X. *International Journal of Heat and Mass Transfer*, *71*, 555–561. doi:10.1016/j.ijheatmasstransfer.2013.12.061

Michel, B., Neveu, P., & Mazet, N. (2014). Comparison of closed and open thermochemical processes, for long-term thermal energy storage applications. *Energy*, *72*, 702–716. doi:10.1016/j.energy.2014.05.097

Mugele, J. (2005). *Optimierung von Speichermaterialien für den Einsatz in geschlossenen thermochemischen Wärmespeichern für gebäudetechnische Anwendungen*. VDI Verlag.

Nagel, T., Beckert, S., Böttcher, N., Gläser, R., & Kolditz, O. (2015). The impact of adsorbate density models on the simulation of water sorption on nanoporous materials for heat storage. *Energy Procedia*, *75*, 2106–2112. doi:10.1016/j.egypro.2015.07.331

Pavlov, K. F., Romankov, N. G., & Noskov, A. A. (1987). Prymeri i zadachi po kursu processov i apparatov khimicheskoy tekchnologiyi. Leningrad: Khimia.

Ruthven, D. M. (1984). Principles of Adsorption and Adsorption Processes. John Wiley & Sons.

Semprini, S., Asenbeck, S., Kerskes, S., & Drück, H. (2017). Experimental and numerical investigations of an adsorption water zeolite heat storage for refrigeration applications. *Energy Procedia*, *135*, 513–521. doi:10.1016/j.egypro.2017.09.492

Sircar, S. (1983). Linear-driving-force model for non-isothermal gas adsorption kinetics, *J. Chem. Soc., Faraday Trans. 1. Phys. Chem. Conden. Phases*, *79*, 785–796.

Sukhyy, K., Belyanovskaya, E., Kovalenko, V., Kotok, V., Sukhyy, M., Kolomiyets, E., Gubynskyi, M., Yeromin, O., & Prokopenko, O. (2018). The study of properties of composite adsorptive materials "silica gel – crystalline hydrate" for heat storage devices. *Eastern-European Journal of Enterprise Technologies*, *91*(1), 52–58. doi:10.15587/1729-4061.2018.123896

Sukhyy, K. M., Belyanovskaya, E. A., & Kolomiyets, E. V. (2018). Design and performance of adsorptive transformers of heat energy. Riga, Latvia: LAP Lambert Academic Publishing.

Sukhyy, K. M., Belyanovskaya, E. A., Kozlov, Ya. N., Kolomiyets, E. V., & Sukhyy, M. P. (2014). Structure and Adsorption Properties of the Composites 'Silica Gel – Sodium Sulphate', obtained by Sol – Gel Method. *Applied Thermal Engineering*, *64*(1-2), 408–412. doi:10.1016/j.applthermaleng.2013.12.013

Sukhyy, K. M., Kolomiyets, E. V., & Belyanovskaya, E. A. (2016). Algoritm rascheta ekspluatacionnyh harakteristik adsorbcionnogo akkumulyatora teplovoj energii dlya sistemy decentralizovannogo otopleniya. *Komp'yuterne modelyuvannya: analiz, upravlinnya, optimizaciya: zbirnik naukovih prats*, *1*(1), 65 – 69.

Sukhyy, K.M., Kolomiyets, O.V., Prokopenko, O.M., Gavrilko, A.V., & Belyanovskaya, E.A. (2017). Algoritm rozrahunku ob'yemu adsorbcijnogo teplovogo akumulyatora dlya sistemi decentralizovanogo opalyuvannya. *Komp'yuterne modelyuvannya, analiz, upravlinnya, optimizaciya. Zbirnik naukovih prac'*, *2*, 72 – 76.

Sun, L., Amar, N. B., & Meunier, F. (1995). Numerical study on coupled heat and mass transfers in an absorber with external fluid heating. *Heat Recov. Syst. CHP*, *15*(1), 19–29. doi:10.1016/0890-4332(95)90034-9

Thomas, W. J., & Crittenden, B. D. (1998). Adsorption Technology and Design. Butterworth-Heinemann.

Vučelic, V., & Vučelic, D. (1985). Heat capacities of water on zeolites. *Surface Science and Catalysis: Zeolites Synthesis, Structure, Technology and Application: Proceedings*, 475 – 480.

Wakao, N., & Funazkri, T. (1978). Effect of fluid dispersion coefficients on particle-to-fluid mass transfer coefficients in packed beds: Correlation of Sherwood numbers. *Chemical Engineering Science*, *33*(10), 1375–1384. doi:10.1016/0009-2509(78)85120-3

Zhang, L. Z., & Wang, L. (1999). Effects of coupled heat and mass transfers in adsorbent on the performance of a waste heat adsorption cooling unit. *Applied Thermal Engineering*, *19*(2), 195–215. doi:10.1016/S1359-4311(98)00023-4

Chapter 6
Technology of Heat and Moisture Regeneration for Ventilation Systems

ABSTRACT

The chapter is focused on technology of heat and moisture regeneration for ventilation systems. In the first sub-division recent progress in adsorptive technologies for air dehumidification, heating and conditioning is analyzed. In the next sub-divisions results of original researches of authors on adsorptive heat and moisture regeneration are given. The design of adsorptive heat-moisture regenerator for ventilation systems is shown. Its operation and the results of field tests are described. The technology of regeneration of low-potential heat and moisture by composite sorbent 'silica gel – sodium sulphate' is suggested. Experimental plots of temperature, absolute and relative humidity at the inlet and the outlet of the apparatus and between cassettes with the composite are given. Correlation of flows switch-over time, airflow rate and temperature drop is stated. The relationships temperature efficiency factor vs. dimensionless temperature drop and moisture efficiency factor vs. absolute humidity dimensionless drop are derived with fair accuracy for engineering calculation. Ability of purposeful modification of the above-mentioned characteristics within broad ranges by changing the half-cycle

DOI: 10.4018/978-1-7998-4432-7.ch006

time, the size of the granules of the adsorbent and its amount is revealed. The mathematical model and algorithm for determining the basic parameters of adsorptive regenerator operating processes are developed. The proposed algorithm involves calculating the volume of air passed through the layer of adsorptive heat-storage material, the concentration of water in the airflow at the outlet of the regenerator, the adsorption, the heat of adsorption, the final temperature of the cold air, the air temperature after mixing the cold air from the street and the warm air in the room at the warm end of the regenerator during inflow, calculation of the final concentration of water in the flow at the cold end of the regenerator, the volume of air passing through the layer of heat-accumulating material, adsorption and heat of adsorption, the final temperature of the air at the cold end of the regenerator, the air temperature after mixing of the cold air from the street and the warm air from the room at the cold end of regenerator during outflow, determining the temperature efficiency coefficient, summarized adsorption and maximal adsorption time. The correlation of air temperatures near the warm and cold end of the regenerator, as well as the temperature efficiency factors calculated according to the proposed algorithm and obtained by experimental way is confirmed. The mathematical modeling of the processes of operation of adsorption regenerators based on composites 'silica gel – sodium sulphate' and 'sodium acetate' in the conditions of the typical ventilation system of residential premises is carried out. The dependences of the temperature efficiency factor vs. the time of switching air flows and the velocity of air flow, as well as the temperatures of external and internal air under stationary conditions are shown. An optimal composition of composite adsorbents is stated to be 20% of silica gel and 80% of salt, that is, sodium sulphate or sodium acetate. Due to higher value of maximal adsorption composite 'silica gel – Na_2SO_4' is shown to be required in half as much as compared with 'silica gel – CH_3COONa'. The results of the research can be used in the development of energy-efficient ventilation systems and devices for residential and warehouse premises.

NOMENCLATURE

A_{lim} – maximal adsorption, kg/kg;
c – air special heat, kJ/(kg·°C);
! ' – volumetric specific heat of air, kJ/m^3 ·°C;

$C_{0.r}$ – is initial absolute humidity at the warm end of the regenerator, kg/m^3;
$C_{0.str}$ – initial absolute humidity at the cold end of the regenerator, kg/m^3;
$!_{fin.r}$ – final absolute humidity when inflowed, kg/m^3.
$!_{fin.str}$ – final absolute humidity at the outflow, kg/m^3;
d_{hyd} – hydraulic diameter of the packed bed;
$d_{particle}$ – particle diameter;
F_s – area of the section, m^2.
k – mean dehumidification rate;
K_H – experimental value of heat regeneration factor (heat efficiency factor);
K_W – experimental value of humidity regeneration factor (humidity efficiency factor);
L_i – inhaust airflow rate, m^3/h;
M_{ads} – mass of adsorbent, kg;
Q_i – thermal load for inhaust airflow heating, W;
$t_{0.r}$ – initial air temperature at the room outlet (warm end of regenerator), °C;
$t_{ext.}$ – temperature of external air, °C;
$t_{fin.r}$ – final temperature of the inflowing cold air, °C;
$t_{fin.r.aft.\ mix.}$ – temperature after mixing the cold air from the street and the warm air in the room during inflow, °C;
t_{0str} – temperature of outdoor air (near the cold end of regenerator), °C;
$t_{fin.str}$ – final warm air temperature during outflow, °C;
$t_{fin.str.aft.\ mix.}$ – temperature after mixing the cold air from the street and the warm air from the room at the outflow, °C;
t_{in} – temperature of the internal air, °C;
t_{inf} – temperature of inflowing air, °C;
t_{out} – temperature of outflowing air, °C;
V – airflow rate,
V_{inf} – volume of the inflow air passed through a layer of the adsorptive heat storage material, m^3;
V_{outf} – volume of air passed through the layer of heat storage material during outflow, m^3;
V_r – volume of the ventilated premise, m^3;
V_{str} – volume of air at the outside end of the regenerator, m^3;
w – speed of humid air, m/s;
z – a height of the bed;
Δd – differential of the air absolute humidity;

Δd_{max} – maximal air absolute humidity differential;
Δh – heat adsorption of water vapor, kJ/ mole;
ΔH_{ads} – heat of adsorption, kJ/kg;
$\Delta H_{ads.inf}$ – heat of adsorption during inflow, kJ/kg;
$\Delta H_{ads.outf}$ – heat of adsorption at the outflow, kJ/kg;
Δm_{uptake} – difference between the measured mass at the start and end of adsorption;
Δp – pressure drop across a bed of spherical and granular particles;
$\Delta t_{adsorption}$ – time between adsorption and desorption equilibrium;
ΔT – temperature differential;
ΔT_D – dimensionless temperature differential;
ΔT_{max} – maximal temperature differential;
δ – resistance coefficient;
δ_{Ergun} – resistance coefficient by Ergun Equation;
η_{tem} – temperature efficiency factor (calculated value);
η_{vent} –efficiency factor of ventilator;
μ_{H_2O} – the molar mass of water, g/mole;
ρ – density;
ρ_a – is air density in a premise, kg/m^3;
τ – time of inflow or outflow, s;
ν – kinematic viscosity of air;
ν_{f0} – velocity of the empty adsorber;
ψ – void ratio of the bed.

RECENT PROGRESS IN ADSORPTIVE TECHNOLOGIES FOR AIR DEHUMIDIFICATION, HEATING AND CONDITIONING

The most important and serious problems in the operating the district heating systems of the housing and utilities sector, as well as energy-efficient decentralized heating systems is a securing the load upon heat and ventilation, in particular warming of the inflow air, as well as the periodic increase of the concentration of carbon dioxide in the space air (Őstin, 2017). According to the sanitary-hygienic requirements, periodic ventilation of residential premises is necessary, which results in significant heat loss as a consequence the relatively low temperatures in Ukraine in the winter. To support an acceptable value of humidity in residential premises is obviously one of the urgent options of ventilation systems in the countries not only with cold but with hot and humid climate.

Adsorbents are widely used as a dessicants to dry air. A promising technology for air drying is obviously desiccant dehumidification, because it it can confines to low-potential heat thermal energy from waste heat or renewable energy source [Jia, Dai, Wu, & Wang, 2006, Zheng, Ge, & Wang, 2014). The properties the used adsorbents are crucial for performance of desiccant dehumidification systems (Sultan, El-Sharkawy, Miyazaki, Saha, Koyama, 2015). Currently, in practice drying applications are limited by the mostly commercially available adsorbents such as silica gels or zeolites [Zheng et al., 2014]. These adsorbents exhibit high hydrothermal stability but lower dehumidification rate. Several experimental studies have been focused on identifying the most promising adsorbents which fulfill all desirable properties for drying applications (Jia et al., 2006, White et al, 2011, Ramzy, Kadoli, & Ashok Babu, 2012). However, these experimental studies have been limited to less than 3 commercially available adsorbents. As a rule, studies compare silica gel or zeolite as reference adsorbent to a novel promising adsorbent (Zheng, Ge, Hu, & Wang, 2015, Yu, Wang, Lu, Wang, 2014, Hong et al., 2016).

Erdogan, Bau and Bardow (2019) considered a wide range of commercially available adsorbents at the process conditions which are similar to the operating mode. The adsorbents were characterized experimentally by three indicators which were selected according to the most important objectives:

1. working capacity to reflect the necessary amount of adsorbent,
2. pressure drop across the adsorbent bed to reveal the necessary auxiliary energy for ventilation and
3. dehumidification rate which results in reflects the process duration.

Erdogan et al. [2019] evaluated 12 commercially available adsorbents: silica gels SG 125, SG 127, SG 127H, SG 125B, SG 127B, ProSorb, ArtSorb, SG E (2–4 mm), SG E (3–6 mm), SG M) and zeolites (Zeolite 13X, AQSOA Z02) for conditions derived from an adsorption dishwasher application. Notably, an adsorption temperature of 36 °C, a desorption temperature of 150 °C and a relative humidity of 71% were used.

Results for the working capacity, i.e. water uptake or adsorption, are given on Figure 1. At first, results for the volumetric capacity are represented in the Figure 1a. The highest volumetric capacity is exhibited by SG 127B and SG

E (3 – 6mm) with 289 g/l and 271 g/l, respectively. The lowest volumetric capacity is shown by SGM with 35 g/l. SGM is the only adsorbent with a wide pore size and the lowest inner surface of 300 – 400 m²/g which leads to a low working capacity. The second lowest working capacity of about 129 g/l is stated for Zeolite 13X which resulted from the lower inner surface as compared to the other adsorbents and the low desorption temperature, zeolite 13X being the adsorbent with the second lowest inner surface area of 488

Figure 1. Working capacity of adsorbents between adsorption and desorption equilibrium (adsorption temperature of T_{ads} = 36 °C and φads = 71% for a regeneration temperature of T_{des} = 150 °C). (a) Volumetric capacity and (b) mass-specific capacity. Due to the low measurement error of the scale, the error bands are neglectable (Erdogan et al., 2019).

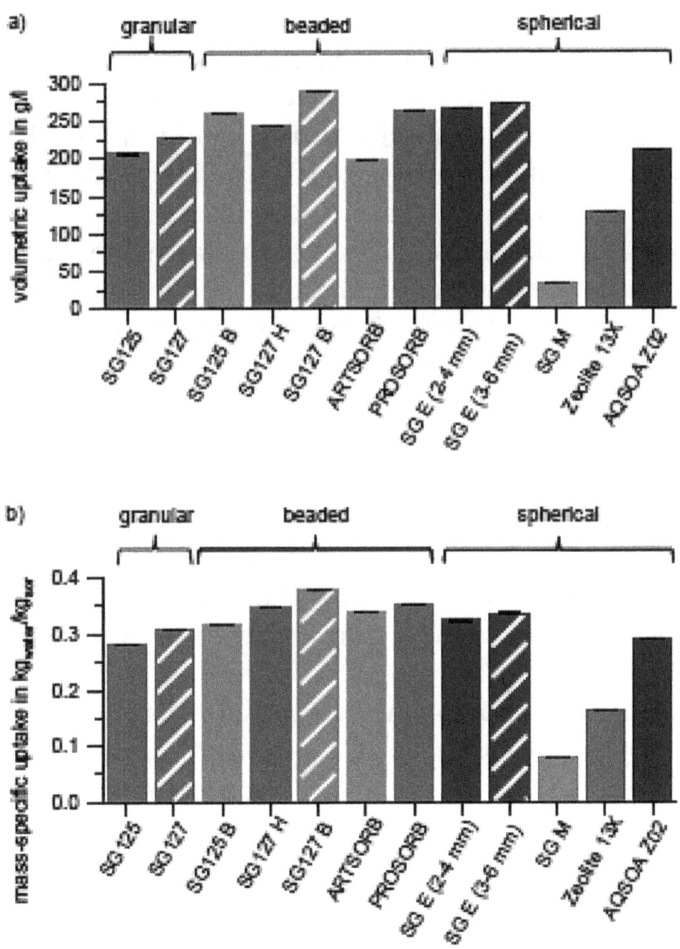

m^2/g. Moreover, the desorption temperature of 150 °C is not sufficient to fully desorb Zeolite 13X. ArtSorb, SG 125 and AQSOA Z02 shows similar working capacities of about 200 g/l. ProSorb, SG E (2–4mm), SG E (3–6mm), SG 125B and SG 127H are stated to have similar working capacities of about 260 g/l.

Results of the mass-specific capacity of adsorbents are represented in Figure 1(b). The mass-specific capacity is utterly important feature for applications with a mass limitation (e.g. mobile applications) or with expensive adsorbents. A qualitative difference between volumetric- and mass-specific capacity is noticeable only for ArtSorb and SG 125B. As compared with their volumetric capacity, the mass-specific capacity of SG 125B is lower and the mass-specific capacity of ArtSorb is higher. Such difference between volumetric and mass specific capacity is resulted from the bulk density of the adsorbents. Specifically, Art-Sorb has a lower bulk density than the other adsorbents which results in a higher mass-specific capacity. By contrast, SG 125B has a higher bulk density as compared to the other adsorbents which results in a lower mass specific capacity. SG 127B also shows the best results with a mass-specific capacity of 0.38 kg of water per kg adsorbent. The lowest mass-specific capacity is again stated for SGM with 0.08 kg of water per kg of adsorbent and Zeolite 13X with 0.16 kg of water per kg of adsorbent. AQSOA Z02 and SG 125 reveal similar mass-specific capacities of about 0.29 kg of water per kg of adsorbent. The other adsorbents exhibit a mass-specific capacity of around 0.33 kg of water per kg of adsorbent. The different working capacities of the adsorbents are assumed to result from the structural differences of the adsorbents.

Generally, a low pressure drop across the bed is favorable in order to reduce the necessary fan power and thus, the auxiliary energy consumption. One way to reduce the pressure drop across the packed bed is to change the particle shape or increase the grain size. We studied different particle shapes and grain size distributions and determined the pressure drop across the bed for the selected adsorbents. The pressure drop over the median grain diameter is shown in Figure 2.

The measured pressure drop can be compared to correlations that describe the pressure drop across a packed bed. For this purpose, we used correlations for the pressure drop Δp across a bed of spherical and granular particles based on Kast (1988) as follows:

Figure 2. Determined pressure drop across the packed bed over measured median grain size. Solid and dotted black lines indicate pressure drop across the bed for granular and spherical particles respectively, based on Kast (1988). Error bars of pressure drop refer to the pressure drop sensor used for measurements. Grain size given by the supplier is shown as error bars
(Erdogan et al., 2019).

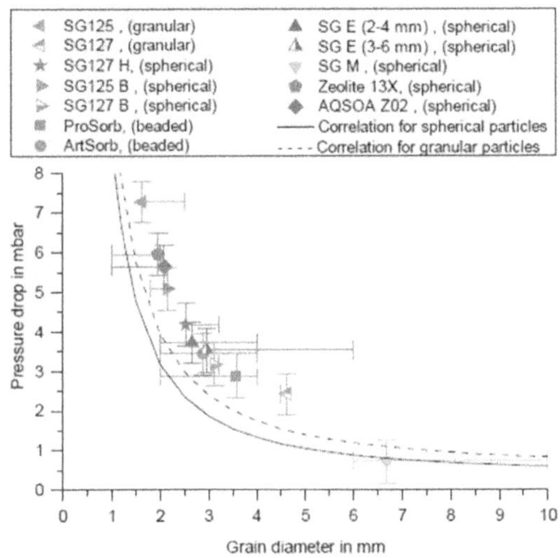

$$\Delta p = \delta \frac{z}{d_{hyd}} \frac{\rho}{2} \left(\frac{\nu_{to}}{\psi} \right) \tag{1}$$

with

$$\delta_{Ergun} = \frac{133.3}{\mathrm{Re}_{\psi,hyd}} + 2.33 \tag{2}$$

with

$$\delta_{spherical} = 2.2\left(\frac{64}{\mathrm{Re}_{\psi,hyd}} + \frac{1.8}{\mathrm{Re}_{\psi,hyd}^{0.1}}\right) \tag{3}$$

or

$$\delta_{granular} = 2.2\left(\frac{64}{\mathrm{Re}_{\psi,hyd}} + \frac{2.6}{\mathrm{Re}_{\psi,hyd}^{0.1}}\right) \tag{4}$$

where

$$d_{hyd} = \frac{2}{3}\cdot\frac{\psi}{1-\psi}\cdot d_{particle}\,, \tag{5}$$

and

$$\mathrm{Re}_{\psi,hyd} = \frac{v_{f0}}{\psi}\frac{d_{hyd}}{v} \tag{6}$$

where δ is a resistance coefficient, z is a height of the bed, d_{hyd} corresponds to the hydraulic diameter of the packed bed, ρ refers to the density, ν_{f0} is the velocity of the empty adsorber, ψ is the void ratio of the bed, d_{particle} corresponds to the particle diameter and ν is the kinematic viscosity of air. Inserting the resistance coefficient δ_{Ergun} results in the well-known Ergun Equation (Ergun, 52).

The Ergun equation is a general equation for calculation the pressure drop across a randomly packed bed covering laminar and turbulent flow conditions. However, as the pressure drop being affected by particle shapes, the Ergun Equation is modified with empirical resistance coefficients which consider the particle shape with $\delta_{\mathrm{spherical}}$ for spherical particle and $\delta_{\mathrm{granular}}$ for granular particle.

The first component of the resistance coefficients (Eqs. (6.2)–(6.4)) corresponds the laminar flow conditions. The second component of the resistance coefficients (Eqs. (6.2)–(6.4)) describes the turbulent flow through the packed bed.

According to Kast (1988), the void ratio for packed beds is variated within a range of about 0.37–0.4. For calculations Erdogan et al. (2019) are chosen an average void ratio of 0.385. The calculated and measured pressure drops through the bed for spherical and granular particles are also represented in Figure 2. Error bars indicate the grain size given by supplier. The supplier information can be considered as inadequate, because the measured median diameter can significantly deviate from the given values.

It should be noted that the correlations for the pressure drop generally underestimate the measured pressure drop across the bed. As compared to the pressure drop correlation for spherical particles, the pressure drop for spherical and beaded particles is stated to be underestimated by 60–90%, with one exception: SGM shows an about 12% lower pressure drop across the bed in reality than the correlation predicts for the corresponding grain size. However, this deviation of 12% corresponds to the range of the uncertainty (Erdogan et al., 2019).

For granular particles, the calculated pressure drop is observed to deviates less from the measured values. For granular particles with small median grain diameter of 1.6 mm (SG 125), the pressure drop is underestimated by 26% (Erdogan et al., 2019).

Furthermore, for granular particles with a large median grain diameter of 4.6 mm (SG 127), the pressure drop is underestimated by 56%. The deviation between calculated and measured values can be resulted from that the particles do not have a unique grain size but a grain size distribution which affects the void ratio and as a consequence the pressure drop. Eventually, correlations of pressure drop across the bed can provide only a lower bound on the actual value and only if the actual grain diameters are employed.

The dehumidification rate is an important criterion to estimate the adsorbents, because the dehumidification rate determines the necessary adsorption time. The mean dehumidification rate is calculated as a ratio of the water removal per time between adsorption and desorption equilibrium, as follows:

$$k = \frac{\Delta m_{uptake}}{\Delta t_{adsorption}} \tag{7}$$

where Δm_{uptake} equals the difference between the measured mass at the start and end of adsorption, $\Delta t_{adsorption}$ refers to the time between adsorption and desorption equilibrium.

Figure 3. Mean dehumidification rate (a) and mean specific dehumidification rate (b) over pressure drop across the packed bed by Erdogan, et al. (2019). Dotted black lines indicate the Pareto frontiers. There are no error bars in the y-axes included, since the uncertainty is mostly below the size of the symbols.

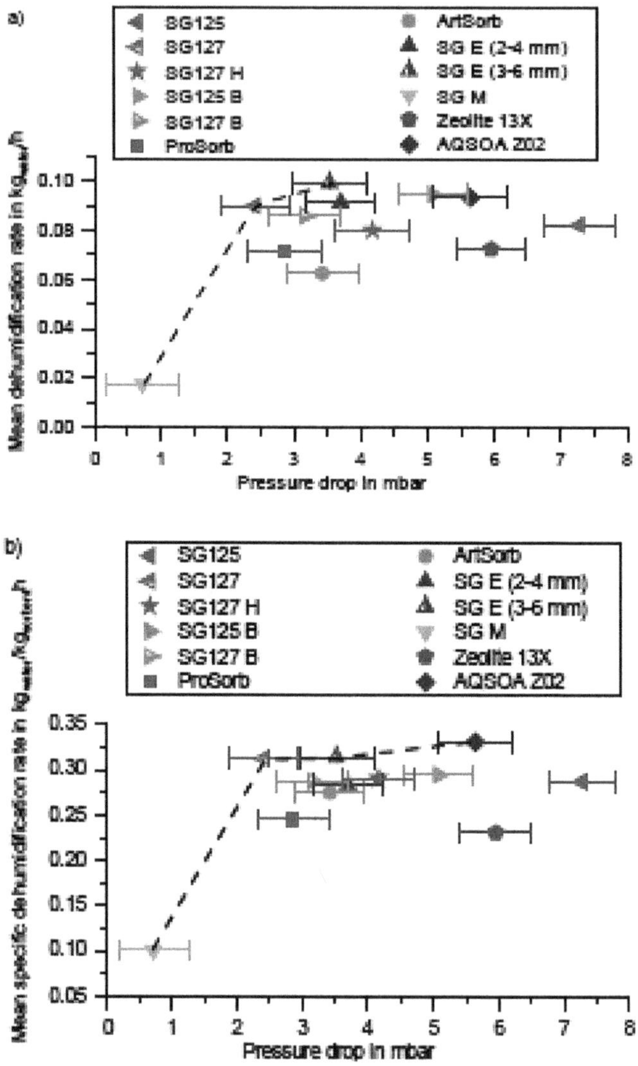

Obviously, the dehumidification rate is affected by heat and mass transfer. As a rule, heat and mass transfer characteristics are enhanced when grain sizes decreased (Bathen & Breitbach, 2001). However, the pressure drop also increases for small grain sizes (Erdogan et al., 2019). This compromise is illustrated by the plot the mean dehumidification rate over the pressure drop (Figure 3).

The Pareto frontier describes the dominant solutions for mean dehumidification rate and pressure drop across the bed. So, the best adsorbents for the selected process conditions are indicated as SG M, SG 127 and SG E (3–6 mm).

The highest mean dehumidification rate is observed for SG E (3–6 mm) with almost 0.1 kg water/h but inducing a pressure drop of about 3.5 mbar. SG 127 decreases the pressure drop by 30% to 2.4 mbar with only a small decrease in the mean dehumidification rate by about 9% to 0.09 kg water/h. SGM also corresponds to the Pareto frontier due to its very low pressure drop of 0.7 mbar but also it exhibits a very low mean dehumidification rate of 0.018 kg water/h. The reason for the low dehumidification rate is the lowest working capacity of SGM (cf. Figure 1).

These 3 adsorbents dominate over all other adsorbents. The mass-specific mean dehumidification rate vs. the pressure drop is given in Figure 3b. Here, again, SG M, SG 127 and SG E (3–6 mm) correspond to the Pareto frontier but AQSOA Z02 can be also one of the most suitable adsorbents for the chosen process conditions. AQSOA Z02 represents the highest mean specific dehumidification rate (0.33 kg water/kg adsorbent·h), followed by

Figure 4. Mean specific dehumidification rate over mass-specific capacity by Erdogan et al. (2019). Dotted black line indicates the Pareto frontier.

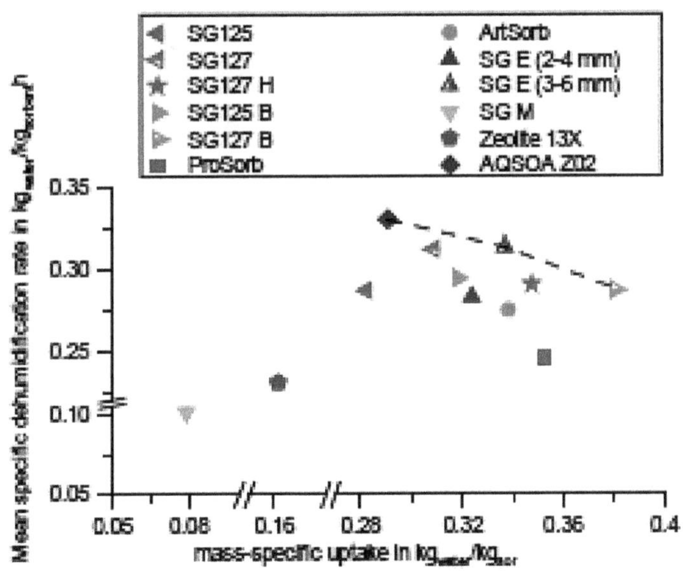

SG E (3 – 6 mm) and SG 127 with about 0.31 kg water/kg adsorbent·h and SGM with 0.1 kg water/kg adsorbent·h.

The best adsorbents according to the mean specific dehumidification rate and working capacity are shown in Figure 4. The Pareto frontier emphasize on AQSOA Z02, SG E (3–6 mm) and SG 127B as most suitable adsorbents as a consequence their mean specific dehumidification rate and mass-specific capacity.

AQSOA Z02 and SG 127B are the key points of the Pareto frontier (Figure 4), because AQSOA Z02 reveals the highest mean specific dehumidification rate (0.33 kg water/kg adsorbent·h) and SG 127B exhibits the highest mass-specific capacity (0.38 kg water/kg adsorbent·h). SG E (3 – 6 mm) is the third adsorbent of the Pareto frontier with 0.31 kg water/kg adsorbent·h and 0.33 kg water/kg adsorbent·h. So, Pareto optimal adsorbents for dehumidification of air are identified.

So, the compromise between the specific dehumidification rate and the working capacity (both, mass-specific and volumetric) reveals that AQSOA Z02, SG E (3 – 6 mm) and SG 127B correspond with the Pareto frontier. According to the specific dehumidification rate and working capacity, the

Figure 5. Comparison between the operation of DAC and VAC system (Sultan et al., 2015)

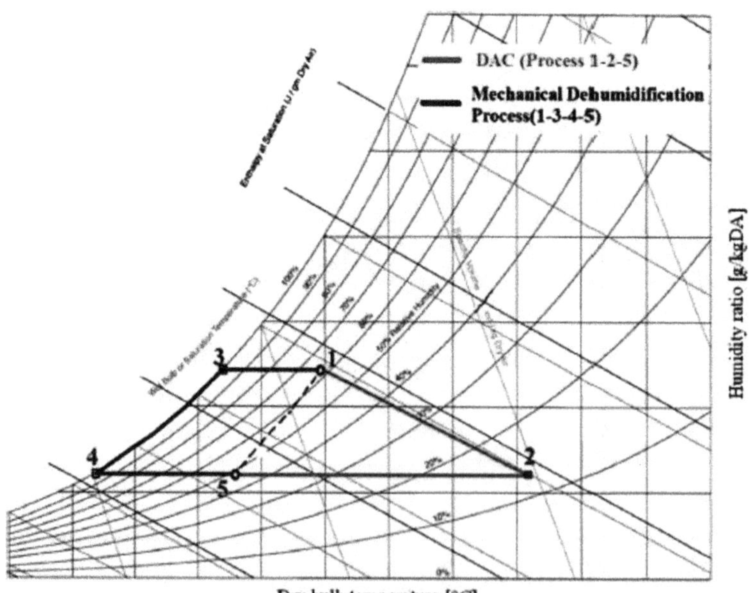

adsorbents of choice ought to be one of these Pareto-optimal 3 adsorbents: AQSOA Z02, SG E (3–6 mm) and SG 127B.

The water adsorption affinity of desiccants make them able to be used in the applications where the main objective is not only to remove the moisture from air in the residential or storage premises but maintain the required temperature. Humidity control is obviously needed in everyday life but it is essential in some cases e.g. hospital buildings where the patients and staff members are under the risk of infection through germs, bacteria or virus and these issues are directly associated with humidity. Relative humidity range associated with different health issues such as allergic rhinitis, asthma and respiratory infections.

The desiccant air-conditioning (DAC) involves not only humidity control with desiccant dehumidification, but also temperature control by evaporative cooling. The desiccant air conditioning is a promising technology because it is free from chlorofluorocarbons, hydro-fluorocarbons and hydrochlorofluorocarbons. Moreover, desiccants can be regenerated by low grade thermal energy. Besides the conventional room air-conditioning (AC), DAC can be of use for industrial processes, hospitals, museums, marine ships, greenhouses, drying grains, product preservation and storage etc. As a rule, the conventional vapor compression air-conditioning (VAC) systems apply to obtain the desired conditions. They use environmentally harmful refrigerants and consume primary energy in an inefficient way.

The VAC controls the humidity by supporting the temperature below the inflow air dew point so the heating is required to obtain the desired conditions of temperature and humidity as shown in process (1–3–4–5) in Figure 5 (Sultan et al., 2015). As a result the VAC system cannot deal the sensible and latent load of air-conditioning distinctly. However, for desiccant air-conditioning the latent heat load of air-conditioning is transformed into sensible heat load due to desiccant dehumidification according to the process (1–2–5) in Figure 5. Moreover, quality of the internal air can be enhanced significantly by utilizing high ventilation, air flow rate and the capability of removing airborne pollutants.

The basic principle of DAC is given in Figure 6.

The air dehumidifying results from natural sorption process between the desiccant and humid air. Obviously, the water vapor pressure difference between the desiccant and process air is the driving force. As the equilibrium condition between the desiccant and air achieved, the water sorption process is stopped until external forces are applied i.e. increasing air pressure or decreasing the temperature. Since the adsorption is an exothermic process, the dehumidified

Figure 6. Principle of desiccant dehumidification and air-conditioning
By Sultan et al. (2015).

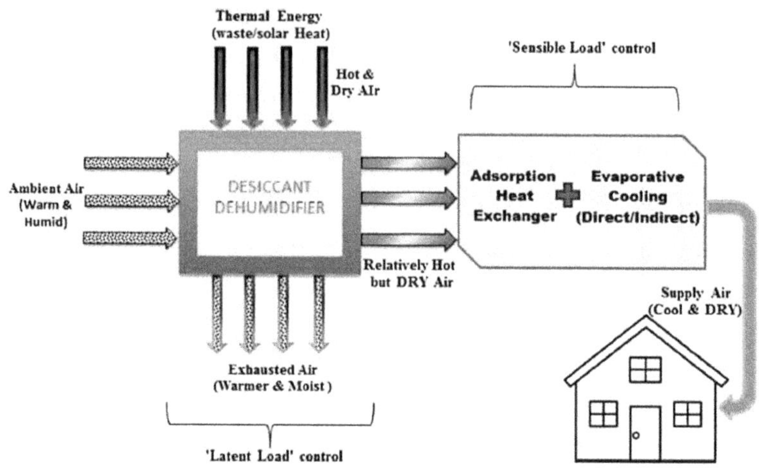

air temperature is increased. The value of this increasing clearly depends on the characteristics of adsorbent as discussed in Chapter 4. Then, the dehumidified air is then cooled at constant humidity using heat exchanger and meanwhile there covered heat is used for the desiccant regeneration which increases the coefficient of performance of the system (Sultan et al., 2015). Then the process air is pushed into the thermal comfort zone by following cooling techniques: (i) direct/indirect evaporative cooling, (ii) cold coil, (iii)M-cycle cooling, and (iv) evaporator of the vapor air-conditioning. In case of direct evaporative cooling the addition of water vapors is an isenthalpic process, the rest of cooling being performed at constant humidity. M-cycle cooling technique is a kind of indirect evaporative cooling. The air is cooled near to

Figure 7. Rotary desiccant dehumidifier
By Rambhad, et al. (2016)

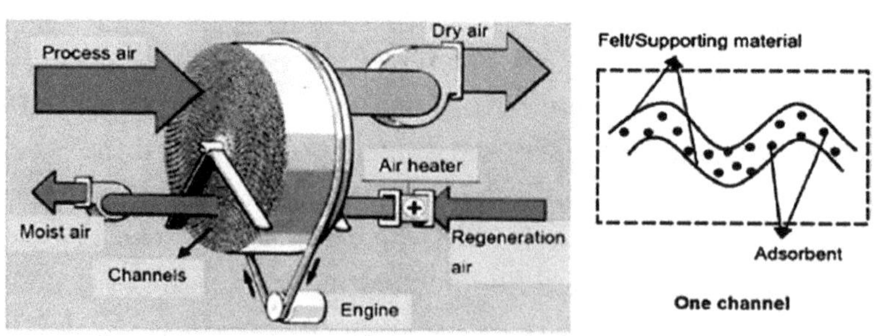

the dew point rather than wet bulb temperature. The desiccant regeneration can be resulted from heating the desiccant or reducing its pressure. A pressure reduction is adopted in industrial processes, while heating is preferred in commercial applications (Sultan et al., 2015).

A typical desiccant air conditioning system involves a desiccant unit, heat recovery component, a cooling unit and a regeneration unit. The available solid DAC systems can be classified as: (a) adsorbent bed or rotating wheel type, (b) single-stage or multi-stage dehumidification type, (ci) single-rotor or multi-rotor type, (d) hybrid system supported by vacuum air-conditioning, and (d) electricity/gas/solar/waste energy operated desiccant air-conditioning system.

In rotary desiccant wheel, an air to air heat and mass transfer occurs at low rotation speed. A thin layer of desiccant material is set in a frame of a wheel. The channels of desiccant wheel frame have a various structures such as honeycomb, triangular, sinusoidal etc. (Rambhad, Walke, & Tidke, 2016). The basic operating principle of rotary desiccant dehumidifier are represented by Figure 7. The cross-section of wheel is divided into air sides of moist(process) and regeneration. When the wheel constantly rotated through two separate sections, the process air is dehumidified by the desiccant because of the adsorption effects of the desiccant material. At the same time,

Figure 8. Schematic figure of solar powered desiccant wheel evaporative cooling system (SDECS)
(Ge et al., 2014).

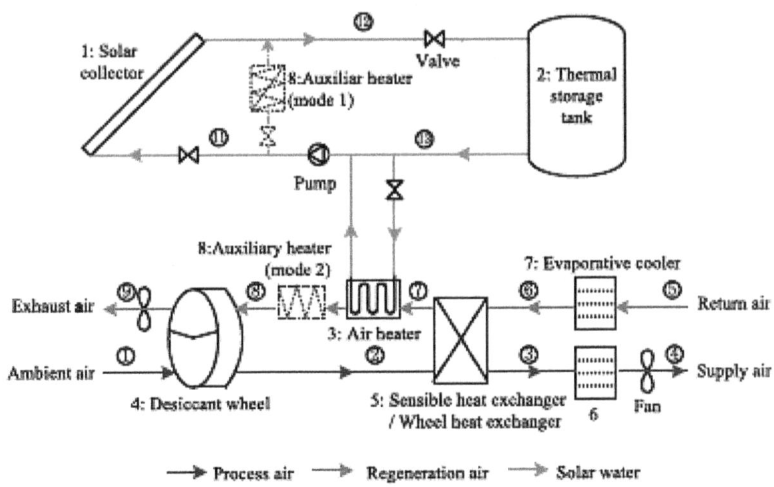

the regeneration air is humidified after heating by a heater and desorbing the water from the wheel.

The widely adopted basic type of solar rotary desiccant wheel cooling system is Solar powered Desiccant Wheel Evaporative Cooling System (SDECS) which is represented by Figure 8 (Ge, Dai, & Wang, 2014). This system mainly involves two subsystems: solar subsystem and desiccant cooling subsystem. The main components of the solar subsystem are similar to conventional solar hot water system, i.e. solar collector, auxiliary heater and water tank. Solar collector (1) absorbs solar radiation and heats up the water which is utilized in air heater (3) to heat up regeneration air. Water tank (2) acts like a heat storage unit which also helps to adjust the flux of water. The desiccant cooling subsystem is operated according to basic ventilation mode: process air from ambient condition is dehumidified and heated in desiccant wheel (4). Further the air is cooled in sensible heat exchanger

Figure 9. Schematic diagram of experimental set-up for solar thermal and electric driven desiccant air-system
(Enteria et al., 2010).

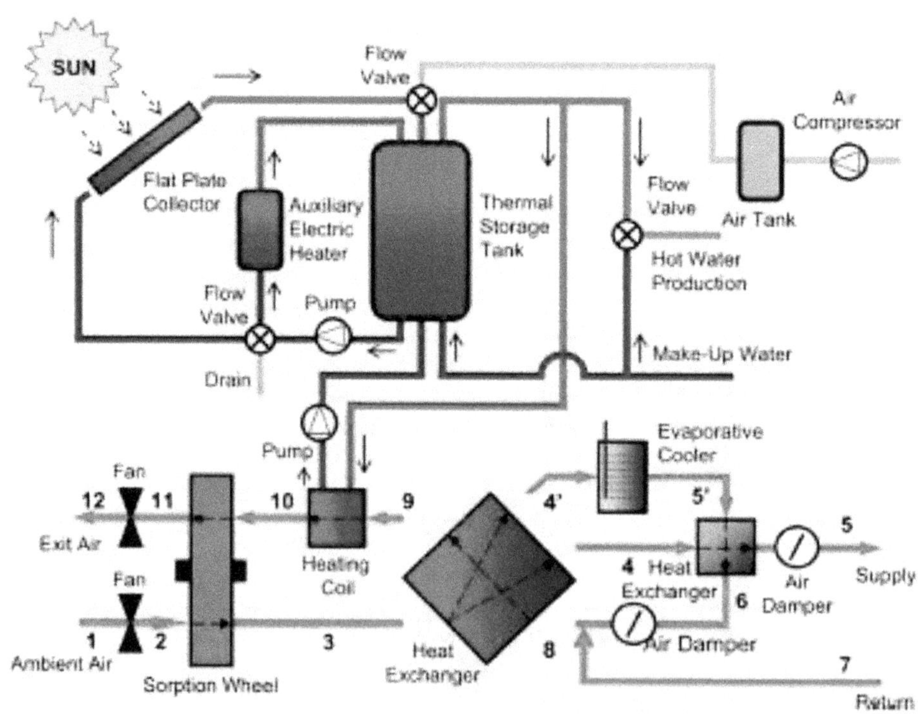

(5), heat exchanger wheel or plate type heat exchanger and is then cooled in evaporative cooler (6), at last the air is delivered to conditioned premise. Regeneration air from indoor condition passes through evaporative cooler (7), sensible heat exchanger (5), air heater (3) and desiccant wheel (4) in series. These two subsystems are connected by air heater (3), in which hot water from solar subsystem is utilized to heat up regeneration air in desiccant cooling subsystem. It should be noted that auxiliary heat source (8) such as gas heater or electricity is installed in some systems to compensate daily or season inequability of solar radiation and ensure the continuous operation. Auxiliary heater can be mounted in solar subsystem (mode 1) or installed directly in regeneration air side (mode 2).

According to data in (Ge T.S., Zhang, Dai, & Wang, 2017), a novel solid desiccant configuration called desiccant coated heat exchanger, which is fabricated by directly coating solid adsorbent to metal outer surface of conventional fin tube heat exchanger. Due to the heat exchanger can provide an inner cooling source for outer coating, descant coated heat exchanger can realize isothermal even temperature-decreasing dehumidification process as compared with desiccant wheel, which also leads to enhanced adsorptive capacity of desiccant with lower adsorption temperature and as a consequence to lower regeneration temperature.

A solar thermal and electric driven desiccant cooling system was developed at Tohoku University, Japan by Enteria et al. (2009). Schematic diagram of the experimental set-up is shown in Figure 9 (Enteria et al., 2010). Storage of the thermal energy during night time for daytime initial operation enhances efficiency of the system by using relatively lower price of electricity hence was preventing total reliance on solar energy. The system was preliminarily tested in winter season. It was shown to be economically efficient when operated in day and night time. The same authors expanded the study the aim to develop a novel desiccant heating, ventilation and air-conditioning system (Enteria et al., 2010). The external air conditions were set to the standard summer condition in Japan i.e. 30 °C and relative humidity of 60% RH with flow rate of 200 m^3/h. There turn air was kept at standard internal air condition in Japan i.e. 26 °C and relative humidity of 55% RH with flow rate of 100 m^3/h.

Dehumidification performance of the system was improved with the increasing regeneration temperature but the supply air temperature was also increased. Furthermore, the system coefficient of performance started decreasing from 0.44 to 0.35 as regeneration temperature increased from 60 to 75 °C. So, the thermal heat supplied to the desiccant air conditioning system is not exactly proportional to the final cooling effect, an optimum regeneration temperature being required to obtain efficient performance of a desiccant air-conditioning system.

A single-stage solar operated hybrid desiccant cooling system was designed and installed at Central Queensland University, Australia (Baniyounes, Rasul, & Khan, 2013). The schematic diagram of the system is represented by Figure 10. The system mainly involves a desiccant wheel, a heat exchanger, an evaporative cooler, flat plate solar collector of 10 m^2, two backup heaters, and a vapor compression cooling unit. The system was analyzed for a metrological year with process air and regeneration air flow ratio of 3 to 1. The cooling load for the reference building was estimated by the same authors which was found to be 6428 kWh with a maximum of 809 kWh in December (Baniyounes, Rasul, & Khan, 2013).

Figure 10. Schematic diagram of single-stage hybrid desiccant air-conditioning system (Baniyounes, Rasul, & Khan, 2013).

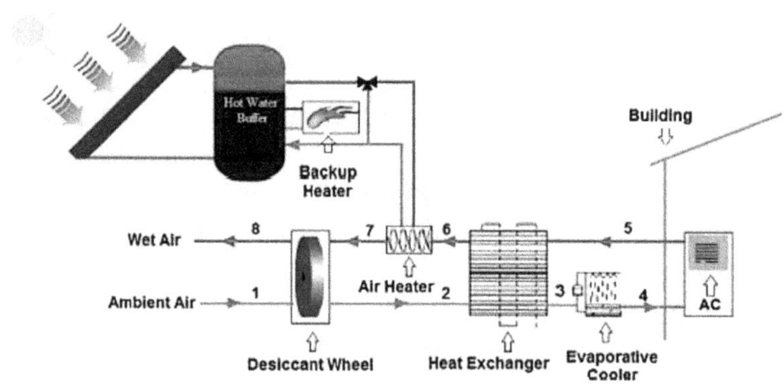

Obviously, performance of the desiccant air-conditioning is strongly affected by operational properties of the adsorbent used. As a rule, designed systems for air-conditioning are based on conventional adsorbents, such as: silica gels of various grades or zeolite 13X (Erdigan et al., 2019), (Silico)aluminophosphates or [115], special Y-type zeolites (Freni, Maggio, Sapienza, Frazzica, Restuccia, & Vasta, 2016), molecular sieves 3A, 4A, 5A, 13x (Rambhad et al., 2016). Regeneration temperatures of the most of these materials are higher than 100°C that strongly limits the driving temperatures. Moreover, their water uptake or adsorptive capacity does not exceed 0.2 – 0.3 kg of water per kg of adsorbent that results in high volumes of adsorptive units.

Another way to improve of heat and moisture conditions in residential premises and to enhance thermal efficiency of the ventilation systems is associated with heat and moisture regeneration. It is of current interest for countries with a cold climate (typical of Ukraine, Russia, Canada, the North Europe and USA), where the difference between indoor and outdoor temperatures can reach in winter up to 30°C or even more (Aristov, 2015). It results in enormous heat losses. Furthermore, during cold winter moisture is continuously lost through ventilation system. Indeed, the absolute humidity of supplied (external) air is extremely low (e.g. at most 0.29 g/m^3 at -30 °C), that leads to falloff the internal relative humidity (down to 10–20% or even below) in winter season and, as a consequence, to a great disbalance the indoor thermal and moisture comfort. So, efficient exchange of heat and moisture between the inflow and outflow air fluxes is urgent for reducing the heat losses. Therefore, in cold countries it is extremely important to organize not only heat exchange, but moisture exchange between the fluxes as well to maintain the indoor humidity within the reasonable range.

Equipment for heating the inflow air in modern ventilation systems of heat supply involve the next units: recuperative heat exchangers (plate heat exchangers with cross flow, plate heat exchangers with counterflow, counter flow channel heat exchangers, partially recirculated system) and regenerative (rotary heat exchanger, static heat exchanger) (Danilevskij, 2014). The disadvantages of the first are caused by a significant volume, thermodynamically unfavorable flows, insignificant surface of heat transfer, the use of liquid thermal fluid and expensive additional equipment. Furthermore, the performance of such devices is substantially complicated by the formation of ice on the cold end of the device. The disadvantage of others (Danilevskij, 2014) is the low efficiency coefficient of ventilation of apartments, the lack of separate air control; fresh air pollution resulted from constant change of air direction. Conventional ventilation and air conditioning systems based on vapor compression machines also provide air cooling to temperatures lower than dew point and heating to acceptable temperatures. Operating of such devices requires significant energy consumption, which requires fossil fuels input and complicates the utilization of heat energy.

In order to fill these gaps of the current technology, the following steps must be taken: to create an efficient heat exchange between outflow and inflow air streams for reducing heat losses; to provide moderate drying of exhaust air in order to avoid the formation of ice at the outlet; to increase the humidity of the air supplied to the room to sustain the necessary thermal and humidity comfort conditions.

A new solution (the so called VENTIREG) was suggested by Aristov (2015) for simultaneous regeneration of both heat and moisture in ventilation systems for cold climate countries. This approach was aimed to ensure an efficient heating and moisturizing of the inflowing air to sustain internal conditions of human thermal and humidity comfort.

The main principle of the VENTIREG method is given by Figure 11.

Figure 11. Scheme of the regeneration process: the temperature and moisture profiles at various times $t_1 < t_2 < t_3$
By Arictov (2015).

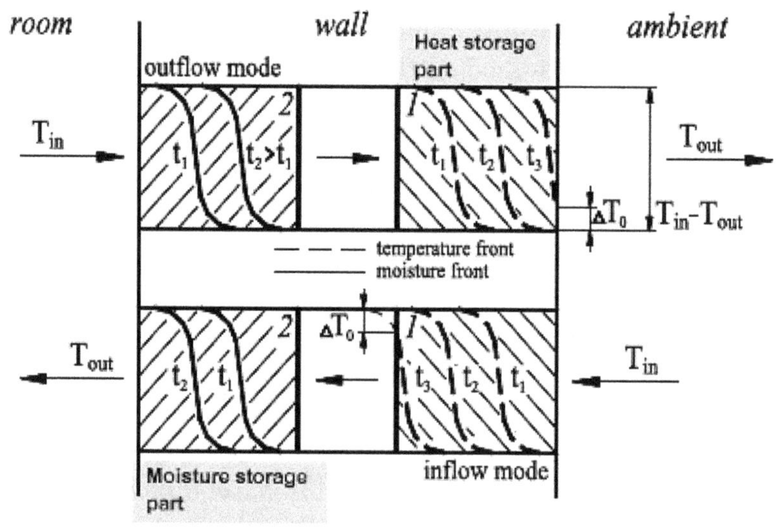

The regenerator includes two layers of heat storage materials (Aristov, 2015). The granulated layer of the sensible heat storage material (1) is maintained closer to the unit outlet, i.e. it is placed closer to the ambient air. Before this layer, layer 2 of water adsorbing material is located closer to the room side. The sensible heat storage material is designated to exchange the sensible heat between the inlet (fresh) and outlet (exhausted) air fluxes. The adsorbent layer works as a water buffer.

The regenerator operates in two modes: outflow and inflow. At the outflow mode a warm and humid indoor air is blown by an ejective fan through the relatively dry adsorbent, which captures and retains the indoor moisture.

Dried and warm air enters layer 1 and heats it up. Then, the air flow switches and inflow mode begins. On inflow mode, a dry and cold outdoor air is inhausted by a supplied fan through the warm layer 1 and is heated up to the temperature close to that in the premise T_{in} and thus, recovers the stored heat. Passing through the layer of the humid adsorbent, warm and dry air causes desorption of the retained water and come back to the room, maintaining the indoor moisture balance. As a consequence of the finite heat capacities of layers 1 and 2, the temperature of inflowing air is slowly decreasing, and the air flow switches when the temperature difference between premise and sensible heat storage material (T_{in}– T) reaches a predetermined value ΔT_0, and so on.

As buffers of water, both common (silica, alumina) and novel (SWS-1A, i.e. alumina impregnated with $CaCl_2$) adsorbents were used (Aristov, 2015). The composite sorbent SWS-1A exhibited better performance as compared with the common commercial adsorbents. Due to higher adsorption capacity of this composite, the adsorbent loading is less by a factor of 2–3 that allows to use smaller unit size and to reduce hydrodynamic resistance of the unit. Obviously, this could allow the using of cheap blade-type fans instead of centrifugal ones and significantly decrease the electricity consumption.

Despite a careful attention devoted to ventilation systems based on adsorptive unit as an ecofriendly replacement of conventional vapor compression system, they still could not break the industrial barriers. Many kinds of desiccant systems are working all over the world with satisfactory performances. Along with the absence and insufficient implementation of global rules for the use of environmentally harmful technologies, an active commercialization of adsorptive ventilation and air-conditioning systems are strongly inhibited by the high initial cost and the long payback period. The running cost of the system generally depends on the adsorbent regeneration. As a consequence, adaptation of adsorptive ventilation and air-conditioning system is complicated where thermal energy sources are limited. As a rule such systems are designed for the regions having enough thermal energy in terms of solar or waste heat. Harvesting there low-grade heat energy is itself expensive which ultimately increases the initial cost of the adsorptive ventilation and air-conditioning system.

Some of the researchers suggest applying desiccant air-cooling for central cooling as compared to split type AC. The vapor air-conditioning industry has a grip on the market which results in the social barriers for the adsorptive technology. Reinforcement of the vapor air-conditioning industry is caused by impetuous urbanization in fast-paced countries such as China and India due to the high demand because of big population and business interests. Another barrier for commercialization of adsorptive ventilation and air-conditioning is also lack of knowledge and familiarity with adsorptive technologies between designers, developers, architects and end users. In this circumstance it is difficult for adsorptive ventilation and air-conditioning technologies to penetrate in the market easily unless having advanced ventilation or air-conditioning system and/or a push from the governments and end users.

The desiccant air-conditioning is based on moisture dehumidification which limits its field of application to relatively moderate in high humid regions. Most perhaps a certain desiccant air-conditioning system design could be practicable for particular region and/or conditions. However, it requires more research to establish such standard systems for each condition. At this stage the size of the system and adsorptive unit becomes also an obstacle for practical application of the technologies. Up to now they are reduced still not enough compete with vapor air-conditioning system. Nevertheless, most of developed ventilation and air-conditioning systems are based on conventional adsorbents with rather low adsorptive capacity, such as silica gels or zeolites. In these conditions, application of advanced composite adsorbents of the type 'salt inside porous matrix' with enhanced adsorptive capacity becomes of urgent necessity. Devices based on them allow maintaining an acceptable temperature and humidity in domestic environment. But the active introduction of the ventilation technology based on new composite adsorbent alongside with mostly complicated techniques for their obtaining is strongly retarded by the lack of an algorithm for their calculation, which allows to simulate and to study the processes of their operation in the conditions of real ventilation systems.

ADSORPTIVE REGENERATOR OF HEAT AND MOISTURE

According to previous chapter, adsorptive heat storage devices are promising for decentralized heating systems. Another field of application of adsorptive energy conversion is adsorptive heat and moisture regeneration.

However, the introduction in practice of such technologies is suspended by the lack of an algorithm for their calculation, which allows simulating the processes of their performance and to determine the design and optimal operational parameters in real conditions of ventilation systems.

Design of the Laboratory Prototype of Adsorptive Regenerator of Heat and Moisture

The adsorptive regenerator of heat and moisture for conditioning of premises (temperature 20ºC and relative degree of humidity 50 – 60% according to State Standard of Ukraine B EN 15251:2011) is developed both to heat cold inhaust air and to sustain its humidity on the comfort level when oxygen concentration kept to be constant. This device is proposed to maintain in ventilation channel. Its scheme is given in Figure 12.

The suggested adsorptive heat regenerator consists of the next basic units: cylindrical case (pipe) (1), cassettes (4) filled with adsorptive heat storage material, external (2) and inner (3) ventilators. Length of laboratory prototype of heat regenerator is 0.6 m. The device operational phases are alternated 'inflow' and 'outflow'. As 'outflow' phase applied, air is aspirated by ventilator 3 and released into the ambient medium. During this process adsorptive heat storage material 4 is warmed with 'outflow' air up to necessary temperature of inner space.

Figure 12. Heat regenerator construction: 1 – pipe (case); 2 –inlet ventilator; 3 – air exhauster; 4 – cassettes with adsorptive heat storage material; 5 – temperature detector; 6 – operating console
(Kolomiyets, Sukhyy, Belyanovskaya, & Tomilo, V. I., 2016).

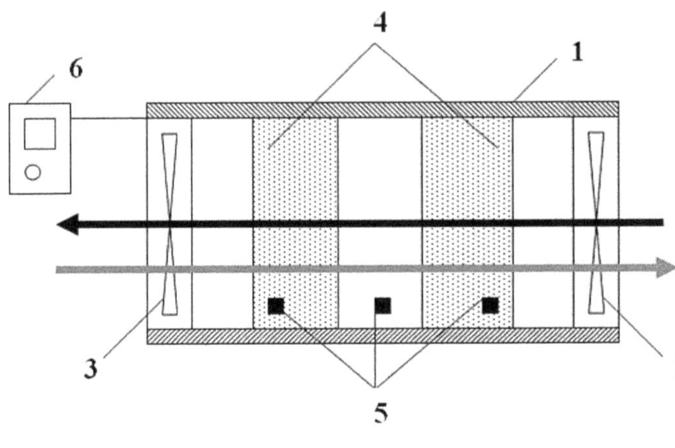

During the 'inflow' phase, air exhauster 3 is switched off, ambient air is intaken by the inlet ventilator 2 and passed through the adsorbent layers into aired premise. Air is passed through the layer of adsorbent which adsorbed the excess of moisture and heated. So, ventilation of premise is performed, both temperature and humidity being supported to be constant.

The composite adsorbents 'silica gel – sodium sulphate' and 'silica gel – sodium acetate' are used as heat storage materials. They are synthesized by sol – gel method from tetraethoxysilane, Na_2SO_4 or CH_3COONa, ethanol (as a solvent) and hydrochloric acid (as a catalyst) according to technique described in (Sukhyy et al., 2018). Adsorption capacities of the applied composites are used to be 0.85 g/g for 'silica gel – sodium sulphate' and 0.42 g/g for 'silica gel – sodium acetate'.

Temperatures of air and adsorbents are determined by group thermoelements Pt1000. Meteorological parameters (air velocity, air flow direction, temperature, relative degree of humidity, atmospheric pressure) are registered by ultrasonic meteorological station Coastal Environmental Systems C–5 S1019RB.

To study laboratory prototype of heat adsorptive regenerator (Figure 12) is installed into ventilation channel. The outer end of studied device is considered to be cold. The inner end is placed in ventilated premise, it being called as warm.

To register temperature-time curves and to determine time of achieving the equilibrium temperatures on 'inflow' or 'outflow', the air temperature is

Figure 13. Typical curves of cooling and heating of air in the adsorption regenerator: 1 – the temperature of the warm end of the regenerator, 0°C; 2 – temperature between cassettes with adsorbent, 0°C; 3 – temperature of the cold end of the regenerator, 0°C (Kolomiets et al., 2016).

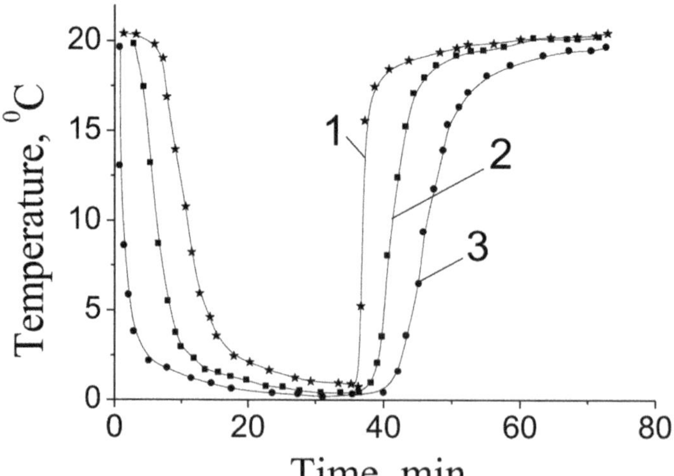

measured after each of the cassettes when inlet ventilator 2 or air exhauster 3 switched on.

Performance of the Laboratory Prototype of the Adsorption Regenerator of Heat and Moisture Based on Composite Adsorbents 'Silica gel – Sodium Sulphate' and 'Silica Gel – Sodium Acetate'

The operation of adsorptive heat-moisture regenerators is based on the alternating the processes of adsorption and desorption, which depends on the deviation of air humidity from equilibrium values.

In order to select an effective operating regime a times of equalization of temperatures on cold and warm ends of regenerator for 'inflow' and 'outflow' phases are determined. Temperature dependences for the warm and cold ends of the regenerator and air between cassettes filled with composite adsorbent occur to be extreme (Figure 13).

The first part of curves corresponds to 'inflow' regime (cooling) and proceeds during about of 30 min. In this period temperature appears to be decrease from 20 °C to 0 – 2 °C. The last values occur to be constant during about of 5 min. Within 35 min. after regenerator start, the phase is changed from 'inflow' to 'outflow'. Temperature is observed to increase during 1 – 5 min. up to 18 – 20°C. Then temperature is observed to be almost constant

Figure 14. Flows switchover time ($\Delta\tau$) vs. airflow rate graphs: 1 – results calculated by Eq. (6.8); 2 – ΔT = 10 °C; 3 – ΔT = 7,5 °C; 4 – ΔT = 5 °C; 5 – ΔT = 2,5 °C (Sukhyy, Belyanovskaya, & Kolomiyets, 2018).

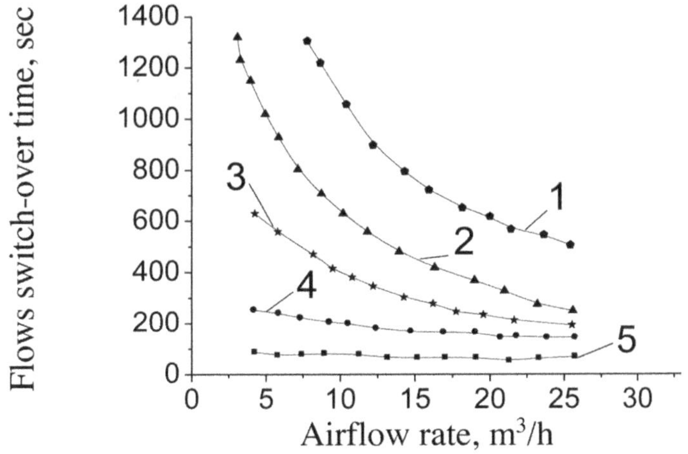

during the next 25 – 30 min. So, switchover time ($\Delta\tau$) of 35 min. is sufficient to complete cooling both ends of the regenerator device cooling to 0°C and heating them up to 20°C. Nevertheless, complete cooling and heating occur to be unsatisfactory for residential premises, as a consequence temperature fall in the premise because of prolonged cold air inflow.

So, sustain comfort heat and moisture conditions switchover time ought to be changed according to differential temperatures of regenerator warm and cold ends. It can establish a convenient regime for each of consumers.

Figure 15. Change of absolute humidity during exploitation of adsorptive regenerator in the warm end (1) and in the cold end (2) of the device when composite 'silica gel – sodium sulphate' (a) and 'silica gel – sodium acetate' (b) used
(Kolomiyets et al. 2016, pp. 108 – 113).

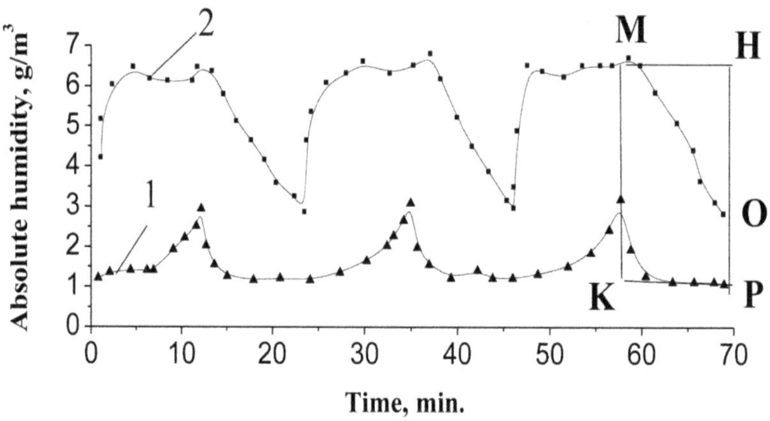

a

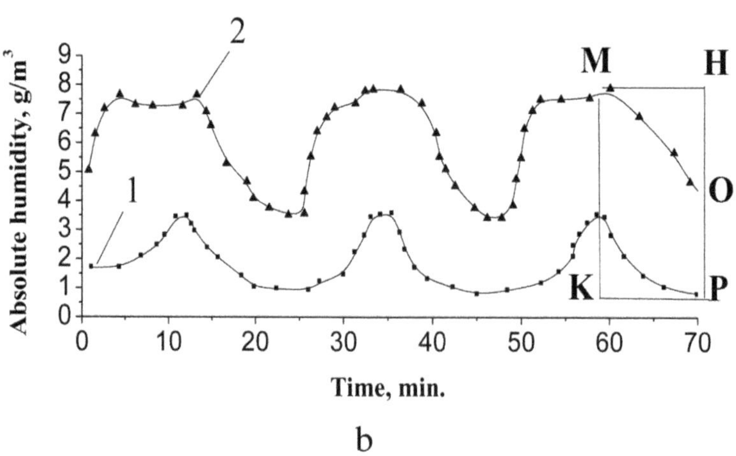

b

The laboratory prototype of the regenerator was tested in situ over period from the 10 January to 25 February 2015. Their averaged results are represented by Figure 14.

As a comparative curve is given (Figure 14, curve 1), it being calculated for step heat front as follows:

$$\Delta\tau = \frac{cm}{c_B' V} \tag{8}$$

Figure 16. Change of relative humidity during exploitation of adsorptive regenerator in the warm end (1) and in the cold end (2) of the device when composite 'silica gel – sodium sulphate' (a) and 'silica gel – sodium acetate' (b) used *(Kolomiyets et al., 2016).*

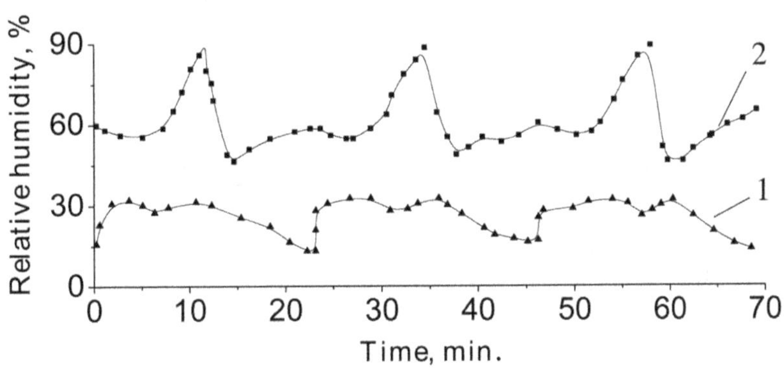

a

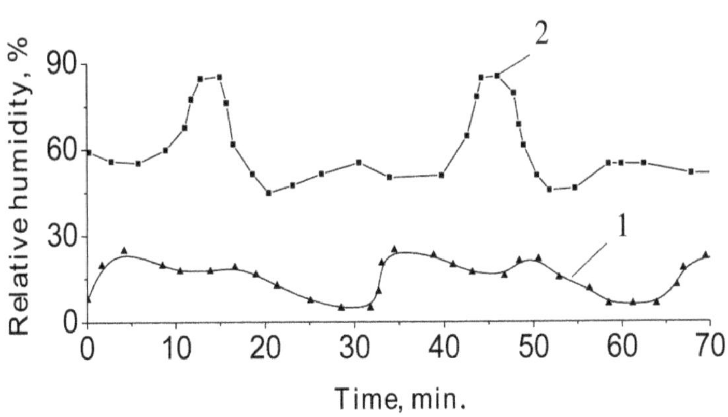

b

where c is specific heat per unit mass of adsorptive material, J/kg•K; c_B' corresponds to specific heat per unit volume of air, J/m^3•K ; m refers to mass of checkerworks, kg; V is airflow rate, m^3/h.

Flows switchover time is revealed to be affected by airflow rate and temperature differential ΔT between warm and cold ends of regenerator in steady-state conditions. As temperature differentials ΔT = 5 – 10°C and airflow rate increased, flows switchover time occurs to be decreased (Figure 14, curves 2 – 4). For ΔT = 2,5°C Δτ is weakly dependent to airflow rate, supposing low slope of heat front. When temperature differentials increased, flows switchover time is observed to increase, too. This effect is the most

Figure 17. Temperature dependences of the adsorptive regenerator in the warm end (1), between cassettes with adsorbent (2) and in the cold end (3) of the device when composite 'silica gel – sodium sulphate' (a) and 'silica gel – sodium acetate' (b) used *(Kolomiyets et al., 2016, pp. 108 – 113).*

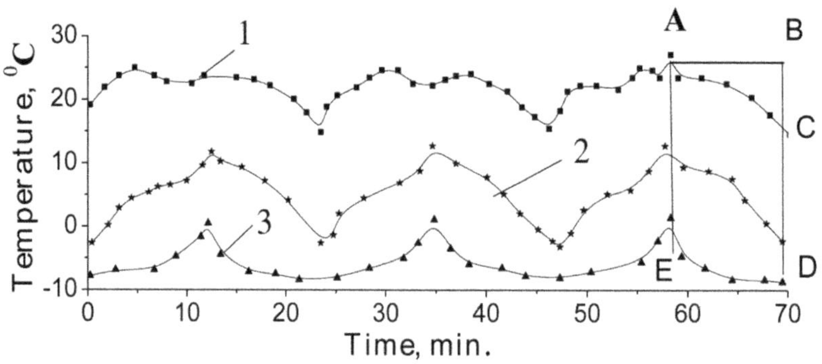

a

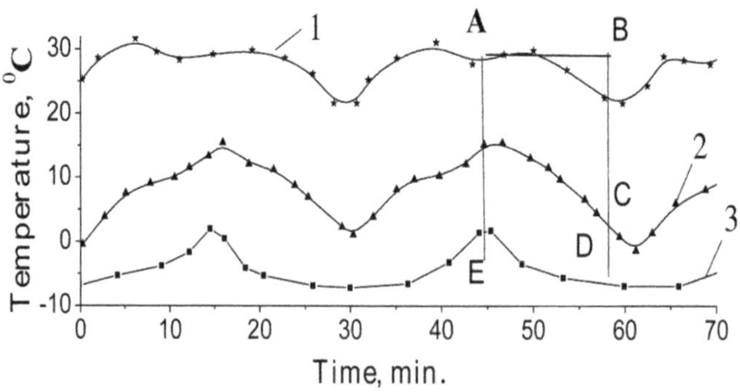

b

conspicuous at low speeds. For example, at V =5 m^3/h $\Delta\tau$ is observed to be equal to 80 s for ΔT = 2.5 °C, $\Delta\tau$ being nearly eighteen times as much as it for ΔT = 10 ^{0}C. Nevertheless, when airflow rate increased to 25 m^3/h, this differential becomes considerably lower, e.g. ΔT = 2,5 °C corresponds to $\Delta\tau$ = 90 c, and when ΔT increased to 10 °C, $\Delta\tau$ is 380 s (Kolomiyets et al., 2016).

As the temperature differential increased and most of the adsorbent involved to the process of heat exchange, plots of flows switchover time vs. airflow rate reaches after the hyperbolic curve calculated by Eq. (6.8).

When the laboratory prototype of the regenerator tested apart from flows switch-over time, temperature variation, air absolute and relative humidity are measured at the warm and cold ends of the device and between cassettes with composite materials. Averaged data of testing are represented by Figure 15 – 17.

The adsorptive regenerator is operated in steady-state conditions with constant airflow rate of 30 m^3/hour and temperature differential of ΔT = 10°C. Each of given dependences is observed to be periodic, which resulted from alternating the flow direction. Increasing of absolute and relative humidity and air temperature on the regenerator warm end is observed when airflow directed from premise outwards (Kolomiyets et al., 2016, pp. 108 – 113).

When airflow directed from ambient media into premise, these values are observed to decrease. On the cold end of the regenerator absolute and relative humidity and air temperature are varied in similar manner, but their variation interval occurs to be much less. The highest amplitude is observed on a plot of temperature between cassettes with adsorptive heat storage material. It can be a consequence of complication of heat and mass exchange with adsorption and desorption processes.

For both adsorbents registered dependences reveal the comparable periodic nature. Nevertheless, as the composite 'silica gel –sodium acetate' used (Figure 15 –17, b), the amplitude of time dependences of absolute and relative humidities and temperature is appeared to be less than observed for 'silica gel – sodium sulphate' (Figure 15 –17, a). Presumably, this distinction can be attributed to the lower value of adsorption capacity of the composite 'silica gel – sodium acetate' as compared to 'silica gel – sodium sulphate'.

The most important characteristics are temperature and moisture efficiency factors, i.e. regeneration factors of heat and humidity corresponded to ratios of actual quantities of heat and water to their maximal values, respectively.

Similarly to (Kolomiyets et al., 2016, pp. 108 – 113), heat regeneration factor K_H is estimated as ratio of surface areas S_{ACDE}/S_{ABDE} (Figure 17), humidity regeneration factor K_W being calculated as ratio S_{MOPK}/S_{MHPK} (Figure 15), and

Since differentials of inner and ambient air temperatures varied during experiments, heat regeneration factor is conveniently expressed in terms of dimensionless temperature differential ΔT_D:

$$\Delta T_D = \frac{\Delta T}{\Delta T_{\max}} \tag{9}$$

where ΔT refers to the temperature differential, ΔT_{max} corresponds to the maximal temperature differential.

Humidity regeneration coefficient is expressed as a dimensionless absolute humidity differential:

Figure 18. Composite layer hydrodynamic resistance vs. airflow rate characteristic when as heat storage material used glass spheres in diameter of 2 mm, composite 'silica gel – sodium sulphate' granules in diameter of 2,0 – 2,5 mm (2) and 3 – 3,5 mm (3)
(Kolomiyets et al., 2016, pp. 108 – 113)

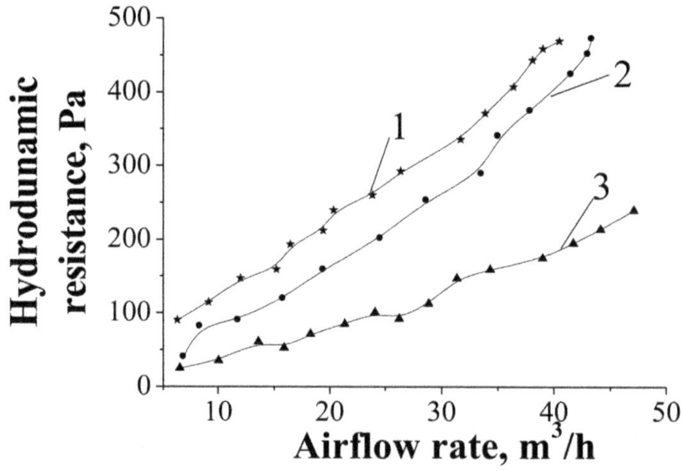

Table 1. Operational parameters of adsorptive regenerator of heat and moisture

Regime	T_{out} °C	V, m³/hour	Δτ, s	W, W	M_{ads}, kg
1	10 – 0	15	500	3	1.5
2	0 – (-5)	10	400	5	1.5
3	(-5) – (-10)	5	200	1	1.5
4	(-10) – (-20)	2	100	0.4	1.5

$$\Delta d_D = \frac{\Delta d}{\Delta d_{max}} \tag{10}$$

where Δd is the differential of the air absolute humidity, Δd_{max} corresponds to the maximal air absolute humidity differential.

Having analyzed plots of heat and humidity regeneration factors vs. dimensionless differential of heat and absolute humidity, all experimental data on heat and humidity regeneration factor occurs to be sufficiently approximated for engineering evaluations by a linear polynomials of degree 2 (Kolomiyets et al., 2016, pp. 108 – 113). The polynomial for heat regeneration coefficient K_H over the range of dimensionless temperature differential $0 \le \Delta T_D \le 0.35$ is estimated to be

$$K_H = 1 - 0.31 \cdot \Delta T_D - 0.13 \cdot \Delta T_D^2 . \tag{11}$$

The similar polynomial is evaluated for humidity regeneration factor K_W on interval $0 \le \Delta d_D \le 0.8$ as

$$K_w = 1 - 0.31 \cdot \Delta d_D - 0.13 \cdot \Delta d_D^2 . \tag{12}$$

Resemblance of these dependences K_H vs. ΔT_D and K_W vs. Δd_D confirms close affinity between heat and mass transfer processes. Indeed, values of these coefficients are affected by regenerator construction and nature and quantity of adsorbent.

Increasing of dispersiveness is stated to lead to rising switchover time of flows. Nevertheless, it results in significant rise of hydrodynamic resistance of adsorbent layer ΔP (Figure 18) and hence increasing of energy consumption for air pumping through the adsorbent layer:

$$W = \frac{W \cdot \Delta P}{\eta_{vent}} . \tag{13}$$

where V is airflow rate, η_{vent} is efficiency factor of ventilator.

For example, when η equals to 0.7, airflow rate is 30 m^3/h, hydrodynamic resistance of the layer of composite 'silica gel – sodium sulphate' with granules in diameter of 2,0 – 2,5 mm is 260 Pa, energy consumption for air pumping through the layer of the composite equals to 3.1 W. As diameter of

Figure 19. Control-flow chart for calculation of efficiency factor of adsorptive regenerator
(Belyanovskaya, Lytovchenko, Sukhyy, Yeremin, Sukha, & Prokopenko, 2019).

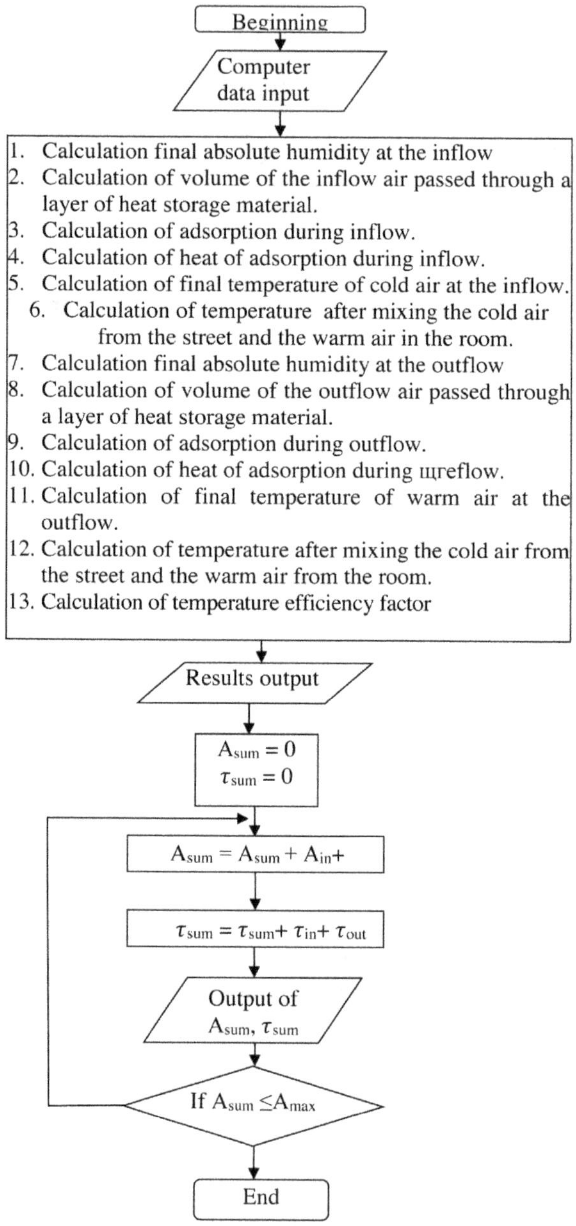

composite granules increased to 3 – 3,5 mm, hydrodynamic resistance is of 100 Pa, and energy consumption is decreased to 1.2 W.

In situ tests of the described laboratory prototype of the adsorptive heat-moisture regenerator over the 10 January to 25 February 2015 period exhibited the ability to support convenient temperatures (20 – 22°C) and relative humidity (50 – 60%) by this device. Operating parameters of this device are given in the Table 1.

Furthermore, the heat and moisture regeneration factors can be purposively varied in rather wide ranges by changing the switchover time, amount of the adsorbent and size of its granules. Consequently, the suggested adsorptive regenerator can be used not only as a buffer of moisture in residential premises and warehouses, but also to heat the supply air, that allows to significantly reduce energy inputs in ventilation systems.

MATHEMATICAL SIMULATION OF OPERATING PROCESSES OF AN ADSORPTIVE HEAT-MOISTURE REGENERATOR

Algorithm for Estimation of the Performance of an Adsorptive Regenerator of Heat and Moisture

Despite the promising characteristics of adsorptive heat-moisture regenerators for ventilation and conditioning systems the active commercialization of such devices is strongly hurdled not only because of the low quality of the used adsorptive materials but with lack of an algorithm for their calculation, which allows simulating and studying their operation in the real conditions.

According to operational phases described, the next algorithm is proposed (Figure 19).

The temperature coefficient of efficiency (efficiency factor) is determined by the formula (Belyanovskaya et al., 2019):

$$\eta_{tem} = \frac{t_{inf} - t_{ext}}{t_{out} - t_{ext}}. \tag{14}$$

where t_{inf} corresponds to temperature of inflowing air, °C; $t_{ext.}$ refers to the temperature of external air; t_{out} is the temperature of outflowing air.

Temperature of inflowing air is calculated as a temperature after mixing the cold air from the street and the warm air in the room during inflow (Belyanovskaya et al., 2019):

$$t_{fin.r.aft.\ mix.} = \frac{V_r \cdot t_{0r} + V_{inf} \cdot t_{fin.r}}{V_r + V_{inf}} \tag{15}$$

where V_r is volume of the ventilated premise, m^3; $t_{0.r}$ refers to the initial air temperature at the room outlet (warm end), °C; $t_{fin.r}$ corresponds to the final temperature of the inflowing cold air, °C.

Final temperature of cold air-in is estimated by equation of thermal balance (Belyanovskaya et al., 2019):

$$t_{fin.r} = \frac{\left(! ' \cdot t_{0str} \cdot V_{inf} + 4.19 \cdot t_{0str} \cdot V_{inf} \cdot C_{0.str} + \cdot H_{ads.inf} \cdot M_{ads}\right)}{\left(! ' \cdot V_{inf} + 4.19 \cdot V_{inf} \cdot !_{fin.r}\right)} \tag{16}$$

where ! ' refers to the volumetric specific heat of air, $kJ/m^3 \cdot °C$; t_{0str} refers to the temperature of outdoor air (near the cold end of regenerator), °C; V_{inf} is volume of the inflow air passed through a layer of the adsorptive heat storage material, m^3; $C_{0.str}$ corresponds to the initial absolute humidity at the cold end of the regenerator, kg/m^3; $\Delta H_{ads.inf}$ is the heat of adsorption during inflow, kJ/kg; M_{ads} refers to the mass of adsorbent, kg; $!_{fin.r}$ is the final absolute humidity when inflowed, kg/m^3.

The volume of air passed through the layer of heat storage material at the inflow or outflow, V_{inf} or V_{outf}, m^3 is computed as follows:

$$V = F_s \cdot w \cdot \tau \tag{6.17}$$

where w is speed of humid air, m/s; τ is time of inflow or outflow, s; F_s is area of the section, m^2.

The temperature of the exhaust air is estimated as the temperature after mixing the cold air from the street and the warm air from the room at the outflow (Belyanovskaya et al., 2019):

Table 2. Initial data for verifying mathematic model of adsorptive regenerator

Designation	Symbol	Unit of Measurement	Initial Data
Initial absolute humidity at the warm end of the regenerator (outflow)	$!_{0.r}$	kg/m³	$1 \cdot 10^{-3}$
Initial absolute humidity at the cold end of the regenerator (inflow)	$!_{0.str}$	kg/m³	$6.5 \cdot 10^{-3}$
Final absolute humidity at the inflow	$!_{fin.r}$	kg/m³	$3.9 \cdot 10^{-4}$
Final absolute humidity at the outflow	$!_{fin.str}$	kg/m³	$1.8 \cdot 10^{-4}$
Initial temperature of the outside air when ejected from the room (inflow)	$t_{0.str}$		-9
Temperature of indoor air (near the warm end of regenerator) (outflow)	$t_{0.r}$		25
Adsorbent mass	M_{ads}		3,5
Area of the section	F_s	M²	0.0314
Time of inflow	τ_{in}	s	600
Time of inflow	τ_{out}	s	600
Mass transfer coefficient	β	s^{-1}	$8.12 \cdot 10^{-8}$
Diffusion coefficient	D_y		0.001
Adsorbent pore content	E	–	0.2

$$t_{fin.str.aft.\ mix.} = \frac{V_{str} \cdot t_{0.str} + V_{outf} \cdot t_{fin.str}}{V_{str} + V_{outf}} \tag{6.18}$$

where V_{str} refers to the volume of air at the outside end of the regenerator, m³; $t_{0.str}$ is the initial temperature of the external air when ejected from the room, °C; $t_{fin.str}$ corresponds to the final warm air temperature during outflow, °C.

Final temperature of the warm air during exhaust is computed by thermal balance equation (Belyanovskaya et al., 2019):

$$t_{fin.str.} = \frac{!\,'\cdot t_{0r}\cdot V_{outf} + 4.19\cdot t_{0r}\cdot V_{outf}\cdot C_{0r} + \Delta H_{ads.outf}\cdot M_{ads}}{!\,'\cdot V_{outf} + 4.19\cdot V_{outf}\cdot !_{fin.str}} \qquad (6.19)$$

where $!\,'$ is volumetric specific heat of air, kJ/m^3 ·°C; t_{0r} is temperature of indoor air (near the warm end of regenerator), °C; V_{outf} is the volume of air passed through the layer of heat-accumulating material during outflow, m^3; $C_{0.r}$ is initial absolute humidity at the warm end of the regenerator, kg/m^3; $\Delta H_{ads.outf}$ is heat of adsorption at the outflow, kJ/kg; M_{ads} is adsorbent mass, kg; $!_{fin.str}$ is final absolute humidity at the outflow, kg/m^3.

Adsorption processes during inflow and outflow are suggested to estimate by Eq. (5.47), (5.52) – (5.54) (Chapter 5).

To validate the adequacy of the suggested mathematical model of the operation of the adsorptive regenerators based on the composite 'silica gel – sodium sulphate' and 'silica gel – sodium acetate' described in [219, 220] were performed.

Initial data for calculation are given in the Table 2.

The results of the calculations reveal the alternation of the temperatures on the cold and warm ends of the adsorptive regenerator (Figure 20, curves 1 and 2), which qualitatively conforms to the experimental data (Figure

Figure 20. Time-temperature dependences on the cold and warm ends of the adsorptive heat-moisture regenerator based on 'silica gel – sodium sulphate' (a) and 'silica gel – sodium acetate' (b): 1, 1' – temperatures at the cold end of the regenerator; 2, 2' – temperatures at the warm end of the regenerator; 1, 2 – calculated data; 1', 2' – experimental data
(Belyanovskaya et al., 2019)

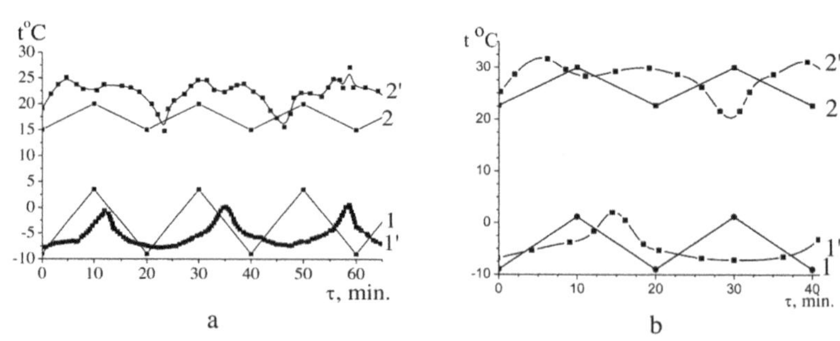

Figure 21. Ventilation plan of three-room apartment with electric hot plate: 1 – outflow air line on the regulator outlet; 2 – inflow air line on the regulator inlet; 3 – filter for inflow air; 4 – regulator; 5 – filter for outflow air; 6 – air line for air distribution inside of the apartment; 7 – affluent diffuser; 8 – sound extinguish insertion; 9 – exhaust diffuser; 10 – outflow air duct; 11 – electric fan with backdraft damper; 12 – air closer conduit; 13 – outflow airshaft; 14 – inflow airshaft; 15 –adsorptive regenerator
(Sukhyy, Belyanovskaya, & Kolomiyets, 2018).

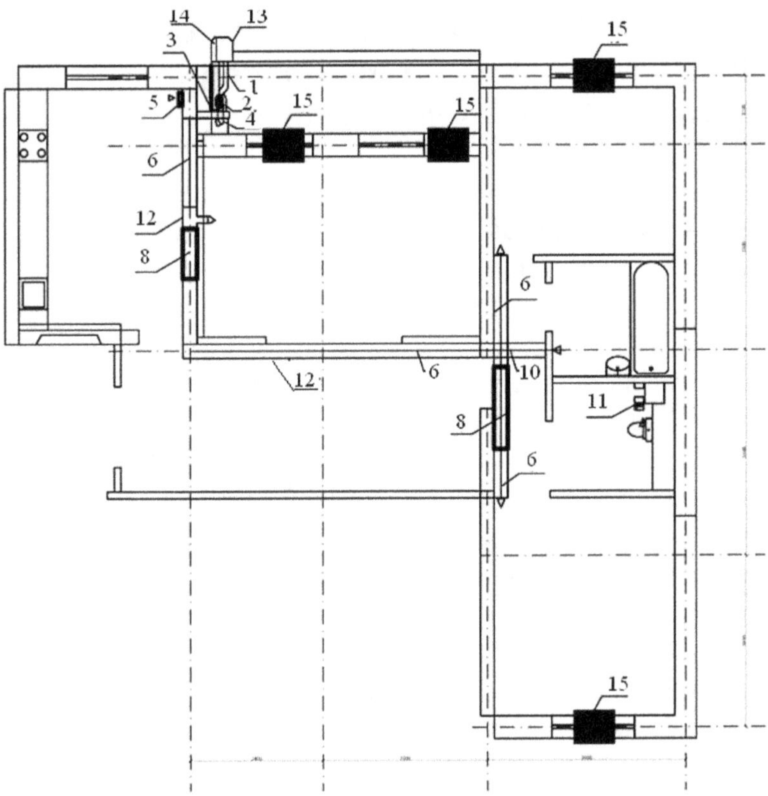

20, curves 1 'and 2'). For regenerator based on the composite 'silica gel – sodium sulphate', the difference between the calculated and experimental temperatures is at most 2 – 3 ° C at the cold end of the regenerator, and 1 – 5 ° C at the warm end.

For regenerator based on the composite 'silica gel – sodium acetate', the deviation of the calculated temperatures from the experimental does not exceed 2 – 2.5 ° C at the cold and warm ends of the regenerator.

According to the experimental data, the temperature efficiency factor of the adsorptive regenerator based on 'silica gel – sodium sulphate' is approximately 95%, and the calculated value being about of 91%. When the composite 'silica gel – sodium acetate' used, experimental and calculated values of the temperature efficiency factor are about of 85% and 82%, respectively. So, the proposed mathematical model allows estimating the performance of adsorptive regenerators in ventilation systems.

Determination of Design Characteristics of Adsorptive Regenerators for the Heating the Air-in for the Ventilation System

An adsorptive regenerators based on composite 'silica gel – sodium sulphate' or 'silica gel – sodium acetate' are suggested to use for the heating of the incoming air in the conditions of a typical three-room apartment with electric hot plate, a total area of 103 m^2, a height of 2.5 meters. Ventilation plan is given in Figure 21.

Adsorptive regenerators are suggested to be fixed in the rooms and in the kitchen to ventilate premises during heating season. This technical solution can reduce the inputs associated with warming by supply an inhaust air heated up to convenient temperature into residential spaces. As a rule, aeration heat losses occur to be significant and crucial for amenities due to temperature-rule drop. The procedure of calculation the design characteristics of the adsorptive heat-moisture regenerator is proposed as follows.

The estimation of the design characteristic of the adsorptive heat-moisture regenerator is based on determining of the thermal load upon airflow heating which can be calculated e.g. according to the conventional procedure by the Sanitary Regulations 2.04.05-91:

$$Q_i = 0.28 \cdot L_i \cdot \rho_a \cdot A \cdot \left(t_{in} - t_{ext}\right) \tag{20}$$

where Q_i is the thermal load for inhaust airflow heating, W; L_i is the inhaust airflow rate, m^3/h; ρ_a is air density in a premise, kg/m^3; c is the air special heat, 1.005 kJ/(kg·°C); t_{in} is the temperature of the internal air, °C, t_{ext} is the temperature of the external air, °C.

According to the results of the calculation by the Sanitary Regulations 2.04.05-91, the thermal load for heating the supply air is about 327.9 MJ per a day when the space aerated during 5 hours per a day.

Then mass of adsorbent M_{ads}, kg is calculated as (Belyanovskaya et al., 2019)

$$M_{ads} = \frac{Q_i}{\Delta H_{ads}} \tag{21}$$

where ΔH_{ads} is heat of adsorption, kJ/kg which estimated as (Belyanovskaya et al., 2019)

$$\Delta \mathrm{H}_{ads} = \Delta \mathrm{h} \cdot \mathrm{A}_{lim} \cdot \frac{1000}{\mu_{\mathrm{H_2O}}} \tag{22}$$

where $\Delta h \cong 60$ is heat adsorption of water vapor, kJ/ mole; A_{lim} is the maximal adsorption, kg/kg; μ_{H_2O} is the molar mass of water, g/mole.

The value of A_{lim} can be determined according to adsorbent characteristics or estimated as linear superposition for mechanical mixture (Belyanovskaya et al., 2019). Volume of an adsorbent is calculated as a ratio of adsorbent mass and adsorbent density according to (Belyanovskaya et al., 2019). Results of calculation of mass of adsorbents are shown in Table 3.

Table 3. Adsorption heat and adsorbent mass and volume for heat load to inflow air heating of 327.9 MJ per a day

Adsorbent	Maximal Adsorption A_{lim}, kg/kg	Heat of Adsorption, kJ/kg	Adsorbent Mass, kg	Adsorbent Volume, m³	Regeneration Temperature, °C
Silica gel KCM (State Standard 3956-76)	0.20	666.67	491.85	0.68	150
Silica gel Type 3A (Ng et al., 2001)	0.32	1066.67	307.41	0.42	100
Silica gel Type RD (Ng et al., 2001)	0.36	1200.00	273.25	0.35	100
Silica gel (20%) – sodium sulphate (80%)	1.046	3486.67	94.0	0.13	90
Silica gel (20%) – sodium acetate (80%)	0.557	1856.10	176.6	0.25	60

(Belyanovskaya, Lytovchenko, Sukhyy, Prokopenko, Yeromin, & Sukha, 2019).

Calculation results confirm that according to the heat of adsorption silica gels of domestic and foreign grades underperform to composite adsorbents by a factor of 1.5 – 20. So, to supply the same heat load the more mass of silica gel is required as compared with composite adsorbents.

The same ratio could be predicted for inputs of heat transfer agents. Apparently, potential of waste energy source is determined with regeneration temperatures of adsorbents. Regeneration temperatures of composite are stated to be substantially lower by contrast to conventional adsorbents. The lowest value of regeneration temperature is revealed for composite 'silica gel – sodium acetate'. However, its maximal adsorption or water capacity is 0.557 kg/kg, when A_{lim} for composite 'silica gel – sodium sulphate' surpasses it in nearly 2 times. The same ratio occurs for adsorption heats of studied composites. So, mass and volume of composite 'silica gel – sodium sulphate' required to supply the same heat load is estimated to be half as much as compared with 'silica gel – sodium acetate' with the same content of salt.

So, as an adsorptive material, the composites containing 20% silica gel and 80% sodium sulphate or sodium acetate are exhibited to be the most efficient.

Table 4. Initial data for calculation of apartment ventilation system

Designation	Symbol	Unit of Measurement	Initial Data
1	2	3	4
Initial absolute humidity at the warm end of the regenerator (outflow)	$!_{0.r}$	kg/m³	$1 \cdot 10^{-3}$
Initial absolute humidity at the cold end of the regenerator (inflow)	$!_{0.str}$	kg/m³	$6.5 \cdot 10^{-3}$
Final absolute humidity at the inflow	$!_{fin.r}$	kg/m³	$3.9 \cdot 10^{-4}$
Final absolute humidity at the outflow	$!_{fin.str}$	kg/m³	$1.8 \cdot 10^{-4}$
Initial temperature of the outside air when ejected from the room (inflow)	$t_{0.str}$	°C	-9
Temperature of indoor air (near the warm end of regenerator) (outflow)	$t_{0.r}$	°C	25
Time of inflow	T_{in}	s	600
Time of outflow	T_{out}	s	600

Simulation of the Operating Processes of Adsorptive Regenerative Device in Residential Premises

Figure 22. Temperature-time dependences of adsorptive heat-moisture regenerator on the cold (1) and warm ends of the device (2). Air-flow switchover time, min.: a – 5; b – 10, c – 15, d – 20. Temperature efficiency coefficient vs. air-flow switchover time (e)
(Sukhyy, Belyanovskaya, & Kolmiyets, 2018).

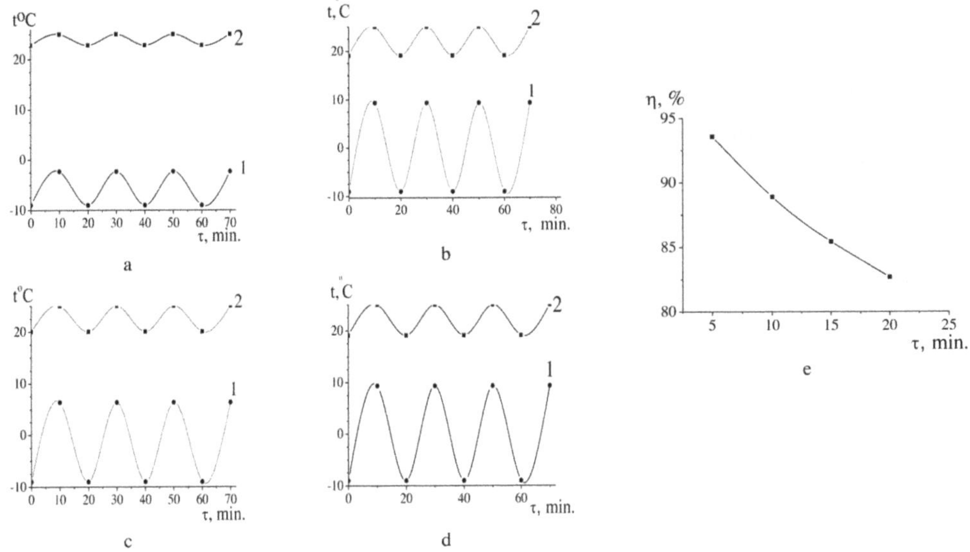

Figure 23. Temperature efficiency factor vs. switchover time for adsorptive regenerator of heat and moisture based composite 'silica gel – sodium sulphate' (a) and 'silica gel – sodium acetate' (b). Air flow speed, m/s: 1 – 0.22; 2 – 0.32; 3 – 0.42; 4 – 0.52
(Belyanovskaya et al., 2019).

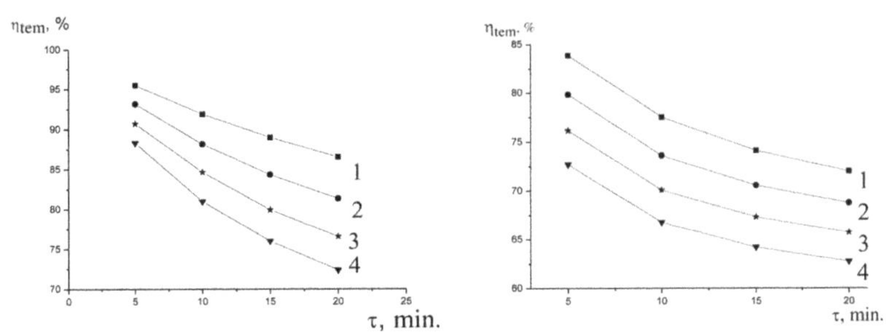

Then operation processes of the adsorptive regenerators based on composites 'silica gel sodium – sulphate' and 'silica gel – sodium acetate' were simulated for ventilation system in the conditions of a typical three-room apartment with electric cookers, a total area of 103 m^2, a height of 2.5 meters. Since an optimal option is to install in the scheme of 3-bedroom apartments four adsorptive regenerators, masses of adsorbents 'silica gel – Na_2SO_4' and 'silica gel – CH_3COONa' on the regenerator are 24 kg and 45 kg, respectively.

Having considered calculation results, simulation of operational processes of adsorptive heat-moisture regenerator was performed on the suggested algorithm. Initial data for calculation are shown in Table 4.

Figure 24. Time of achieving the maximal adsorption vs. switchover time of adsorptive heat moisture regenerator based on composite 'silica gel – sodium sulphate' (A) and 'silica gel – sodium acetate' (B). Air flow speed, m/s: 1 – 0.22; 2 – 0.32; 3 – 0.42; 4 – 0.52
(Belyanovskaya et al., 2019).

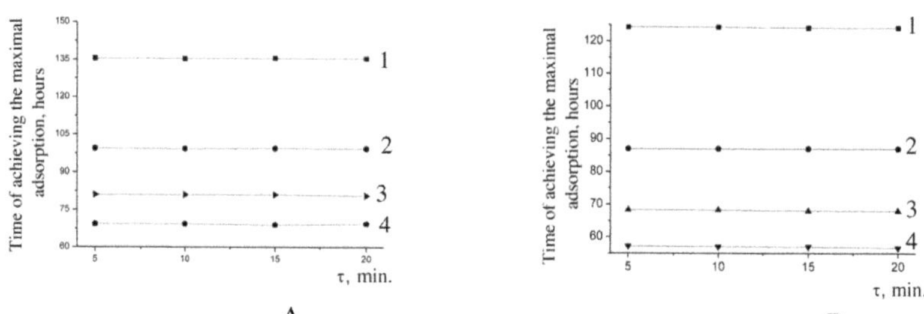

Figure 25. Temperature efficiency factor vs. switchover time of adsorptive heat moisture regenerator based composite composite 'silica gel – sodium sulphate' (A) and 'silica gel – sodium acetate' (B) at the temperatures of external air, °C: 1 – -3; 2 – -9; 3 – -15; 4 – -25
(Belyanovskaya et al., 2019).

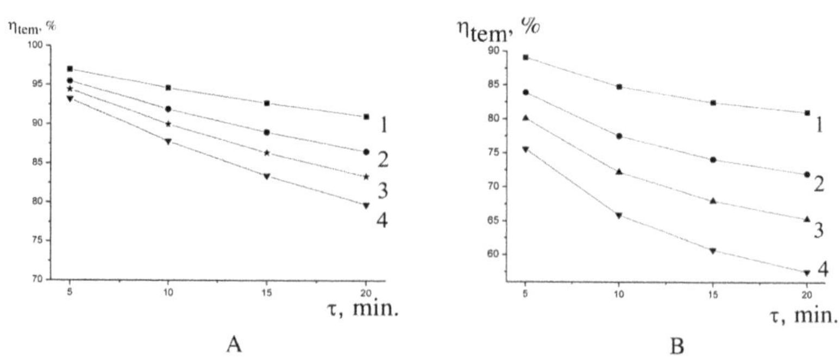

The results of the calculations are shown in Figure 22 – 28. Under typical residential conditions, the temperatures on the warm and cold ends of the regenerator are stated to be periodically changed.

Results of mathematical modelling exhibit that the temperature efficiency factor η_{tem} of both heat-moisture regenerators strongly depends on airflow speed (Figure 23). When the speed of airflow increased, the temperature efficiency factor is decreased. It is explained by increasing of volume of inflowed air. Similarly, as switchover time (that is changing direction of flows) increased, amplitudes of time-temperature dependences on both cold and warm ends of

Figure 26. Temperature efficiency factor vs. switchover time of adsorptive heat moisture regenerator based composite 'silica gel – sodium sulphate' (A) and 'silica gel – sodium acetate' (B) at the temperatures of internal air, °C: 1 – 15; 2 – 20; 3 – 25 *(Belyanovskaya et al., 2019).*

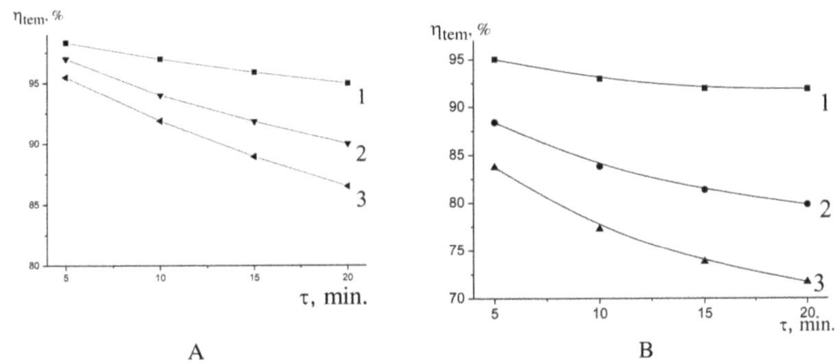

Figure 27. Influence of the absolute humidity of the external air on the temperature efficiency factor (a) and time to achieve the maximal adsorption (b) for adsorptive heat-moisture regenerator based on composite 'silica gel – sodium sulphate'. Absolute humidity of the external air, g/m³: 1 – 4.5; 2 – 6.5; 3 – 8.5 *(Belyanovskaya et al., 2019).*

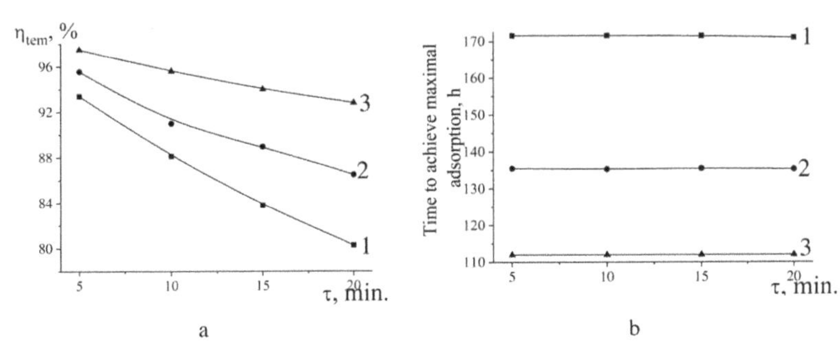

the device are stated to be magnified. This results in decreasing values of the temperature efficiency factors (Figure 23). Maximal values of temperature efficiency factors are observed at the airflow speeds 0.22 – 0.32 m/s and switchover time at most of 5 – 10 min. Time of achieving of maximal adsorption is negligibly affected by switchover time of flows (Figure 24). However, it is observed to strongly depend on airflow speeds. Maximal time to achieve the maximal adsorption is registered at airflow speed of 0.22 m/s, it being stated of 125 and 135 hours for composite 'silica gel – sodium acetate' and 'silica gel – sodium sulphate', respectively.

Also, meteorological conditions strongly affect the efficiency of adsorptive heat-moisture regenerator. For example, the external air temperature and temperature efficiency factor are increased simultaneously (Figure 25). Obviously, time of achieving of the maximal adsorption is negligibly affected by the external air temperature. Decreasing the internal temperature leads to increasing the temperature efficiency factor (Figure 26).

Figure 28. Temperature efficiency factor vs. internal and external temperatures for adsorptive heat moisture regenerator based on composite 'silica gel – sodium sulphate'
(Belyanovskaya et al., 2019).

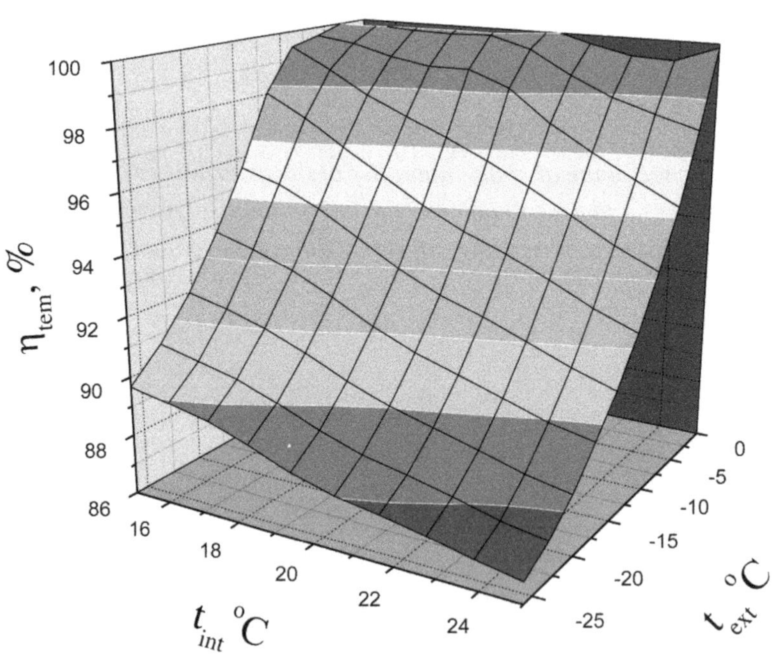

Similarly, the temperature efficiency factor is affected by the increasing of the absolute humidity of the external air (Figure 27a). It is accompanied by the increasing of the adsorption rates and reducing the time to achieve the maximal adsorption (Figure 27b). Temperature of external air does not affect the time to achieve the maximal adsorption.

It should be noted that temperature efficiency factors of regenerators based on composites 'silica gel – sodium sulphate' are stated to exceed η_{tem} for devices based on 'silica gel – sodium acetate' by at least 9 – 10%. This difference becomes more significant at higher airflow speeds, switchover times or internal temperature (Figures 23, 25, 26). It appears to increase when external temperature decreased (Figure 26).

Therefore, the correlation of the temperature efficiency factor and meteorological conditions were considered for regenerator based on composite 'silica gel – sodium sulphate' (Figure 28). Maximal values of temperature efficiency factor are observed at the external air temperature of -5 – 0°C and internal air temperature of 15 – 16°C.

REFERENCES

Aristov, Yu. I. (2015). Current progress in adsorption technologies for low-energy buildings. *Future Cities and Environment*, *1*(10), 1–13. doi:10.118640984-015-0011-x

Baniyounes, A. M., Rasul, M. G., & Khan, M. M. K. (2013). Experimental assessment of a solar desiccant cooling system for an institutional building in subtropical Queensland, Australia. *Energy and Building*, *62*, 78–86. doi:10.1016/j.enbuild.2013.02.062

Bathen, D., & Breitbach, M. (2001). *Adsorptionstechnik*. Springer. doi:10.1007/978-3-642-18235-8

Belyanovskaya, E.A., Lytovchenko, R.D., Sukhyy, K.M., Prokopenko, O.M., Yeromin, O.O., & Sukha, I.V. (2019). Choice criteria of adsorbents for heat energy converters in ventilation systems. *Scientific Works*, *83*, 3 – 9.

Belyanovskaya, E. A., Lytovchenko, R. D., Sukhyy, K. M., Yeremin, O. O., Sukha, I. V., & Prokopenko, E. M. (2019). Operating regime of adsorptive heat-moisture regenerators based on composites «silica gel – sodium sulphate' and 'silica gel – sodium acetate'. *J. Chem. and Chem. Technology*, *27*(2), 158–168. doi:10.15421/081917

Danilevskij, L. N. (2014). *Sistemy prinuditel'noj ventilyatsii s rekuperatsiej teplovoj ehnergii udalyaemogo vozdukha dlya zhilykh zdanij.*

Enteria, N., Yoshino, H., Mochida, A., Takaki, R., Satake, A., Yoshie, R., Mitamura, T., & Baba, S. (2009). Construction and initial operation of the combined solar thermal and electric desiccant cooling system. *Solar Energy, 83*(8), 1300–1311. doi:10.1016/j.solener.2009.03.008

Enteria, N., Yoshino, H., Satake, A., Mochida, A., Takaki, R., Yoshie, R., & Baba, S. (2010). Development and construction of the novel solar thermal desiccant cooling system incorporating hot water production. *Applied Energy, 87*(2), 478–486. doi:10.1016/j.apenergy.2009.08.026

Erdogan, M., Bau, U., & Bardow, A. (2019). Benchmarking commercial adsorbents for drying air in a packed bed. *Applied Thermal Engineering, 160*, 113942. doi:10.1016/j.applthermaleng.2019.113942

Ergun, S. (1952). Fluid flow through packed columns. *Chemical Engineering Progress, 48*, 89–94.

Freni, A., Maggio, G., Sapienza, A., Frazzica, A., Restuccia, G., & Vasta, S. (2016). Comparative analysis of promising adsorbent/adsorbate pairs for adsorptive heat pumping, air conditioning and refrigeration. *Applied Thermal Engineering, 104*, 85–95. doi:10.1016/j.applthermaleng.2016.05.036

Ge, T. S., Dai, Y. J., & Wang, R. Z. (2014). Review on solar powered rotary desiccant wheel cooling system. *Renewable & Sustainable Energy Reviews, 39*, 476–497. doi:10.1016/j.rser.2014.07.121

Ge, T. S., Zhang, J. Y., Dai, Y. J., & Wang, R. Z. (2017). Experimental study on performance of silica gel and potassium formate composite desiccant coated heat exchanger. *Energy, 141*, 149–158. doi:10.1016/j.energy.2017.09.090

Hong, S. W., Ahn, S. H., Chung, J. D., Bae, K. J., Cha, D. A., & Kwon, O. K. (2016). Characteristics of FAM-Z01 compared to silica gels in the performance of an adsorption bed. *Applied Thermal Engineering, 104*, 24–33. doi:10.1016/j.applthermaleng.2016.05.058

Jia, C. X., Dai, Y. J., Wu, J. Y., & Wang, R. Z. (2006). Experimental comparison of two honeycombed desiccant wheels fabricated with silica gel and composite desiccant material. *Energy Conversion and Management, 47*(15-16), 2523–2534. doi:10.1016/j.enconman.2005.10.034

Kast, W. (1988). *Adsorption aus der Gasphase: Ingenieurwissenschaftliche Grundlagen undtechnische Verfahren*. Wiley-VCH.

Kolomiyets, E. V., Sukhyy, K.M., Belyanovskaya, E.A., & Tomilo, V. I. (2016). Tekhnologiya regeneratsii niz'kopotentsijnogo tepla ta vologi kompozitnim sorbentom «Silikagel'/natrij sul'fat». *Vopr. khimii i khim. tekhnologii, 5 – 6*(109), 68 – 73.

Kolomiyets, O. V., Sukhyy, K. M., Belyanovskaya, E. A., Tomilo, V. I., & Prokopenko, O. M. (2016). Operating characteristics of adsorptive regenerator of low-potential heat and moisture based on composite sorbents 'silica gel – sodium sulphate and silica gel – sodium acetate' synthesized by sol – gel method. *Scientific Works*, *80*, 108 – 113.

Őstin, R. (2017). Evaluation of a Single Family Low Energy Building in Cold Climate. *Energy Procedia*, *132*, 9–14. doi:10.1016/j.egypro.2017.09.623

Rambhad, K. S., Walke, P. V., & Tidke, D. J. (2016). Solid desiccant dehumidification and regeneration methods – A review. *Renewable & Sustainable Energy Reviews*, *59*, 73–83. doi:10.1016/j.rser.2015.12.264

Ramzy, A. K., Kadoli, R., & Ashok Babu, T. P. (2012). Performance studies on the desiccant packed bed with varying particle size distribution along the bed. *International Journal of Refrigeration*, *35*(3), 663–675. doi:10.1016/j.ijrefrig.2011.11.016

Sukhyy, K., Belyanovskaya, E., Kovalenko, V., Kotok, V., Sukhyy, M., Kolomiyets, E., Gubynskyi, M., Yeromin, O., & Prokopenko, O. (2018). The study of properties of composite adsorptive materials "silica gel – crystalline hydrate" for heat storage devices. *Eastern-European Journal of Enterprise Technologies*, *91*(1), 52–58. doi:10.15587/1729-4061.2018.123896

Sukhyy, K. M., Belyanovskaya, E. A., & Kolomiyets, E. V. (2018). Design and performance of adsorptive transformers of heat energy. Riga, Latvia: LAP Lambert Academic Publishing.

Sultan, M., El-Sharkawy, I. I., Miyazaki, T., Saha, B. B., & Koyama, S. (2015). An overview of solid desiccant dehumidification and air conditioning systems. *Renewable & Sustainable Energy Reviews*, *46*, 16–29. doi:10.1016/j.rser.2015.02.038

White, S. D., Goldsworthy, M., Reece, R., Spillmann, T., Gorur, A., & Lee, D. Y. (2011). Characterization of desiccant wheels with alternative materials at low regeneration temperatures. *International Journal of Refrigeration*, *34*(8), 1786–1791. doi:10.1016/j.ijrefrig.2011.06.012

Yu, N., Wang, R. Z., Lu, Z. S., & Wang, L. W. (2014). Development and characterization of silica gel–LiCl composite sorbents for thermal energy storage. *Chemical Engineering Science*, *111*, 73–84. doi:10.1016/j.ces.2014.02.012

Zheng, X., Ge, T. S., Hu, L. M., & Wang, R. Z. (2015). Development and characterization of mesoporous silicate–LiCl composite desiccants for solid desiccant cooling systems. *Industrial & Engineering Chemistry Research*, *54*(11), 2966–2973. doi:10.1021/ie504948j

Zheng, X., Ge, T. S., & Wang, R. Z. (2014). Recent progress on desiccant materials for solid desiccant cooling systems. *Energy*, *74*, 280–294. doi:10.1016/j.energy.2014.07.027

Chapter 7
Design and Performance of Adsorptive Heat Pumps

ABSTRACT

The chapter is devoted to design and performance of adsorptive heat pumps. In the first sub-division, state-of-the-art of the adsorptive heat pumping is analyzed. It involves analysing operating principle of adsorptive heat pumps, comparing of the properties of adsorbents used, bed specifications, and operating conditions. Original construction of the adsorptive heat pump is designed by authors for independent heat supply systems or hot water supply of buildings and other structures for various purposes. The composites 'silica gel – sodium sulphate' or 'silica gel – sodium acetate' were used as adsorbents. Discharging was performed in a daytime, when heat pump supplied heating system with water warmed to 45 – 35°C. The regeneration mode proceeded at night from 0.00 to 8.00 a.m. Efficiency of suggested adsorptive heat pump is estimated by two methods: as ratio of adsorption heat to sum of desorption heat and external heat supplied to sorbent during its heating up to regeneration temperature (coefficient of performance of cycle) and as ratio of heat of adsorption to heat supplied by solar collector (net coefficient of performance). Suggested heat pump coefficients of energy performance of

DOI: 10.4018/978-1-7998-4432-7.ch007

cycle are stated to be 2.084 when composite 'silica gel – sodium sulphate' used and 2.021 when 'silica gel – sodium acetate' used. Seasonal dependence of net coefficient energy performance for suggested adsorptive heat pump based on composites 'silica gel – sodium sulphate' and 'silica gel – sodium acetate' is revealed. Correlation of coefficients of energy performance of adsorptive heat pump and composite sorbents properties (sorption capacity and regeneration temperature) is stated. Insignificant decreasing of coefficients of energy performance when 'silica gel – sodium acetate' used is explained by lower sorptive capacity as compared to 'silica gel – sodium sulphate'. Suggested heat pump application perspectives are shown for heat supply systems to result from traditional energy sources independence and environmental advantages. Adsorptive heat pumps development challenges, major limitations for commercialization of adsorptive heat pumping, and requirements to ongoing innovations are analysed. The present chapter can be useful for energy efficient decentralized heat supply systems based on adsorptive heat pump unit.

NOMENCLATURE

C_{bed} – the specific heat of the adsorbent bed;
$C_{p,a}$ – the constant pressure specific heat of the adsorbate;
C_s – the specific heat of the adsorbent;
COP_h and COP_c – coefficients of performance for heating and cooling;
Q_c – useful heat condensation;
M_{bed} – adsorbent bed mass;
M_s – adsorbent mass;
SHP or *SCP* – specific heating or cooling power;
T_c – condensation temperature;
T_e – evaporation temperature;
X – the adsorbate concentration in the adsorbent, i.e. water uptake or adsorption.
X_{max} – maximal uptake of the adsorbate, i.e. maximal adsorption;
X_{min} – minimal uptake of the adsorbate, i.e. minimal adsorption;
ΔH_{ads} – adsorption heat;
ΔH_v – heat of evaporation.

More than 35% of the Europe's overall energy consumption is used by the building sector alone, from which 75% is supplied domestic hot water production and room heating (Pinheiro, Salústio, Rocha, Valente, & Silva, 2016). As most of this energy is produced by non-renewable and environmentally harmful energy sources, development of new processes and design of devices with low environmental impact becomes a crucial task for renovation heat supply systems.

Adsorptive heat pumps attract attention of researchers and heating market over the last years. As distinct from the conventional vapor compression heat pumps, which apply environmentally harmful refrigerants with high global warming potential, adsorptive heat pumps can use natural refrigerants such as water or ammonia (Dawoud, 2014).

STATE-OF-THE-ART OF THE ADSORPTIVE HEAT PUMPING

Working Principle

Operating principle of heat pumps is based on extracting heat from a low temperature level and delivering it at an intermediate temperature level, driven by a third energy source (Dias, João, & Costa, 2018). In adsorptive heat pumps, as third energy source, for example, waste heat, gas burners, electricity, geothermal or solar energy can be used.

A scheme of a simple adsorptive heat pump system is given in Figure 1. An adsorptive heat pump involves an adsorbent bed (metallic structure where the adsorbent is placed) which filled or coated with an adsorbent (adsorptive heat storage material), an evaporator, a condenser, an expansion valve and a heat transfer system or fluid to supply or withdraw heat to or from the adsorbent bed .

In heating applications, the evaporator is fed by a free of charge low temperature level heat source (ambient, geothermal or waste heat) to vaporize the adsorbate, which is supplied to the adsorbent bed during the adsorption phase. The heat transfer system collects useful adsorption heat normally by a heat transfer fluid. When desorbed, heat has to be provided to the adsorbent bed for releasing adsorbate from the adsorbent in vapour state.

The adsorbate is directed to the condenser, where its condensation occurs before returning to the evaporator. This process leads to releasing the useful heat condensation, Q_c.

Figure 1. Scheme of a common adsorption heat pump. HTF – heat transfer fluid. *(Dias, João, & Costa, 2018)*

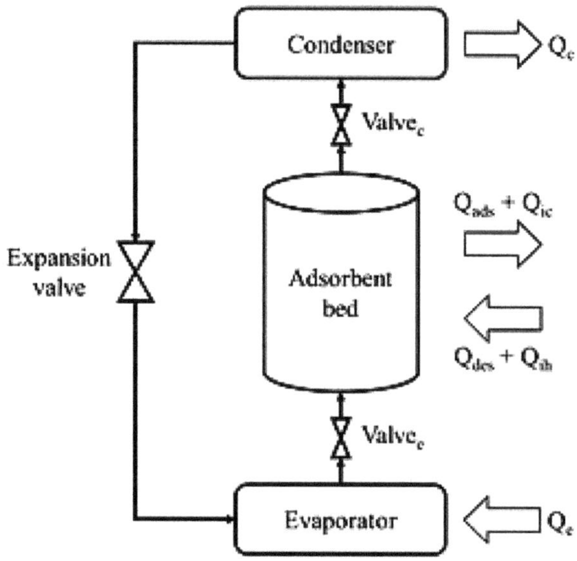

The ideal thermodynamic cycle of an adsorptive heat pump (AHP) is presented in Figure 2 in the Clapeyron diagram. It involves four phases, that is, isobaric ad sorption (1–2), isosteric heating (2–3), isobaric desorption (3–4) and isosteric cooling (4–1).

When the valve between the evaporator and the adsorbent bed is open and the valve between the adsorbent bed and the condenser is closed, the isobaric adsorption phase (1–2) proceeds. Adsorption of the vaporized adsorbate by the adsorbent results in releasing heat Q_{1-2}, which ought to be removed until the bed temperature reaches T_2 (~T_c). The maximal adsorbate uptake in the cycle, X_{max}, is achieved at the end of the phase, and it is determined by the bed pressure, which is normally at the evaporator pressure, P_e. The heat released by the adsorbent bed during the adsorption phase is given by the following equation (Demir, Mobedi, & Ulku, 2008):

Figure 2. Adsorption heat pump thermodynamic cycle
(Dias, João, & Costa, 2018).

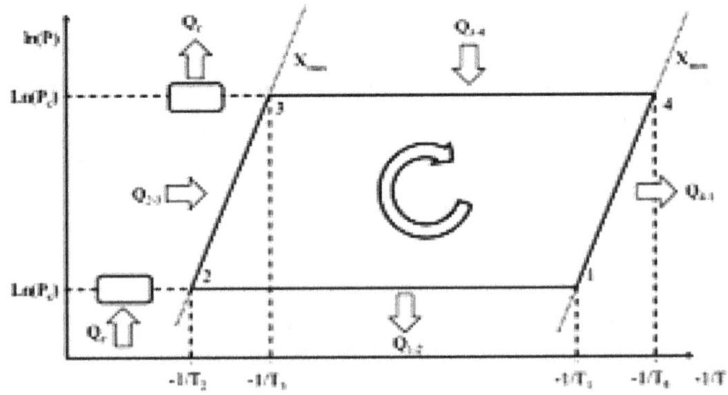

$$-Q_{1-2} \approx \int_{T_1}^{T_2} \left[M_s \left(C_s + X C_{p,a} \right) + M_{bed} C_{bed} \right] dT - \int_{X_{\min}}^{X_{\max}} M_s \Delta H_{ads} dX \tag{1}$$

where M_s corresponds with the adsorbent mass; M_{bed} refers to the adsorbent bed mass; C_s is the specific heat of the adsorbent; $C_{p,a}$ is the constant pressure specific heat of the adsorbate; C_{bed} is the specific heat of the adsorbent bed; ΔH_{ads} refers to the adsorption heat; X is the adsorbate concentration in the adsorbent, i.e. water uptake or adsorption.

Then, the isosteric heating phase (2–3) begins. The valve between the evaporator and the adsorbent bed is closed when heat provided to the bed. The adsorbate uptake remains constant during this phase, as both valves are closed. The pressure inside the bed occurs to increase with temperature until it reaches the condenser pressure, P_c which corresponded to the termination of this phase. The amount of heat supplied during this process can be calculated by (Demir et al., 2008):

$$Q_{2-3} \approx \int_{T_2}^{T_3} \left[M_s \left(C_s + X_{\max} C_{p,a} \right) + M_{bed} C_{bed} \right] dT \tag{2}$$

where X_{max} refers to the maximal uptake of the adsorbate, i.e. maximal adsorption.

The heating process continues during the isobaric desorption phase (3–4) which proceeds when the valve between the bed and the condenser opened. The desorbed adsorbate flows from the bed to the condenser, where it condenses and useful heat of condensation, Q_c is released. The pressure remains constant during this phase, it being corresponded to the condenser pressure.

The next phase begins when desorption temperature is achieved, $T_4 = T_{des}$, the valve is closed. The heat provided to the bed during this phase is given according to (Demir et al., 2008):

$$Q_{3-4} \approx \int_{T_3}^{T_4} \left[M_s \left(C_s + XC_{p,a} \right) + M_{bed} C_{bed} \right] dT - \int_{X_{\max}}^{X_{\min}} M_s \Delta H_{ads} dX \tag{3}$$

where X_{min} refers to the minimal uptake of the adsorbate, i.e. minimal adsorption.

The isosteric cooling (4–1) is the last phase of the cycle. When the valves between the adsorbent bed and the condenser and evaporator closed, the temperature of the bed is decreased. The adsorbate uptake remains constant during the process and the pressure and temperature inside the bed are decreased simultaneously. When the pressure desreased to the evaporator pressure, P_e, the phase is finished and the subsequent isobaric adsorption phase begins. The heat withdrawn from the bed during the isosteric cooling is calculated as (Demir et al., 2008):

$$-Q_{4-1} \approx \int_{T_4}^{T_1} \left[M_s \left(C_s + X_{\min} C_{p,a} \right) + M_{bed} C_{bed} \right] dT \tag{4}$$

The heat required for the vaporization of the adsorbate in the evaporator, Q_e, taken from the low-level heat source during adsorption, and the heat of condensation, Q_c, obtained during desorption phase, can be calculated by [226]:

$$Q_e = M_s \Delta X \Delta H_v + \int_{T_c}^{T_e} M_s \Delta X C_{p,a} dT \tag{5}$$

$$Q_c = M_s \Delta X \Delta H_v \tag{6}$$

where T_e is the evaporation temperature; T_c refers to the condensation temperature; ΔH_v corresponds to the heat of evaporation.

Despite the fact that the main task of adsorptive heat pumps is focused on heating applications, the evaporator withdraws heat from its surroundings and provides cooling effect. The criteria for estimation of performance of an adsorptive heat pump are usually coefficient of performance (*COP*) and specific heating or cooling power (*SHP* or *SCP*). For heating applications, COP_h is determined as the ratio of the heat released by the adsorbent bed during isosteric cooling and adsorption phases (Q_{4-1} $+Q_{1-2}$), plus the heat of condensation Q_c, to the heat required during isosteric heating and desorption (Q_{2-3} $+Q_{3-4}$). For cooling purposes, COP_c is calculated as the ratio of the heat of evaporation to the heat provided to the bed [Dias and Costa, 2018].

$$COP_h = \frac{Q_c + Q_{4-1} + Q_{1-2}}{Q_{2-3} + Q_{3-4}} \tag{7}$$

$$COP_c = \frac{Q_e}{Q_{2-3} + Q_{3-4}} \tag{8}$$

The specific heating power *SHP* is determined as the ratio of useful heat to product of the adsorbent mass and cycle time:

$$SHP = \frac{Q_c + Q_{4-1} + Q_{1-2}}{M_s \tau_{cyc}} \tag{9}$$

The specific cooling power SCP is estimated as the ratio of the evaporation heat to the product of the adsorbent mass and cycle time as follows:

$$SCP_c = \frac{Q_e}{M_s \tau_{cyc}} \tag{10}$$

The specific heating or cooling power are known to be strongly affected by the system design. It allows comparison of different system setups and designs (Demir et al., 2008, Ulku, 1986).

A new adsorptive cycle for upgrading the ambient heat is suggested and briefly analyzed by Aristov (2017). The simplest 3T HeCol cycle also operates between three thermostats with constant temperatures T_L, T_M and T_H. However, it consists of two isosters and two isotherms (Figure 3).

The main feature of this cycle is that regeneration of adsorbent is performed at constant temperature T_M by dropping the vapour pressure over the adsorbent (stage 4–1). This pressure drop is ensured exclusively by the low ambient temperature. So, the regeneration does not require any supply of heat that has commercial value. It is somewhat similar to the Pressure Swing Adsorption process used for gas drying and separation (Ruthven, Farooq, & Knaebel, 1994). The initial adsorbent state (point 1 on Fig. 7.3) corresponds to temperature T_M and pressure of the adsorptive vapour $P_L = P_0(T_L)$, where $P_0(T_L)$ is the saturation vapour pressure at low temperature T_L of the ambient air. Due to the low pressure P_L, the equilibrium adsorbate uptake $w_1 = w(T_M, P_L)$ is small. The isoster 1–2 corresponds with heating the adsorbent up to temperature T_H. At point 2, the adsorber is connected to an evaporator maintained at T_M which generates the constant vapour pressure $P_M = P_0(T_M)$. This pressure jump results in the vapour adsorption and an increasing the equilibrium uptake up to $w_2 = w(T_H, P_M)$ (point 3 on Figure 3).

Figure 3. p – T-diagram for the new cycle for heat upgrading – isothermal version
By Aristov (2017).

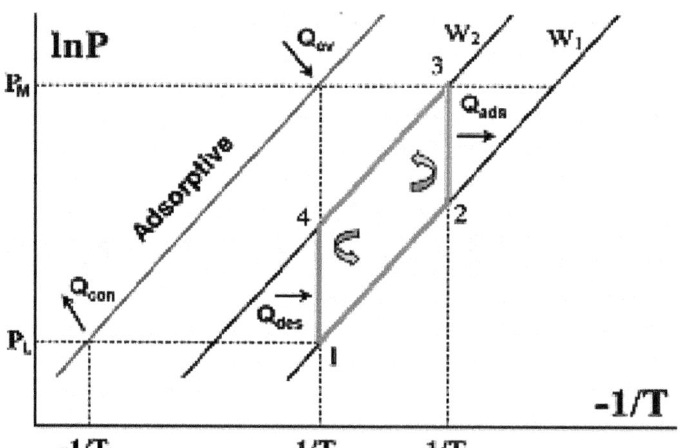

The heat of evaporation Q_{ev} is absorbed in the evaporator at T_M and the useful adsorption heat Q_{ads} is released at constant temperature T_H (isotherm

2–3) in the adsorber and supplied to the heating circuit of a Consumer. Then, the adsorber is isolated from the evaporator and cooled down to temperature T_M (isoster 3–4). At point 4, the adsorber is connected with the condenser maintained at the low temperature T_L and pressure $P_L = P_0(T_L)$. This pressure drop leads to the adsorbate releasing to restore the initial uptake $w^1 = w(T_M, P_L)$ (point 1). The heat Q_{des} required for adsorbate desorption is supplied to the adsorbent from the water basin at temperature T_M (isotherm 4–1). The desorbed vapour is condensed in the condenser releasing the heat Q_{con} to the ambient media, and the cycle is closed.

If neglect all thermal masses, the energy balance and the first law efficiency can be presented as $Q_{des} + Q_{ev} = Q_{ads} + Q_{con}$ and $F_{LT} = Q_{ads}/(Q_{des} + Q_{ev}) \approx 0.5$. The last value, however, has no practical meaning since the heats Q_{des} and Q_{ev} are suggested to be taken for free from an inexhaustible natural heat reservoir at low temperature T_M. Contrariwise, the useful heat Q_{ads} may be of a commercial value and be used for heating.

The heat upgrading may be also accomplished under nonisothermal mode as represented by Fig. 4. At point 2, the adsorber is linked with the evaporator maintained at T_M. As the heat of adsorption released, the adsorbent temperature increases up to $T(2^*) > T_H$, if all thermal masses are equal to zero. If the thermal masses are non-zero, the adsorbent temperature is increased up to $T(2') < T(2^*)$, because a part of the adsorption heat is spent for heating the adsorbent and heat exchanger. Temperature 2' can be evaluated, if the heat

Figure 4. P-T diagram of the non-isothermal HeCol cycle
(Aristov, 2017)

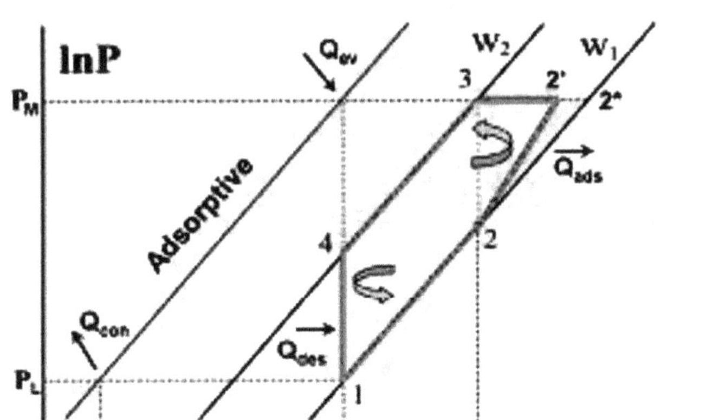

capacities of the adsorbent and heat exchanger are known. After that, the rest of adsorption process (2* – 3) or (2' – 3) is proceeds at $P = P_M$ = const until the equilibrium is set at point 3 corresponding to (T_H, P_M).

This non-isothermal cycle can supply a Consumer with a somewhat lower quantity of heat, but with higher quality, namely, at the temperature between T(2*) or T(2') and T_H (Figure 4) which is higher than in the isothermal cycle (Figure 4).

So, the main feature of this cycle is that regeneration of adsorbent is resulted from dropping the vapour pressure over adsorbent at a constant temperature, rather than by adsorbent heating as usual. This pressure drop does not need the supply of energy that has commercial value. It is of crucial importance that the regeneration occurs to be easier at colder ambient. Therefore, such technology is promising for countries with cold climate, especially for the Arctic zone.

Applied Adsorbents

Traditionally, a working pair 'adsorbent – water' is used in adsorptive heat pumps. Adsorbents are choiced from wide range of standard adsorbents which historically used in adsorptive heat storage devices or chillers driving by low-temperature heat sources.

Microporous silica gel is known (Chahbani, Labidi,& Paris, 2002) to be promising because of its low cost and its sufficient reliability for practical application if the temperature lift between evaporator and condenser is not too large. Nevertheless, silica gel is mainly considered as a less stable material because of its amorphous nature as compared with crystalline zeolites.

Despite the fact that most of research has been done on pair 'zeolite – water', zeolites can be used with all the major chilling agents. The commercial alumino-silicate zeolites of type 4A and 13X are generally used (Pesaran, Lee, Hwang, Radermacher, & Chun, 2016). These zeolites exhibit a strong hydrophilic character and as a consequence a high water adsorptive capacity and high ad/desorption heat. However, commercialization of heat pumps based on such alumino-silicate zeolites is strongly inhibited by high regeneration temperature, which limits their use to applications where a high temperature driving heat source (200 – 300°C) is available.

As alternative to conventional zeolites, a new class of zeolite-like materials denominated as aluminophospate (AIPO) and silico-aluminophospate (SAPO) can be considered. AIPO and SAPO exhibit a moderately low regeneration temperature (60–100°C) which results in high performance (Hassan, Mohamad, Alyousef, Y., Al-Ansary, 2015).. The interesting properties of such materials for adsorption cycles were confirmed by a number of studies for different types of silico-aluminophosphates (AIPO-5, AIPO-18, SAPO-34, etc.) [233, 234].

Novel silico-aluminophosphates 'AQSOA' have been developed and commercialized by Mitsubishi Plastic Inc. (MPI) for desiccant and adsorptive heat pumping systems (Dias and Costa, 2018). Adsorption equilibrium and service life studies of adsorbents AQSOA-FAM-Z01 and AQSOA-FAM-Z02 are given in (Teo, Chakraborty, & Fan, 2017). Both materials present S-shaped isotherms (Boman, Hoysall, Pahinkar, Ponkala, & Garimella, 2017). However, the water uptake for these materials occurs to be within limits 0.25 g/g [Kakiuchi et al., 2005], which obviously require not only large volumes of adsorptive units but high inputs of heat transfer media for regeneration of these adsorbents. Among other factors, a commercialization of AIPO and SAPO zeolite-like materials is strongly inhibited by its the high capital cost, because the synthesis process is more expensive than standard commercial zeolites. So, no industrial mass-production has yet been established.

Y-type zeolites reveal partially hydrophilic character, and can be considered as a cheaper alternative to AIPO and SAPO. A novel zeolite ''DDZ-70'' has been developed by UOP Llc., a Honeywell Company. This material is a high silica content Y zeolite exchanged with cerium. Such modification results in reducing of the hydrophilicity [Lutz, 2014]. Application of DDZ-70 is shown to be promising for low temperature adsorption cycles, because of the lower regeneration temperature (<150°C) and lower water adsorption heat (Freni et al., 2015). Despite the fact that hydrophobic zeolites repel water molecules but they can be used with such refrigerants as methanol or ethanol. Consequently, a very high silica content Y-type zeolite CBV-901 from Zeolyst Int., has been suggested for cooling applications as methanol adsorbent (Freni et al., 2015).

Activated carbons were proposed and closely studied as adsorbents of ammonia and methanol in adsorptive heat pumping systems (Freni et al., 2016). A huge variety of porous carbons is available in granular, expanded and consolidated forms as well as fibers, felts and cloths. As adsorbates, such fluids are suggested: methanol, ethanol and ammonia (Tamainot-Telto, Metcalf, Critoph, Zhong, Thorpe, 2009). A typical adsorbate uptake is reached of 0.1 – 0.6 g adsorbate per g adsorbent within adsorptive heat conversion cycles. Lately, ethanol adsorption onto different carbonaceous adsorbents has been studied for adsorptive cooling systems (Frazzica et al., 2016). The activated carbon exhibits COP of 0.39 – 0.55 and 0.45 – 0.63 for refrigerating and air conditioning cycles, respectively.

According to Freni et al. (2016), one of a major trends of the research in the field of adsorptive heat pumping and cooling is estimation of super-molecular metal-organic frameworks (MOFs). MOFs attract attention, due to their open crystalline structure and very high surface area. Some studies revealed a large adsorptive capacity of water, methanol and ethanol under practically useful operating conditions (Freni et al., 2016). Nevertheless, a significant ad/desorption hysteresis and insufficient stability strongly inhibit practical application of MOFs.

Most promising adsorbents are the composites formed by inorganic salts inserted in porous matrices. These adsorbents can interact with convenient working fluid such as water, methanol, ethanol or ammonia (Veselovskaya &Tokarev, 2011). A wide class of these materials based on mesoporous silica impregnated with $CaCl_2$ (SWS-1L) and LiBr (SWS-2L) exhibits a high water uptake (0.75 kg/kg) and a relatively low temperature of desorption (70–120 °C) (Aristov, Parmo, Restuccia, & Cacciola, 2002). The composite SWS-8L reveals a water uptake of 0.2 kg/kg and an even lower desorption temperature (75–80 °C) (Veselovskaya & Tokarev, (2011). Some studies are focused on the utilization of methanol (Oliveira, Wang & Li, 2010) and ethanol (Gordeeva & Aristov, 2010) as adsorbate. Another works consider water as adsorbate (Gordeeva, Restuccia, Cacciola, & Aristov, 1998).

Another composite material, called EML, was developed for magnesium oxide/water chemical heat pump (Myagmarjav, Zamengo, Ryu, & Kato, 2015). It was obtained by mixing pure $Mg(OH)_2$ with expanded graphite (EG) and lithium bromide (LiBr), which reveals higher thermal conductivity and reactivity, respectively. With the aim to achieve a high energy density, the EML composite was compressed into figure of the EML tablet. Despite the rather high value of heat storage density of 815.4 MJ per m^3 of the tablet, dehydration temperatures for this adsorbent reach of 200 – 300°C. So, such property substantially complicates commercialization of EML in sector of housing and utilities especially for utilization of low-grade heat.

Bed Specifications

Several studies have been focused on the modelling and experimental testing of adsorbent bed and heat exchanger configurations. As a rule, the parameters studied are flat or circular tube configurations, the use of fins, fin height, fin pitch and number of fins.

A prototype of an adsorption chiller and heat pump was designed by Núñez Mittelbach and Henning (2007) for domestic applications. It is shown to be capable of delivering a nominal power of 3–8 kW (COP = 0.5) and 16 kW (COP = 1.5) for cooling and heating modes, respectively. Nevertheless, as an adsorbent silica gel was used. It results in high mass and volume of adsorbent. For example, two beds of silica gel with 35 kg are required to reach this power level. They are corresponded to SCP of 86–229W/kg and SHP of 457W/kg. So, a big machine which cannot be adequate for domestic applications is needed.

A parametric study of a heat exchanger (HEx) to be used as adsorbent bed was carried out by Kowsari, Niazmand & Tokarev (2017).. A numerical model was validated with an experimental set up. Rectangular and trapezoidal bed geometries were studied for different fin heights and pitch and the results were compared based on SCP, VSCP, COP_c, adsorbent mass and metal mass. The rectangular geometry exhibits the best SCP as compared to a similar configuration for the trapezoidal geometry. According to previous studies, higher metal mass results in a lower COP_c and large adsorbent thickness provides poor SCP.

A parametric study of a carbon-ammonia adsorber for a gas fired AHP operating under room heating conditions was performed by Rivero-Pacho, Critoph, and Metcalf (2017). Rivero-Pacho et al. (2017), using two 2-D finite difference simulation models programmed in Matlab™. The adsorber involves 1700 tubes with 1.2mm outer diameter and a pitch of 3mm, with a mixture of carbon grains and powder packed between the tubes. A finned tube and shell and tube geometries exhibit no major differences in the COP. Nevertheless, when the finned tube geometry used, the number of tubes can be decreased seven times, reducing the manufacturing costs.

A novel adsorber structure with aluminium sintered metal fibres was designed by Wittstadt et al. (2016). Results obtained by a driving temperature difference analysis revealed that this new type of heat exchanger reaches a very high volume specific power density for a reasonable adsorbent to HEx mass ratio.

Graf et al. (2016) developed an experimentally calibrated dynamic model to predict the SCP and COP of AHPs. Gravimetric large temperature jump (G-LTJ) experiments were performed on a representative adsorber part in order to calibrate the adsorber model, which is then used to simulate an adsorptive heat pump (AHP) and predict its specific cooling power SCP and coefficient of performance COP. The heat and mass transfer coefficients were calculated by least-square fitting the measured and simulated adsorbate uptake. The model allows a fast evaluation of design and operating parameters, such as fin geometry, cycle times and operating temperatures. It was validated for a small representative adsorber part. However, the model was not corroborated for a real adsorptive heat pump system and the question of whether it allows to predict the performance of a real sized adsorber still remains. The design and testing of AHP systems based on previous parametric analysis of single system components are represented in (San & Tsai,. 2014). Since these approaches being purely experimental, they provide accurate characterizations of a system for a given experimental scenario. Nevertheless, these kind of experimental approaches are restricted to the implemented setup, performing a parametric analysis of the entire system being complicated as the whole setup would have to be dismantled and built over again in order to change some parameters.

Operating Conditions

Sapienza et al. (2016) developed an adsorption chiller prototype with three flat tube heat exchangers coated with the AQSOA-Z02 adsorbent and simultaneously packed with silica gel grains. The tests of the prototype were conducted in a test bench for adsorption chillers at CNR-ITAE Testing Centre under two different operating temperatures sets. The temperatures of the evaporator and condenser in/outlet were registered and the effect of the cycle time on the average cooling power (ACP) and COPc was studied for the two different operational conditions. The cycle times which lead maximum ACP were stated as well as the cycle times which result in maximum COP_c. The results exhibit that maximum ACP and maximum COP_c cannot be obtained simultaneously and the balance between them can be controlled by adjusting cycle times.

An experimental study of an adsorption heat storage system using AQSOA-Z02-water pair was performed by Palomba, Vasta and Freni (2017). The system was tested under different operating regimes for cold storage, daily heat storage and seasonal heat storage working modes. The temperatures in the adsorber, evaporator and condenser in/outlets were controlled. The desorption temperature is revealed to be the parameter with higher repercussions on heat the storage capacity of the system. Moreover, the system's performance is strongly affected by the evaporator temperature ranged from 5 to 10 °C, effect for temperatures above this range being negligible.

A gas-fired adsorption heat pump for heating appliance was designed by Dawoud (2014), from Viessmann Werke Allendorf GmbH. Its performance is divided in four modes. The concept is focused on using adsorption and desorption phases to sustain the low heat demand operation modes. Under the medium heat demand, the adsorption and desorption processes can deliver the required heat to the heating net only at the beginning of each phase. To achieve the desired supply temperature, the gas burner is used to heat backup the adsorption heat pump when the adsorption and desorption phases nearly finished. Even when the processes are close to their ends, they are still used for pre-heating the return cold water, the temperature lift needed in the gas burner being reduced.

During high heat demand mode, the water is heated using solely the gas burner (Dawoud, 2011). The system was implemented in two one-family houses in Germany and exhibits 27% more efficient seasonal gas utilization, which results in 20% less CO_2 emissions as compared to the standalone gas condensing boiler technology. As a result, a commercial product called Vitosorp 200-F has launched been by Viessmann (Dawoud, 2014). The heat pump adsorption module involves a heat exchanger coated with zeolite, it being regenerated by a water circuit heated by a gas-fired boiler. Nevertheless, a performance-predicting model does not exist or is unavailable to outsiders, because it can provide designing and dimensioning details to develop the construction of a similar system. The system's components are not described in detail, which makes it impossible to develop a physical model for representing the system. So, detailed information about Vitosorp 200-F is is unavailable,

Figure 5. Possible AHP system schematics
[Dawoud, 2014]

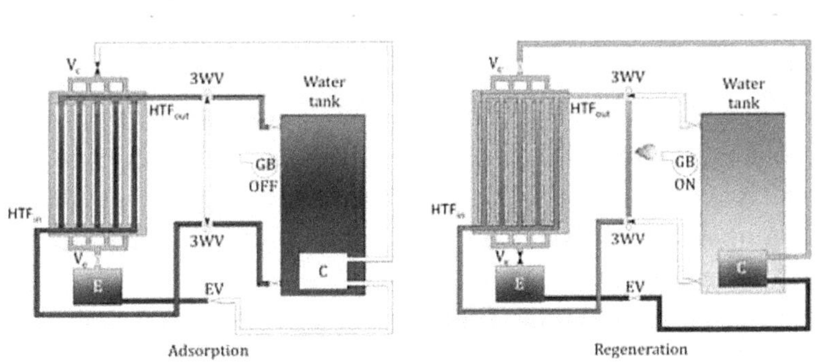

because it is protected by a patent legislation. In this system the adsorber unit is uses to heat the water inside the water tank. The water is used as heat transfer fluid, a gas burner being used for the adsorbent regeneration. The major components of the system involve the adsorber module, evaporator, condenser, gas burner, expansion valve (EV) and the water tank.

A possible configuration of adsorptive heat pump system is given in Figure 5.

During the adsorption phase, adsorbate vapour comes from the evaporator to the adsorber, when the valve to the condenser closed. As the adsorbent material uptaking the adsorbate, heat of adsorption is released. Retrieving this heat occurs as cold water is taken from the bottom of the water tank, or from the supply network, and circulated inside the tubes of the adsorber. After leaving the adsorber, the warmed water is returned to the tank. The process continues until a pre-determined uptake value is reached or a certain time is passed. Then, desorption process starts and water is taken from the water tank, heated by the gas burner and circulated inside the tubes of the adsorber. The released adsorbate circulates through the condenser transferring the heat of condensation to the water inside the tank. The process occurs until a pre-determined uptake value is achieved or a certain time is passed. At the end of the phase of desorption, the hot water which was circulated inside the tubes of the adsorber is returned to the water tank.

This system operated in three different modes: low power demand, medium power demand and high power demand. In the low power demand mode, the gas burner is only used to regenerate the adsorbent material and the heat delivered to the water tank is solely obtained by the adsorption and regeneration processes. In the medium peak demand mode, the gas burner is used not only for the regeneration of the adsorbent but also to heat the water the leaves the adsorber during the adsorption process, boosting its temperature to a higher level. Finally, in the high power demand mode the system works as a gas boiler, using the gas burner to heat the water directly, providing instant hot water.

As a rule, coefficient of performance of adsorptive heat pump is suggested to improve by surface cascade cycle, or multi-bed, or the dual mode cycle (Li, Hou, Zhang, Yuan, 2015).

Another construction of the adsorptive heat pump is designed for independent heat supply systems or hot water supply of buildings and other structures for various purposes.

The adsorptive heat pump (Figure 6) is operated into two modes (Pat. 91481 UA, 2014).

The first mode corresponds to adsorption and the second one associates with adsorbent regeneration. On the first stage (adsorption) through open distributing checker (10) water vapor is supplied from the evaporator (7) to adsorber (1) and then adsorbed by composite material (adsorbent) (4). The adsorption heat transfers from granula to granula and to grid, and then to heat transfer media in the hydraulic circuit and to consumer. Heat is supplied to the evaporator by finned warmer (8).

Figure 6. Solar adsorptive heat pump: 1 –cylindric adsorber; 2 – hydraulic circuit; 3 – fins of hydraulic; 4 – adsorbent; 5 – condenser; 6 – cylindrical rings; 7 – evaporator; 8 – warmer, 9 – narrowed neck; 10 – distributing checker; 11 – lid; 12 – polymeric solar collector; 13 – pump; 14 solar rays; 15 – heated water diversion; 16 – cold water supply; 17 – taps.
(Pat. 91481 Ukraina, 2014)

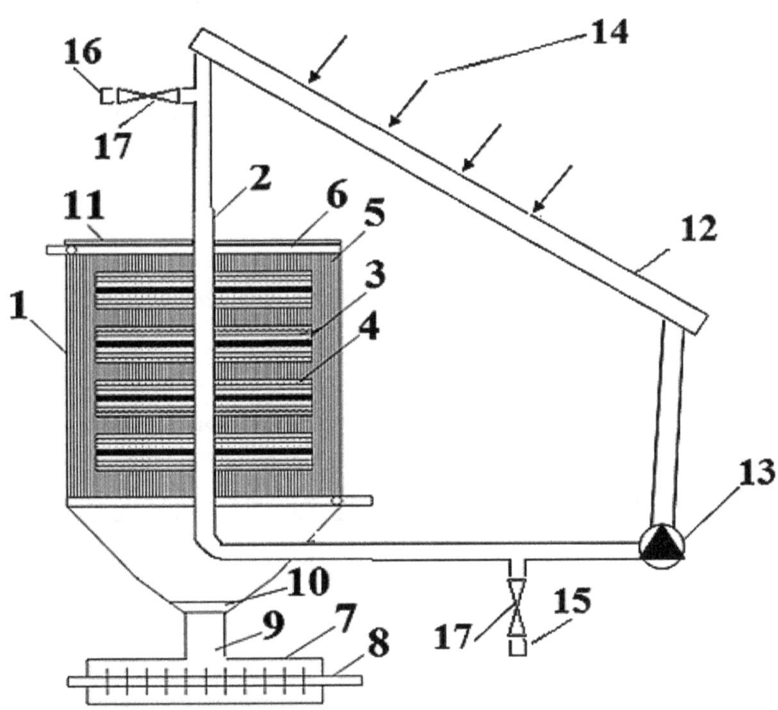

On the second stage distributing checker 10 is closed. Heat from external source (solar collector PSK-VS1-2) is transferred through the hydraulic circuit to one-side finned grid and to adsorbent granules which placed in cylindrical rings. At that, heat of condensation is released. Then heat is supplied to chilling heat carrier which is directed to the consumer or warmed in the hydraulic circuit and used for adsorbent regeneration. The condensate flows down to the evaporator by radial-vertical fins through narrowed neck to the evaporator.

The coefficient of energy performance was calculated according to the formula (Morozyuk, 2006):

$$COP = \frac{T_1}{T_1 - T_2} \tag{11}$$

where T_1 is a temperature in the evaporator, °C; T_2 corresponds to a temperature in a condenser.

Its value equals to 1.55.

Calculation of the mass of the adsorbent and water which are required for performance of the adsorptive heat pump is carried out similarly to the adsorptive heat storage device (subdivision 4.4).

Adsorptive heat pump was tested through February 2017. Period-averaged day changes of temperature of adsorbent, adsorbate and pressure in the system are given in a Figure 7.

Figure 7. Averaged day operating cycle of the experimental adsorptive heat pump (Belyanovskaya, Sukhyy, Kolomiyets, Gubinskyi, 2019).

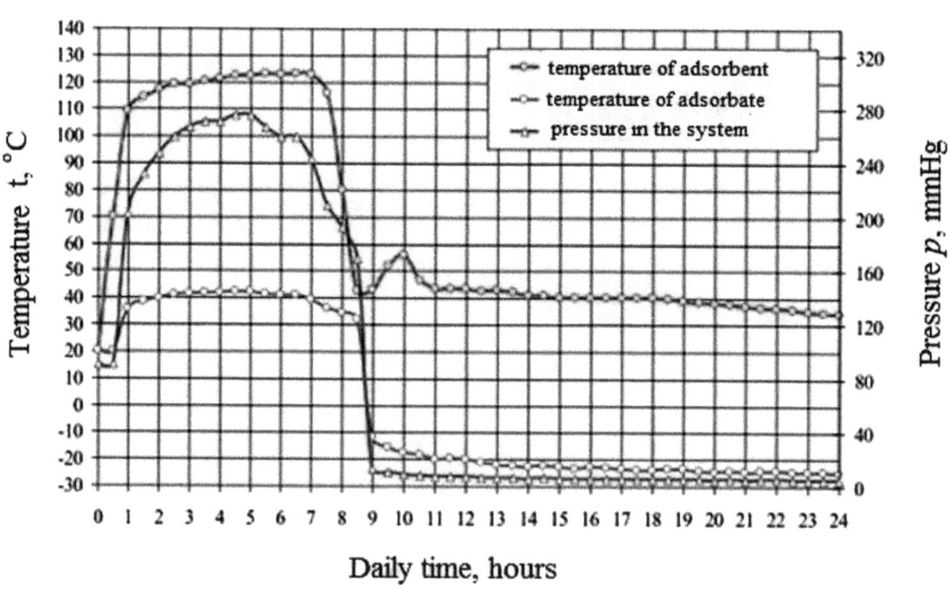

As adsorbent were used the composites 'silica gel – sodium sulphate' or 'silica gel – sodium acetate'. Mass of composite was 10 kg. As a thermal insulation, mineral wool with thickness of 100 mm and heat conduction coefficient 0.035 W/m·K was used. Discharging was performed in a day time, when heat pump supplied heating system with water warmed to 45 – 35°C. The regeneration mode proceeded at night from 0.00 to 8.00 a.m.

The efficiency adsorptive heat pumps is estimated by coefficient of energy performance COP, which is equal to ratio of heat supplied to consumer with a higher temperature to heat expended at that. COP can be calculated by two methods:

1. as ratio of adsorption heat to sum of desorption heat and external heat supplied to adsorbent during its heating upto regeneration temperature (COP_c);
2. as ratio of adsorption heat to heat supplied by solar collector (COP_s(b)) (net).

Results of calculation are represented in Table1.

Table 1. Adsorptive heat pump coefficient of energy performance

Month	The Total Average Daily Value of Solar Radiation (kW-h/ m²)	'Silica Gel – Na_2SO_4'		'Silica Gel – CH_3COONa'	
		COP_c	COP_s(b)	COP_c	COP_s(b)
May	5.99	2.084	1.130	2.021	1.189
June	5.86	2.084	1.133	2.021	1.193
July	6.06	2.084	1.128	2.021	1.187
August	5.66	2.084	1.137	2.021	1.200
September	4.41	2.084	1.176	2.021	1.257

(Sukhyy, Belyanovskaya, Kolomiyets, 2018)

COP_c and COP_s(b) of adsorptive heat pumps based on both composite sorbents are stated to be of the same magnitude. It results from lower regeneration temperature of composite 'silica gel – sodium acetate' (60°C) as compared with 'silica gel – sodium sulphate' (90°C). Negligible decrease of COP_c when composite 'silica gel – CH_3COONa' applied is a consequence of the lower adsorption capacity as compared to 'silica gel – Na_2SO_4'. Higher values of COP_s(b) when composite 'silica gel – CH_3COONa' used results from its lower regeneration temperature.

ADSORPTIVE HEAT PUMPS DEVELOPMENT CHALLENGES

Despite the several studies focused on the adsorption kinetics, adsorbent bed specifications and operating conditions, a system-level approach considering an entire adsorptive heat pump system still remains a valuable contribution towards the development of the technology of adsorptive heat pumping.

The optimization of a single or few parameters, isolated components and working pairs can only slightly affect the performance of the complete system. There is obvious lack of system-level studies which involve the simultaneous optimization of the adsorption kinetics, adsorbent bed specifications, operating conditions and relationship between all system components (Pesaran et al., 2017). Moreover, detailed experimental data for the validation of physical dynamic models that consider an entire AHP system are still unavailable. Future research must integrate the individual optimizations into a practical and commercial adsorptive heat pump in order to identify and distinguish the high sensitive parameters from the more irrelevant ones.

There are the crucial challenges which are yet to be overcome in order to develop an adsorptive heat pump system for heating appliances.

Conveniently, a suggested adsorptive heat pump systems for heating applications are designed basing on several adsorbent-adsorbate working pairs, such as

- Silica gel – Water.
- Activated carbon – Ammonia.
- Zeolite 4A – Water.
- Zeolite 13X– Water.
- AQSOA-Z02 – Water.

The hydrothermal stability and material aging for these few particular materials are already well stated and considered suitable. Despite the interesting characteristics and readiness to market most of these working pairs are not available in the amounts needed for application in a real size adsorptive heat. Moreover, low water uptake with the traditional adsorbents lead to high volumes of adsorptive units that together with high regeneration temperature and significant cost strongly inhibits commercialization of adsorptive heat pumps in heat supply systems because of performance, sanitary regulations and unsustainable costs for production and operating. Nevertheless, the above-mentioned working pairs are still remained the most studied.

Another challenge for operating the adsorptive heat pump is regeneration the adsorbent material. The most commonly it is performed by solar thermal energy, waste heat and gas burners.

Waste heat is generally used in such applications as industrial processes or thermal engines because it is abundant at low cost. However, for domestic applications it is as a rule unavailable in a useful amount.

As a rule, solar thermal energy is used for adsorption heat storage because it can store heat energy during the day. However, regeneration of heat storage material is restricted by the period when solar energy is available. To control the cycle time, the adsorbent material ought to regenerate according to the system requirements, which depend on different operating conditions and system's configuration.

As a solution for cycle time control, gas burners can be considered. Regrettably, as distinct from solar energy, natural gas is not free and its burning results in carbon dioxide to the atmosphere. As a promising solution can be considered combination of the two technologies, since by using solar energy when it is available and the gas burner only when the solar energy is insufficient or unavailable. So, the operational costs and CO_2 emissions intrinsic to the burning of natural gas can be lowered while maintaining the cycle time control. Additionally, the gas burner can be directly applied to provide heat under peak hour of the heat power demands which are too high to be satisfied by the adsorption process alone. It can also be used to elevate the temperature over 50 °C, which is very difficult to achieve through the adsorption process.

The usage of electricity for water heating becomes recently promising [251], because it can be obtained from renewable energy sources and applied for compensation of the night-time valley.

The main areas to improve for enhancing the an adsorptive heat pump system performance are physical and thermodynamic properties of adsorbents, adsorber configuration and operating modes.

Considering physical and thermodynamic properties of adsorbents, the simultaneous enhancement of the adsorbent material thermal conductivity and permeability can result in a major improvement. Nevertheless, according to [226], improving one of these parameters leads to deterioration of the other.

So, improvement the system's performance requires the right balance between these two parameters. Moreover, the convective heat transfer coefficient between the adsorber metal wall and the adsorbent material directly affects the heat flux that is transferred between them, while the cycle time and consequently the specific heat power remain almost constant. This parameter is affected by the effective contact between the adsorbent material and the adsorber metal wall. So, another way to improve the system's performance is focused on enhancement of the effective contact better coating techniques that bind the adsorbent material to the metal wall without compromising the adsorbent permeability. All these engineering solutions are currently being developed at lab scale but their sustainable production for the market is still a challenge.

Obviously, enlargement of the contact areas between the adsorbent material and the metal wall of the adsorber will enhance the heat transfer between them. The common method to increase this heat transfer area is to include fins in the adsorber. Nevertheless, disregarding the adsorber metal mass is common in the performance calculation, which needs to be taken into account in the physical models to obtain accurate results. Recent studies (Frazzica & Freni, 2017) exhibited that the performance of AHP systems are affected by the ratio of the metal to adsorbent mass. Increasing metal to adsorbent mass ratios result in lower performances, this was confirmed for several working pairs. However, application of an adsorber configuration with fins not only allows enhancing the heat transfer in the bed, but substantially increases the metal to adsorbent mass ratio. The added fins also occupy precious space which could be otherwise filled with adsorbent, lowering the adsorbent mass when compared to a similar configuration without fins.

There are several packing techniques and adsorber configurations such as shell packing plate-finned, tube, finned tube, fin plate, flat pipe, spiral plate and coated tube, which are described according to (Li et al., 2015). All the different configurations lead to different adsorbent-metal wall contact areas, metal to adsorbent mass ratios, adsorbent area exposed to the adsorbate and heat losses to the environment. The best adsorber configuration may be different for each application and dependent on the adsorber metal and adsorbent-adsorbate working pair thermodynamic and physical properties. Any adsorber configuration reveals both advantages and disadvantages. For example, shell configuration shows less adsorbent-adsorbate contact area but it has more adsorbent-metal contact area as compared to the tube configuration. Nevertheless, in the shell configuration the heat transfer media circulates around the adsorbent whereas in the tube configuration the adsorbent material is packed around the tube, which will minimize heat losses since the heat can only be transferred from the tube to the adsorbent material. Finned HExs have higher adsorbent-metal contact area but less adsorbent-adsorbate contact area and higher metal to adsorbent mass ratio. When an adsorber designed, the following aspects ought to take into account, such as: maximization of the heat transfer between adsorbent material and the heat transfer fluid, maximization of the adsorbent surface exposed to the adsorbate, minimization of the metal to adsorbent mass ratio, guarantee of a good thermal contact between the adsorbent bed and the metal (for example using thermal interface material paste), and keep a simple design that can be built using vacuum technology elements. Selecting the best configuration and designing the optimum adsorber is a difficult problem, which requires further investigation aiming at the best balance of all these parameters.

MAJOR LIMITATIONS FOR COMMERCIALIZATION OF ADSORPTIVE HEAT PUMPING: REQUIREMENTS TO ONGOING INNOVATIONS

Despite the fact the adsorptive heat pumping is one of the promising technologies of heat supply, its commercialization is still inhibited. The next limitations are still yet to overcome:

- Low COP;
- Low value of maximal adsorption which results in a large mass of adsorbent material is required to produce the needed heating powers for most applications;
- The materials used in the system (valves, joints, welding, etc.) have to be adequate for vacuum operating conditions;
- It is difficult to provide heat at temperatures over 55 °C;
- The adsorbent-adsorbate working pairs with higher COP and capable of delivering higher heating powers require high regeneration temperatures;
- Many cautions have to be taken when regeneration temperatures higher than the heat transfer fluid's boiling temperature are used due to the possible resulting high pressure inside the pipes.

Not many new adsorbent-adsorbate working pairs are ready to market, easily accessed and produced, hydrothermal stable and with a long life cycle. A bridge between the material's level and system's level studies still remains to be built.

The Linear Driving Force model and Darcy's Law are the most common models to estimate the internal and external mass transfer resistances, respectively. Most of numerical simulation are focused on only one operating mode and do not allow to determine integral operating parameters and efficiency of adsorptive heat pump. The similar limitations are exhibited by physical models.

REFERENCES

Aristov, Yu. I. (2017). Adsorptive transformation of ambient heat: A new cycle. *Applied Thermal Engineering*, *124*, 521–524. doi:10.1016/j.applthermaleng.2017.06.051

Aristov, Yu. I., Parmon, V. N., Restuccia, G., & Cacciola, G. (2002). A family of new working materials for solid sorption air conditioning systems. *Applied Thermal Engineering*, *22*(2), 191–204. doi:10.1016/S1359-4311(01)00072-2

Belyanovskaya, E., Sukhyy, K., Kolomiyets, E., & Gubinskyi, M. (2019). *Operating characteristics of heat conversion units in heat supply system, The Actual Problems of the World Today, 2*. Sciemcee Publishing.

Boman, D. B., Hoysall, D. C., Pahinkar, D. G., Ponkala, M. J., & Garimella, S. (2017). Screening of working pairs for adsorption heat pumps based on thermodynamic and transport characteristics. *Applied Thermal Engineering*, *123*, 422–434. doi:10.1016/j.applthermaleng.2017.04.153

Chahbani, M. H., Labidi, J., & Paris, J. (2002). Effect of mass transfer kinetics on the performance of adsorptive heat pump systems. *Applied Thermal Engineering*, *22*(1), 23–40. doi:10.1016/S1359-4311(01)00067-9

Dawoud, B. (2011). Gas-driven sorption heat pumps, a potential trend-setting heating technology. *IEA Heat Pump. Cent. News.*, *29*, 18–22.

Dawoud, B. (2014). On the development of an innovative gas-fired heating appliance based on a zeolite-water adsorption heat pump; System description and seasonal gas utilization efficiency. *Applied Thermal Engineering*, *72*(2), 323–330. doi:10.1016/j.applthermaleng.2014.09.008

Demir, H., Mobedi, M., & Ulku, S. (2008). A review on adsorption heat pump: Problems and solutions. *Renewable & Sustainable Energy Reviews*, *12*(9), 2381–2403. doi:10.1016/j.rser.2007.06.005

Dias, J., & Costa, V. A. F. (2018). Adsorption heat pumps for heating applications: A review of current state, literature gaps and development challenges. *Renewable & Sustainable Energy Reviews*, *98*, 317–327. doi:10.1016/j.rser.2018.09.026

Frazzica, A., & Freni, A. (2017). Adsorbent working pairs for solar thermal energy storage in buildings. *Renewable Energy*, *110*, 87–94. doi:10.1016/j.renene.2016.09.047

Frazzica, A., Palomba, V., Dawoud, B., Gullì, G., Brancato, V., Sapienza, A., Vasta, S., Freni, A., Costa, F., & Restuccia, G. (2016). Design, realization and testing of an adsorption refrigerator based on activated carbon/ethanol working pair. *Applied Energy*, *174*, 15–24. doi:10.1016/j.apenergy.2016.04.080

Freni, A., Dawoud, B., Lucio, B., Chmielewski, S., Frazzica, A., Calabrese, L., & Restuccia, G. (2015). *Characterization of Zeolite-Based Coatings for Adsorption Heat Pumps. Springer*. Advance online publication. doi:10.1007/978-3-319-09327-7

Freni, A., Maggio, G., Sapienza, A., Frazzica, A., Restuccia, G., & Vasta, S. (2016). Comparative analysis of promising adsorbent/adsorbate pairs for adsorptive heat pumping, air conditioning and refrigeration. *Applied Thermal Engineering*, *104*, 85–95. doi:10.1016/j.applthermaleng.2016.05.036

Gordeeva, L., & Aristov, Yu. (2010). Novel sorbents of ethanol—Salt confined to porous matrix for adsorptive cooling. *Energy*, *35*(6), 2703–2708. doi:10.1016/j.energy.2009.04.001

Gordeeva, L. G., Restuccia, G., Cacciola, G., & Aristov, Yu. (1998). Selective water sorbents for multiple applications: 5. LiBr confined in mesopores of silica gel: sorption properties. *Reaction Kinetics and Catalysis Letters*, *63*(1), 81–88. doi:10.1007/BF02475434

Graf, S., Lanzerath, F., Sapienza, A., Frazzica, A., Freni, A., & Bardow, A. (2016). Prediction of SCP and COP for adsorption heat pumps and chillers by combining the large-temperature-jump method and dynamic modeling. *Applied Thermal Engineering*, *98*, 900–909. doi:10.1016/j.applthermaleng.2015.12.002

Hassan, H. Z., Mohamad, A. A., Alyousef, Y., & Al-Ansary, H. A. (2015). A review on the equations of state for the working pairs used in adsorption cooling systems. *Renewable & Sustainable Energy Reviews*, *45*, 600–609. doi:10.1016/j.rser.2015.02.008

Kakiuchi, H., Shimooka, S., Iwade, M., Oshima, K., Yamazaki, M., Terada, S., Watanabe, H., & Takewaki, T. (2005). Novel water vapor adsorbent FAM-Z01 and its applicability to an adsorption heat pump. *Kagaku Kogaku Ronbunshu*, *31*(5), 361–364. doi:10.1252/kakoronbunshu.31.361

Kowsari, M. M., Niazmand, H., & Tokarev, M. M. (2017). Bed configuration effects on the finned flat-tube adsorption heat exchanger performance: Numerical modeling and experimental validation. *Applied Energy*, *213*, 540–554. doi:10.1016/j.apenergy.2017.11.019

Li, X. H., Hou, X. H., Zhang, X., & Yuan, Z. X. (2015). A review on development of adsorption cooling – novel beds and advanced cycles. *Energy Conversion and Management*, *94*, 221–232. doi:10.1016/j.enconman.2015.01.076

Lutz, W. (2014). Zeolite Y: Synthesis, Modification, and Properties—A Case Revisited. *Advances in Materials Science and Engineering*. doi:10.1155/2014/724248

Morozyuk, T. V. (2006). Teoriya kholodil'nykh mashin i teplovykh nasosov. Odessa: Studiya «Negotsiant».

Myagmarjav, O., Zamengo, M., Ryu, J., & Kato, Yu. (2015). Energy density enhancement of chemical heat storage material for magnesium oxide/water chemical heat pump. *Applied Thermal Engineering*, *91*, 377–386. doi:10.1016/j.applthermaleng.2015.08.008

Núñez, T., Mittelbach, W., & Henning, H. M. (2007). Development of an adsorption chiller and heat pump for domestic heating and air-conditioning applications. *Applied Thermal Engineering*, *27*(13), 2205–2212. doi:10.1016/j.applthermaleng.2005.07.024

Oliveira, R. G., Wang, R. Z., & Li, T. X. (2010). Adsorption characteristics of methanol in activated carbon impregnated with lithium chloride. *Chemical Engineering & Technology*, *33*(10), 1679–1686. doi:10.1002/ceat.201000075

Palomba, V., Vasta, S., & Freni, A. (2017). Experimental testing of AQSOA FAM Z02/water adsorption system for heat and cold storage. *Applied Thermal Engineering*, *124*, 967–974. doi:10.1016/j.applthermaleng.2017.06.085

Pat. 91481 Ukraina, MPK F 25 B 30/00, F 25 B 17/00 Adsorbtsijnij teplovij nasos / Sukhyy K.M., Sukhyy M.P., Kolomiyetsts O.V. [ta in.]; zayavnik i patentovlasnik Derzhavnij vishhij navchal'nij zaklad —Ukraïns'kij derzhavnij khimiko/tekhnologichnij universitet.– N° u 2013 15448; zayavl. 30.12.2013; opubl. 10.07.2014, Byul. N°.13, 4 (in Ukrainian)

Pesaran, A., Lee, H., Hwang, Y., Radermacher, R., & Chun, H. H. (2016). Review article: Numerical simulation of adsorption heat pumps. *Energy*, *100*, 310–320. doi:10.1016/j.energy.2016.01.103

Pinheiro, J. M., Salústio, S., Rocha, J., Valente, A. A., & Silva, C. M. (2016). Analysis of equilibrium and kinetic parameters of water adsorption heating systems for different porous metal/metalloid oxide adsorbents. *Applied Thermal Engineering*, *100*, 215–226. doi:10.1016/j.applthermaleng.2016.01.142

Rivero-Pacho, A. M., Critoph, R. E., & Metcalf, S. J. (2017). Modelling and development of a generator for a domestic gas-fired carbon-ammonia adsorption heat pump. *Renewable Energy*, *110*, 180–185. doi:10.1016/j.renene.2017.03.089

Ruthven, D.M., Farooq, S., & Knaebel, K.S. (1994). *Pressure Swing Adsorption*. Wiley.

San, J. Y., & Tsai, F. K. (2014). Testing of a lab-scale four-bed adsorption heat pump. *Applied Thermal Engineering*, *70*(1), 274–281. doi:10.1016/j.applthermaleng.2014.05.014

Sapienza, A., Gullì, G., Calabrese, L., Palomba, V., Frazzica, A., Brancato, V., La Rosa, D., Vasta, S., Freni, A., Bonaccorsi, L., & Cacciola, G. (2016). An innovative adsorptive chiller prototype based on 3 hybrid coated/granular adsorbers. *Applied Energy*, *179*, 929–938. doi:10.1016/j.apenergy.2016.07.056

Sukhyy, K. M., Belyanovskaya, E. A., & Kolomiyets, E. V. (2018). Design and performance of adsorptive transformers of heat energy. Riga, Latvia: LAP Lambert Academic Publishing.

Tamainot-Telto, Z., Metcalf, S. J., Critoph, R. E., Zhong, Y., & Thorpe, R. (2009). Carbon-ammonia pairs for adsorption refrigeration applications: Ice making, air conditioning and heat pumping. *International Journal of Refrigeration*, *32*(6), 1212–1229. doi:10.1016/j.ijrefrig.2009.01.008

Teo, H. W. B., Chakraborty, A., & Fan, W. (2017). Improved adsorption characteristics data for AQSOA types zeolites and water systems under static and dynamic conditions. *Microporous and Mesoporous Materials*, *242*, 109–117. doi:10.1016/j.micromeso.2017.01.015

Ulku, S. (1986). Adsorption heat pumps. *J. Heat Recover. Syst.*, *6*, 277–284. doi:10.1016/0198-7593(86)90113-X

Veselovskaya, J. V., & Tokarev, M. M. (2011). Novel ammonia sorbents "porous matrix modified by active salt" for adsorptive heat transformation: 4. Dynamics of quasi-isobaric ammonia sorption and desorption on $BaCl_2$/vermiculite. *Applied Thermal Engineering*, *31*(4), 566–572. doi:10.1016/j.applthermaleng.2010.10.018

Wittstadt, U., Füldner, G., Laurenz, E., Warlo, A., Große, A., Herrmann, R., Schnabel, L., & Mittelbach, W. (2017). A novel adsorption module with fiber heat exchangers: Performance analysis based on driving temperature differences. *Renewable Energy*, *110*, 154–161. doi:10.1016/j.renene.2016.08.061

Chapter 8
Construction and Operating Parameters of Adsorptive Chillers

ABSTRACT

The chapter is devoted to the design and performance of adsorptive chillers. Basic types of design and operating principle of adsorptive chillers were analyzed. Advantages and disadvantages performance of one-, two-, three-, and four-bed solar power adsorptive chillers are compared. Performance of adsorptive refrigerators based on composite adsorbents was studied. The correlation between the adsorbent composition and the coefficient of energy performance of the adsorptive chiler was revealed. An optimal composition of adsorbent 'silica gel – sodium sulphate' is stated to be of 20% silica gel and 80% sodium sulphate. The maximal values of the coefficient of performance of cycle of studied solar adsorptive chiller about of 1.14 are stated for composites containing about 20 wt. % silica gel and 80 wt% sodium sulphate. As a consequence of decreasing of adsorbent mass, the coefficient of performance is shown to increase when sodium sulphate content in the composite increased. Regeneration process parameters of the composite were shown to strongly affect on the coefficient of performance of the adsorptive chiller. The growth of the coefficient of performance is stated to result from decreasing the difference between adsorbent temperature and regeneration

DOI: 10.4018/978-1-7998-4432-7.ch008

temperature from 85 to 55°C. The basic factors affecting the net coefficient of energy performance of the adsorptive solar refrigerator were stated daily solar radiant flux alongside with composition of the adsorbent and difference between adsorbent temperature and temperature regeneration. Net coefficients of performance of solar adsorptive refrigerator based on composite 'silica gel – sodium sulphate' were stated to change from 0.25 to 0.34 during operating period. Utilization of the adsorption heat is suggested to warm the heat carrier which applied to heat adsorbent during regeneration. The ways to improve the design and performance of adsorptive solar chillers are suggested. The first one involves the introduction of solar collectors made of cellular polycarbonate plastics in the design of adsorptive solar chiller. Instantaneous efficiency coefficient were calculated as special thermal performance-solar radiant flux surface density ratio, optical efficiency factor is determined as special thermal performance-solar radiant flux surface density ratio at the equal temperatures of heat transfer medium and environment, reduced heat loss factor being calculated as the product of solar collector efficiency factor and net heat loss coefficient. The environmental test of developed collectors PSK-AV2-3, PSK-AV1-2, PSK-AV2-1, PSK-VS1-2, PSK-VS2-2, PSK-VS2-3, PSK-ST10-PW were conducted. The correlation of their results with laboratory tests when the thermohydraulic stand applied is shown. Relative accuracy of laboratory and environment tests was shown to be not exceeding 5 – 7%. The optical efficiency factor and the coefficient of thermal losses of polymeric solar collectors were determined. On the basis of the dependencies of the efficiency of the solar collectors vs. the reduced temperature, optimal designs of the polymeric solar collectors for the adsorption chilling solar systems are determined to be depended on the temperature of the regeneration temperature of the sorbents. As the temperatures of the regeneration of composite adsorbent ranged from 50°C to 60°C, appliance of the collectors PSK-AV2-1, PSK-CT10-PW occur to be expedient, and PSK-AB2-3, PSK-VS2-3, PSK-AB1-2, PSK-VS2-2, and PSK-VS1-2 are revealed to be more efficient when regeneration temperatures increased over 80 °C. Thermotechnical characteristics of designed polymeric solar collectors are shown to surpass conventional metal and vacuum collectors. The perspectives of polymeric solar collectors in the design of adsorptive chilling solar plants were shown. Another way to improve the performance of adsorptive solar chillers concerns with equipping it with a photosensitive element and an electric drive, which will allow changing the angle of slope of the adsorber to the horizon depending on the intensity of the solar radiation. The chapter can be useful for design the efficient adsorptive chilling plants.

NOMENCLATURE

A_{lim} – maximal adsorption of composite adsorbent, kg/kg;
C – heat capacity of the structural elements, J/kg·K;
C_k and C_w – heat capacity of the composite and water, respectively, kJ/kg·K.
C_M, C_A and C_S – specific heats of chilling chamber material, air in it and substances placed into chilling chamber;
C_p – specific heat of the liquid chilling agent, i.e. refrigerant,
C_{pr} – heat capacity of the food substances introduced into the refrigerating chamber, J/kg·K;
COP_c – coefficient of performance of the cycle;
COP_n – net coefficient of performance;
COP_n^{exp} – experimental solar COP;
$COP_{s(n)}$ – net coefficient of performance calculated assuming that the cooling effect is used to cool down the water or other substances contained in the chilling chamber;
$COP_{s(n)\eta}$ –net coefficient of performance calculated as quantity of heat received by composite taken into account;
COP_n –experimental solar COP calculated when efficiency of solar collector taken into account;
E is level of efficiency of protection equipment in the door opening;
D_f – factor that corrects for air flow characteristics in the door opening;
D_τ – coefficient that corrects for time interval when door opened;
F_k – surface area of solar collector, m^2;
$F'(\tau\alpha)$ – optical efficiency factor;
h_e –refrigerant latent heat of vaporization;
k corresponds to specific heat of the material of the cold box and air in it, kJ/K;
M is the mass of refrigerant desorbed in one operation cycle, kg;
m, m_{pr} –mass of the cold chamber and introduced products, respectively, kg;
m_k and m_w – the mass of the composite and the adsorbed water, kg;
M_{ads} – adsorbent mass, kg;
M_{H2O} – mass of the water which is necessary for taking off from the chilling chamber, kg;
M_M, M_A, M_S – masses of chilling chamber material, air in it and substances placed into the cold box, kg;

q – total cooling load of chilling chamber for steady-state air flow with regard to differences of densities, heat contents and humidities and size of door opening, kW;
Q_1 – quantity of heat which have to be taken from the cold box, kJ;
Q_2 – heat quantity taken off in the chilling chamber, kJ;
Q_3 – heat amount required to regeneration, kJ;
Q_{d1} – heat applied to the adsorbent when heated, kJ;
Q_{d2} – heat applied to the adsorbent for releasing the refrigerant vapor, kJ;
Q_e – cooling effect produced at the evaporator, kJ;
Q_k – quantity of heat supplied to adsorbent per 1 m^2 of collector surface area, kJ/m^2;
Q_{reg} – the energy that is effectively transferred to the adsorber to promote its regeneration, kJ;
Q_s – all the solar energy received by the solar collector surface;
Q_p – heat leakages through the camera walls itself, kJ;
Q_V – heat leakage at-the-opening per a day, kJ;
S_k – is external surface area of the chamber, m^2;
T_c and T_e – temperatures of condensation and evaporation, K;.
Z_B – coverage factor;
Z_k – coverage factor of solar collector surface area;
α_{v1} is calorific coefficient for air inside of the chamber, $W/m^2 \cdot °C$;
α_{v2} is calorific coefficient for ambient air, $W/m^2 \cdot °C$;
ΣQ_z – sum of heat leakage into the chamber as a result of heat transfer through its walls, floor and ceiling, from infiltration of outside air when the chamber is opened and loads from lighting, kJ.
ΔH_{des} –the heat of desorption of water, kJ/kg;
ΔH_{ev} – heat of water evaporation, kJ/ kg;
ΔT – difference between ambient temperature and average daily temperature in the refrigerating chamber, K;
ΔT_1 – difference between the temperature of the adsorbent and the temperature of regeneration, K;
ΔH_{des} is the heat of desorption of water, kJ/kg;
λ_{iz} – heat conduction coefficient of heat insulation material, $W/m^2 \cdot °C$;
λ_m – heat conduction coefficient of material of the chamber, $W/m^2 \cdot °C$;
δ_{iz} – thickness of heat-retention lagging, m;
τ – exploitation time per a day, s;
τ_{op} –time interval per a day when door is opened, s.

Conventional vapor compression refrigerating systems and air conditioners in a hot season consume a large amount of electricity that results in depletion of non-recoverable natural resources. Moreover, increasing consumption of electricity in the summer period overloads power supply systems and even leads to serious accidents (Morozyuk, 2006). Adsorptive chilling technologies exhibits a promising features in last three decades because of its noiseless, non-corrosive and environment friendly operation by the utilization of low grade heat sources, especially solar energy. Despite the fact that a plethora of adsorption cooling systems had been developed but still these cooling systems cannot compete with the traditional vapor compressive cooling systems.

Figure 1. Main components of a solar adsorption cooling device (Fernandes et al. 2014).

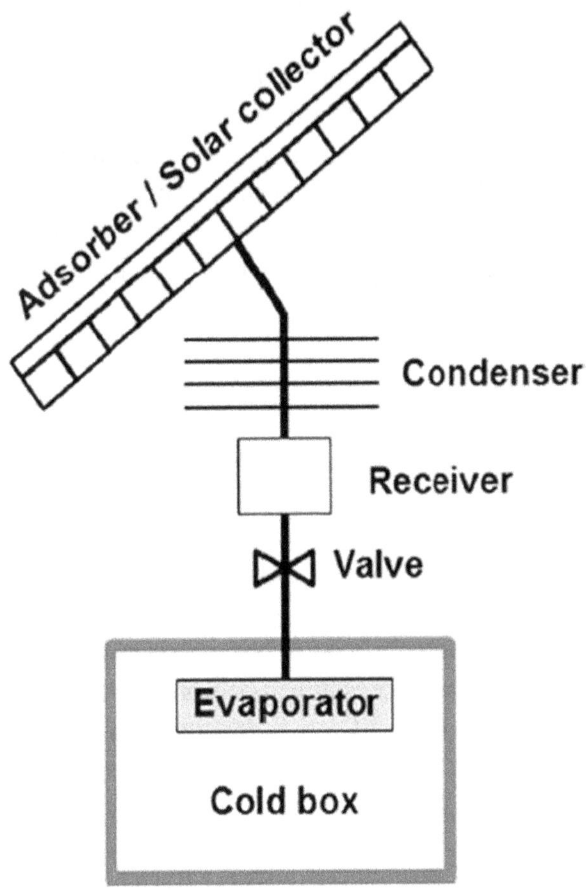

DESIGN AND WORKING PRINCIPLE OF SOLAR ADSORPTIVE CHILLER

Solar energy is the energy source of most adsorption devices operating with the basic cycle. In the remaining cases the components are kept unchanged, the main difference being the heat collection method. A solar adsorption refrigerator based on the basic adsorption refrigeration cycle does not require any mechanical or electrical energy, just thermal energy, and it operates intermittently according to the daily cycle. Like a simple vapor compression system, these adsorption systems are closed systems, comprising a compressor, a condenser and an evaporator (Goyal, Baredar, Mittal, & Siddiqui, 2016). However, in this case, the compressor is an adsorber powered by the thermal energy, and the cooling effect is achieved by the evaporation of a refrigerant while the vapor produced is adsorbed by the adsorbent layer in the adsorber. The adsorbed content of refrigerant varies cyclically, depending on the adsorbent temperature and system pressure, which varies between a maximum limit set by the condensation pressure and a minimum limit imposed by the evaporation pressure. In its simplest form (Fernandes, Brites, Costa, Gaspar, & Costa, 2014), a solar refrigerator is a closed system consisting of a solar collector containing the adsorbent bed (hermetically sealed and painted in black, to optimize the solar radiation absorption), a condenser, a receiver equipped with a 2-way valve and a cold box with the evaporator inside (Figure 1).

A basic adsorption cycle consists of four thermodynamic phases:

1. Heating and pressurization
2. Heating, desorption and condensation
3. Cooling and depressurization
4. Cooling, adsorption and evaporation

The above four phases are explained with the help of Clayperon diagram.

Heating and Pressurization (or isosteric heating). During this phase, adsorber absorbs heat thus increasing the adsorbent's temperature which induces a pressure increase from evaporation pressure to condensation pressure.

This phase is similar to compression in compression cycles. The temperature of adsorbent bed increases from T_{a2} to T_{g1}. After this vapors of adsorbent starts rising and moves towards condenser where they cools down and lowering of temperature takes place (Figure 2).

Heating, desorption and condensation. During this phase, heat is supplied to adsorber. So, its temperature continues to increase. As adsorber connected to condenser, desorption of vapor get is induced and liquefaction of desorbed vapor in condenser further takes place. This phase is similar to condensation in vapor compression cycles. The temperature of the adsorbent bed increases from T_{g1} to T_{g2}. This phase is also known as isobaric desorption. Then, the heated vapor of adsorbate cools down and thus temperature get reduced which could easily be exhibited with the rightward movement in Figure 2.

*Cooling and depressurization (*isosteric cooling*).* This phase is similar to expansion in vapor compression cycles. During this phase, low pressure vapor enters adsorber passing from evaporator. Temperature of adsorbent reduces because of release of heat which induces pressure decrease from condensation pressure to evaporation pressure. The pressure is reduced from P_c to P_e, temperature being also decreased to T_{a1}. This pressure change can

Figure 2. P-T diagram for basic adsorption cycle. (a) Heating and pressurization; (b) Heating and desorption and condensation; (c) Cooling & Depressurization; (d) Cooling, adsorption and Evaporation
(Goyal et al., 2016).

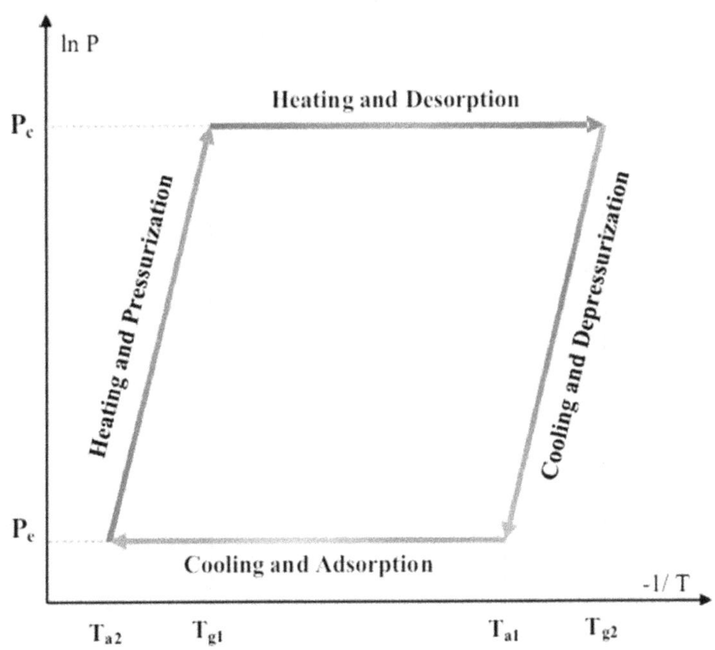

be verified with the downward shift in Figure 2. In this phase, vapors change their state and turns into liquid.

Cooling, adsorption and evaporation. The adsorbent temperature continues to decrease because of further releasing of heat causes adsorption of vapor due to connection with evaporator.

This vapor is formed in the evaporator because of heat supplied by heat source, this heat being withdrawn from cold box or chilling chamber, i.e. refrigerated media. This phase is analogous to evaporation in compression cycles. This phase is also known as isobaric adsorption. The adsorbent temperature is decreased from T_{a1} to T_{a2}. During this phase, cooling is produced which causes temperature of the region to lower (Figure 2).

Heat energy for whole process of adsorptive cooling can be supplied through gas or oil fired or solar energy. So, the need of electricity can be eliminated which is beneficial to areas where electrical networks are not yet installed.

Nevertheless, the adsorbent – refrigerant pair ought to be selected very cautiously, because it is a crucial factor which determines a successful and effective operation of a solid adsorption system this is necessary. Several combinations are available. Each of them exhibits certain advantages and disadvantages which lead to varying performances.

The proper selection is determined by the desired characteristics of refrigeration system, properties of the adsorbent/adsorbate pair, temperature of heat source and their cost, availability and environmental impact. It ought to remember that for any refrigerant application, the adsorbent have to posess low adsorbitivity at high temperature and pressure or high adsorbitivity when accounts to room temperature and low pressure (Alghoul, Sulaiman, Azmi, & Wahab, 2007).

When adsorbent selected, the most important features must be taken into account (Fan, Luo, & Souyri, 2007) to achieve promising cooling characteristics as follows:

- high adsorption and desorption capacity;
- low specific heat;
- good thermal conductivity to shorten cycle time;
- to be nontoxic and non-corrosive;
- to chemically and physically compatible with chosen refrigerant;
- ought to be of reasonable cost;
- should be available in plenty.

For a thermally conductive adsorbtive material a compromise is needed between high porosity to enhance vapor diffusion and high density because of high porosity always results in reducing thermal conductivity and deterioration of the efficiency of refrigeration system.

The next adsorbents are generally used: activated carbon, zeolite and silica gel (Freni et al., 2016). Among these, silica gel satisfies most of the criteria listed above but it is expensive and is not easily available.

To provide an effective performance, the refrigerant properties have to satisfy the following requirements (Choudhury, Saha, Chatterjee, & Sarkar, (2013):

- molecular size should be small for better adsorption;
- high latent heat per unit volume;
- high thermal conductivity;
- good thermal stability;
- low viscosity;
- low specific heat;
- Low toxicity;
- must not be inflammable and corrosive;

Figure 3. Schematic diagram of solar power adsorptive chiller
(Muttakin et al., 2020)

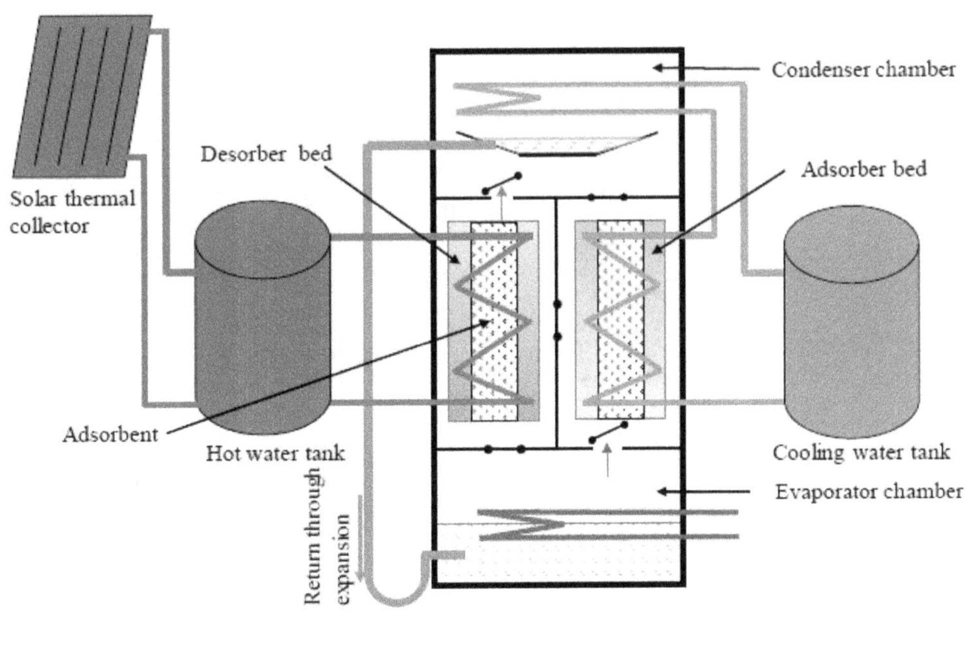

- low saturation pressures;
- must be environment friendly.

The next available refrigerants can considered as promising: ammonia, methanol, water which have relatively high latent heat values and also presents good performances. Among these ammonia is rather capricious because of its flammable and corrosive nature. There are several studies where adsorbent-refrigerant working pairs for adsorption refrigeration systems considered providing their characteristics, performances, advantages and disadvantages. As a rule, prototypes of adsorptive refrigerators are generally based on pairs such as zeolite – water, silica gel – water, carbon – methanol and activated carbon – ammonia (Freni et al., 2016).

Typical designs of adsorptive cooling systems can be divided into one-bed and two-bed configuration. One-bed chilling device or system involves only one adsorptive/desorptive reactor to which intermittently supplies heat from low-potential heat source for regeneration of adsorbent (Fernandes et al., 2014). As a rule, adsorbtive/desorptive reactor is connected with solar collector. Such design is simple. However, cycle time of this refrigerator is strongly limited with 24 hours. When conventional adsorbent with low

Figure 4. Schematic diagram of 3-bed system working in a series (Rouf et al., 2018)

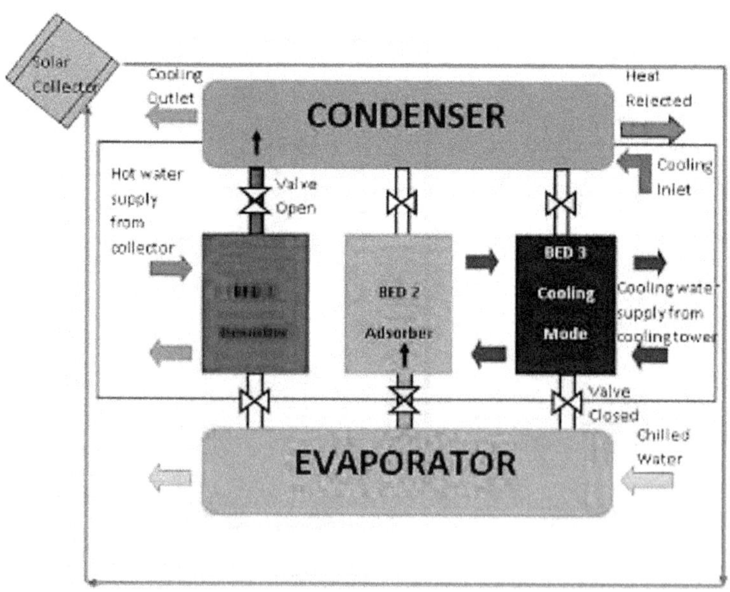

adsorptive capacity used, its performance is complicated because of high volumes of adsorbent.

Rather obvious solution of this problem can be two-bed refrigerator which involves divided at least two adsorber and desorber beds (Figure 3) (Muttakin, Ito, and Saha, 2020). Typical adsorption or desortion bed involves fin tubed heat exchanger where adsorbent is placed between the fins. Operating mode of two-bed refrigerator is similar to one-bed one.

During the adsorption stage, cold water circulates through the tubes of the adsorbent bed. During the desorption phase, hot water from the hot water tanks supplied through those tubes (Muttakin et al., 2020). The difference between the temperatures of hot and cold water is called as a temperature swing. For a typical single stage silica gel—water adsorption chiller, the temperature of the hot water have to be 60–80 °C. A temperature swing for a multistage adsorption chiller can be limited only to 10 °C. This value can be decreased down to 2.2°C for a 10-stage adsorption chiller.

In solar thermal-powered adsorptive cooling devices, heat is supplied primarily by the solar thermal collectors. To compensate a strong dependence of the performance of the solar thermal collector and adsorptive chiller on environmental condition an auxiliary electrical heater is often installed to maintain the temperature in the hot water tank.

As working pairs in two-bed adsorptive chillers, 'activated carbon – ethanol' and 'zeolite – water' are suggested to use (Sangwan, & Chakraborty, 2020). Such systems allow expanding durability of cycle more than 24 hours. They are necessary when adsorbents with low maximal adsorption used. Nevertheless, their performance is rather complicated not only by complex design, but with high masses of adsorbents. For example, for adsorptive chiller based on

Table 1. Operational process of the three bed system in a cycle

Mode	Bed 1	Bed 2	Bed 3
1st 1/3 cycle	desorption/ condensation	adsorption/ evaporation	precool
Mode change time	Precool	Preheat	adsorption/ evaporation
2nd 1/3 cycle	Precool	desorption/ condensation	adsorption/ evaporation
Mode change time	adsorption/ evaporation	Precool	Preheat
3rd 1/3 cycle	adsorption/ evaporation	Precool	desorption/ condensation
Mode change time	Preheat	adsorption/ evaporation	Precool

(Rouf et al., 2018)

working pair 'zeolite – water' mass of adsorbent is 360 kg and solar collector surface is 20 m^2, COP being 0.1. So, adsorbent properties are crucial factors which determine the design of adsorptive refrigerator.

A new 3-bed cycle working in a series had been suggested in (Rouf, Alam, Saha, & Kabir (2018). (Figure 4). The performance of this cycle was encouraging when low heat input had been considered.

The operating principle of three bed adsorptive chiller is similar to that of the conventional two-bed adsorption chiller, but instead of having two half cycles there are three one -third cycles (Table 1). The bed works as the desorber in the 1st on-third cycle. In the 2nd and the 3rd one-third cycles, in the precooling mode it works as an adsorber, and in the next one-third cycle, operates as a desorber again. So, the three beds occur to be desorber/adsorber once in a cycle and continue to work in a cyclic order.

For the 3-bed adsorption cooling cycle, the optimum operating condition is 28 solar thermal collectors each of area 1.72 m^2. As collector area increased, cyclic average cooling capacity (CACC) increases. Nevertheless, very little improvement of the cooling capacity after increasing collector number from 26

Figure 5. Schematic diagram of the 4-bed parallel system
Rouf et al. (2020)

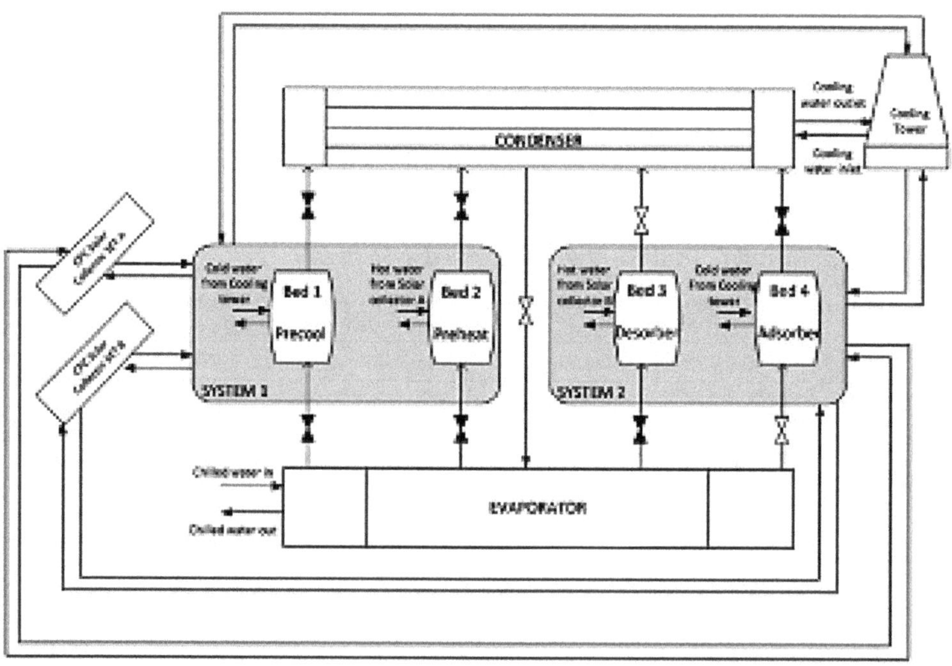

to 30 is observed. Maximum cooling capacity was stated for collector number 28. Evidently, the cost of the cooling system can be reduced by choosing longer cycle time and a smaller collector area, because for extensive heat input temperature of the heat transfer unit gets heated in a shorter cycle time.

However, in optimum operating conditions, the CACC for 3-bed chiller 10.3 kW occurs to be lower as compared with that of the conventional 2-bed basic system (CACC 11.19 kW) under identical operating conditions (Rouf, Jahan, Alam, Sultan, Saha, & Saha, 2020).

The cooling capacity of 3-bed chiller occurs to be less as compared with a 2-bed chiller because of the less amount of adsorbent in each smaller bed. At maximal heat input, for 3-bed chiller bed temperature increases rapidly. It results in decreasing precooling time, the thermal gradient of the bed being not large enough which lead to a small entropy of the adsorber and comparatively lesser cooling capacity. By contrast, at less heat input (34.4 m^2, 20 collector area) and longer cycle time, the result appears to be just opposite since this time the thermal gradient is large. So, the cooling capacity 8.47 kW is gained with 3-bed distributed mass, whereas it is only 7.91 kW with a 2-bed conventional system. These results allows to conclude that with a small amount of heat input, better output could be achieved when the total amount of adsorbent distributed into multiple adsorption beds at a longer precooling time.

As a promising system, 4-bed parallel system was suggested by Rouf et al. (2020) (Figure 5). This new cycle is considered to be effective for not only low-grade heat input but also for a limited amount of adsorbent.

Table 2. The working rule of the 4-bed parallel scheme

Mode of the Bed	Bed 1	Bed 2	Bed 3	Bed 4
Mode 1, phase 1 (120s)	Pre-cool	Preheat	Desorber	Adsorber
Mode 1, phase 2 (120s)	Adsorber	Desorber	Pre-cool	Preheat
Mode 2, phase 1 (120s)	Preheat	Pre-cool	Adsorber	Desorber
Mode 2, phase 2 (120s)	Desorber	Adsorber	Preheat	Pre-cool

Rouf et al. (2020)

Two working modes of this system involve two phases Rouf et al. (2020) which are given in Table 2.

The proposed 4-bed chiller involves two conventional 2-bed solar heat assisted adsorption chiller joined. The system involves four adsorption beds, one evaporator, and one condenser. According to Figure 5, system one is in the pre-heat/precool mode, system two being in desorption/adsorption mode and connected with both condenser and evaporator. 4-bed parallel system provides better specific cooling production as compared with 3-bed system. Despite the enchancing operating parameters of multibed chilling systems,

Figure 6. Adsorption chiller: 1 – adsorber; 2 – coil (hydraulic accumulator); 3 – adsorptive material; 4 – evaporator; 5 – condenser; 6 – cold box; 7 – water cold accumulator; 8 – pipe; 9 – tap; 10 – transparent cellular polycarbonate *(Sukhyy et al., 2015)*

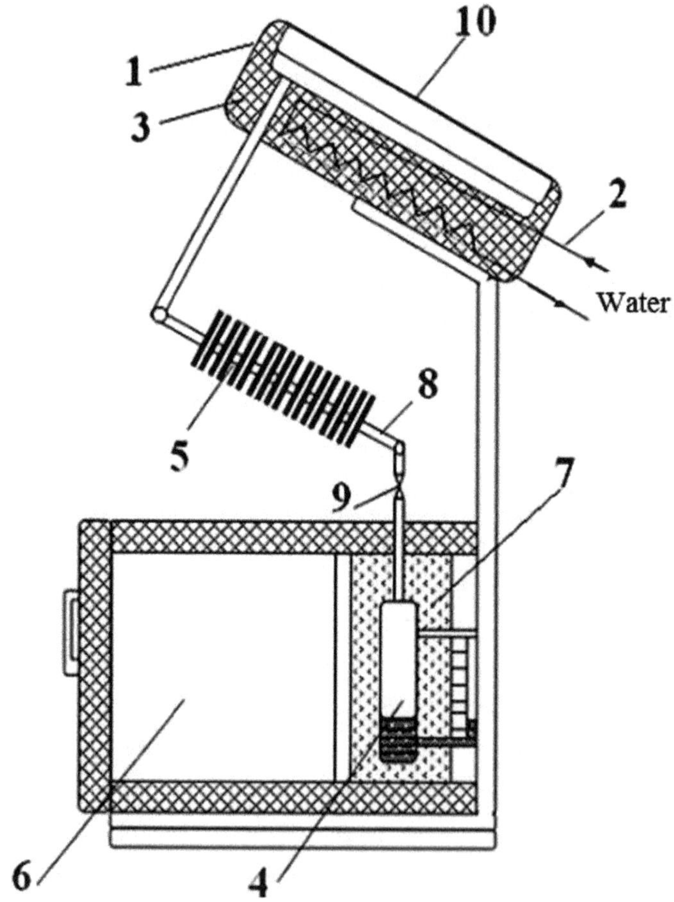

their performance is rather complicated. Efficient one-bed configuration of adsorptive chiller can occur when adsorbent with maximal adsorption more than 0.5 kg/kg used, i.e. composite adsorbent 'salt inside porous matrix'.

PERFORMANCE OF ADSORPTIVE REFRIGERATORS BASED ON COMPOSITE ADSORBENTS

Design of the adsorptive refrigerator is given in Figure 6 (Sukhyy, Belyanovskaya, Sukha, Kolomiyets, Gavrilko, & Sukhyy, 2015). Its main units involve an adsorber (1), made in the form of a solar collector, a condenser (5) and an evaporator (4) in the cold box (6).

In order to reduce the influence of weather conditions on the time of regeneration of the adsorbent a water accumulator (2) is maintained. A cold box in volume of 3 m^3 is made of steel grade 30X thickness of 0.5 mm. As heat isolation foam polystyrene was used.

The adsorptive refrigerator works in accordance with the thermodynamic cycle shown in Figure 7 (Kolomiyets, Kozlov, Belyanovskaya, Sukhyy, & Sukhyy, 2014). The operation proceeds in two-phase mode. The first phase corresponds to the reception of a cold (Figure 4, line 3-4-1- adsorption and evaporation of water).

Figure 7. Operational cycle of adsorptive chiller
(Kolomiyets, 2014)

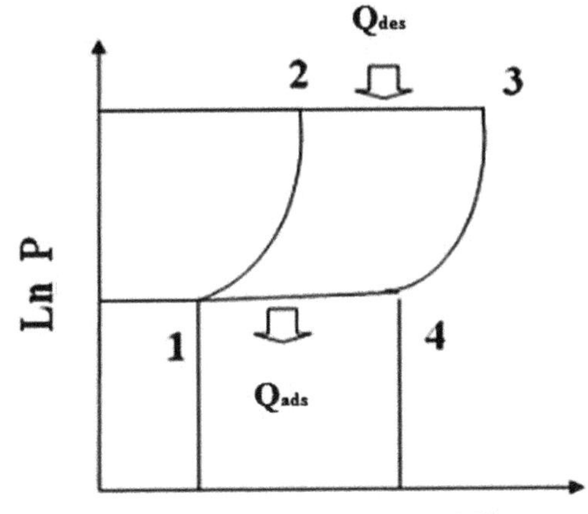

When the tap (9) opened, water vapors begin to diffuse through the condenser to the adsorber. Because of the adsorption by adsorptive material, water is evaporated in the evaporator (4) that leads to a cooling effect in the cold box (6). Due to a large volume of water which containing in the walls of the cold box, the temperature in the cold box (6) is maintained at 5 – 10 °C for 10 – 20 hours, until the next cycle begins. The temperature in the adsorber (1) increases significantly due to adsorption of water by adsorbent (3). To extract this heat, water is delivered to the coil (2). Hot water is supplied to the consumer in a hot water fed or heating system or used to regenerate the adsorptive material at the second stage of the refrigerator operation.

The second phase corresponds to regeneration of the adsorbent (Figure 4, line 1-2-3 – desorption and water condensation). The tap (9) is closed. The adsorptive material (3) is warmed using solar energy or by passing through the coil (2) the hot water heated in the first phase. So, the influence of weather conditions on the time of material regeneration is reduced. The water is collected in the condenser (5) and flows down to the evaporator (4). Then, the process of getting cold starts.

DETERMINATION OF THE WATER MASS AND COMPOSITE ADSORBENT REQUIRED FOR ADSORPTIVE REFRIGERATOR OPERATION

The quantity of heat which have to be taken from the cold box can be calculated as according to (Kolomiyets, Belyanovskaya, Sukhyy, Prokopenko, Kozlov, & Sukhyy, 2015) as the sum of the heat going to cool the chamber itself and the introduced food substances and to cover the heat inleaks both of the chamber and when opening the chamber when adding products:

$$Q_1 = C \cdot m \cdot \Delta T + C_{pr} \cdot m_{pr} \cdot \Delta T + \Sigma Q_z \quad (1)$$

C is the heat capacity of the structural elements, J/kg·K; C_{pr} is the heat capacity of the food substances introduced into the refrigerating chamber, J/kg·K; ΔT is the difference between ambient temperature and average daily temperature in the refrigerating chamber, K; m, m_{pr} is mass of the refrigerating chamber and introduced products, respectively, kg; ΣQ_z is the sum of heat leakage into the chamber as a result of heat transfer through its walls, floor and ceiling, from infiltration of outside air when the chamber is opened and loads from lighting, kJ.

The heat leakages through the camera walls itself Q_p can be determined according the conventional procedure, as follows (Ostrov, 1999), kJ:

$$Q_p = \frac{\Delta T \cdot S_k \cdot \tau}{\left(\frac{1}{\alpha_{v1}} + \frac{\delta}{\lambda_m} + \frac{\delta_{iz}}{\lambda_{iz}} + \frac{1}{\alpha_{v2}}\right)} \tag{2}$$

where $S_k = 2\cdot(a\cdot b+a\cdot c+b\cdot c)$ is external surface area of the chamber, m^2; α_{v1} is calorific coefficient for air inside of the chamber, $W/m^2 \cdot °C$; α_{v2} is calorific coefficient for ambient air, $W/m^2 \cdot °C$; λ_m corresponds to heat conduction coefficient of material of the chamber, $W/m^2 \cdot °C$; λ_{iz} is heat conduction coefficient of heat insulation material, $W/m^2 \cdot °C$; δ_{iz} corresponds to thickness of heat-retention lagging, m; τ is exploitation time per a day, s.

Heat leakage at-the-opening per a day Q_V is suggested to calculate according to (Ostrov, 1999) as a product of a heat load and time interval when door opened:

$$Q_V = q \cdot D_\tau \cdot D_f \cdot (1 - E) \cdot \tau_{op} \tag{3}$$

where q is total cooling load of chilling chamber for steady-state air flow with regard to differences of densities, heat contents and humidities and size of door opening, kW; D_τ is the coefficient that corrects for time interval when door opened; D_f corresponds to the factor that corrects for air flow characteristics in the door opening; E is level of efficiency of protection equipment in the door opening; τ_{op} is time interval per a day when door is opened, s

Heat leakages resulted from work of lightning devices are determined as product of quantity of luminaires, lamp working efficiency and running period per a day (Ostrov, 1999).

Cooling in the chilling chamber proceeds due to evaporation of water in the evaporator. Heat quantity taken off in the chilling chamber can be calculated with regard to the heat of water evaporation (ΔH_{ev}, kJ/ kg) (Ostrov, 1999):

$$Q_2 = \Delta H_{ev} \cdot M_{H2O}\text{, kJ} \tag{4}$$

So, the mass of the water which is necessary for taking off from the chilling chamber (M_{H2O}, kg):

$$Q_2 = Z_B \cdot Q_1,\ M_{H2O} = \left(Z_B \cdot Q_1\right) / 2420\text{, kg} \tag{5}$$

where Z_B is coverage factor. To support continuous running of the device it is taken as 1.05 (Sukhyy, Belyanovskaya & Kolomiyets, 2018).

In order to compensate of daily variability of meteorological conditions mass of working fluid is increased by 50% (Belyanovskaya, Pustovoy, Sukhyy, Kolomiyets & Sukhyy, 2018). So, water mass in the evaporator will be 1.5· M_{H2O}, kg.

Adsorbent mass can be calculated as:

$$M_{ads} = M_{H2O}/A_{lim} \tag{6}$$

where A_{lim} corresponds to maximal adsorption of composite adsorbent, kg/kg.

The heat amount required to regeneration can be calculated as follows (Kolomiyets et al., 2015):

$$Q_3 = m_k \cdot C_k \cdot \Delta T_1 + m_w \cdot C_w \cdot \Delta T_1 + m_w \cdot \Delta H_{des} \tag{7}$$

where ΔT_1 is the difference between the temperature of the adsorbent and the temperature of regeneration, K; ΔH_{des} is the heat of desorption of water, kJ/kg; m_k and m_w are respectively, the mass of the composite and the adsorbed water, kg; C_k and C_w are the heat capacity of the composite and water, respectively, kJ/kg·K.

Adsorbent can be regenerated by the heat supplied with solar energy. Required surface area of solar collector is calculated by the formula:

Table 3. Basic design characteristics of the solar adsorptive chiller

Adsorbent	V_k, m³	M_{H2O}, m³	M_{ads}, kg	Q_3, kJ	F_k, m²
'silica gel –sodium sulphate'	0.181	6.13	13.93	18496.04	5.6
	0.125	4.50	6.2	13466.4	4.1
'silica gel – sodium acetate'	0.181	6.13	6.13	17710.95	5.4
	0.125	4.50	4.50	12894.8	3.9

(Sukhyy,Belyanovskaya, & Kolomiyets, 2018)

$$F_k = \left(Z_k \cdot Q_3\right) / Q_k \text{, m}^2 \tag{8}$$

where Z_k corresponds with coverage factor of solar collector surface area, it being introduced support continuous performance of the chiller at an adverse weather conditions. Depending on region coverage factor can be taken as 1.5 – 2.5. Q_k refers to quantity of heat supplied to adsorbent per 1 m^2 of collector surface area, kJ/m^2.

Basic design characteristics of solar collector calculated by the Eq. (8.1) – (8.8) for various volumes of cold boxes are presented in the Table 3. Coverage factor of the solar collector and filling rate of chilling chamber are taken as 2 and 0.5, respectively.

PERFORMANCE OF SOLAR ADSORPTIVE REFRIGERATOR

The temperature dependences of different modules of the adsorption refrigerator vs. time for average daily cycle are given in Figure 8.

Figure 8. Averaged daily temperature plots in July 2014: 1 – environment temperature; 2 – absorber; 3 – transparent insulation; 4 – in the chilling chamber; 5 – adsorbent. a – when composite 'silica gel – Na_2SO_4' used; b – when composite 'silica gel – CH_3COONa' used
(Sukhyy et al., 2018)

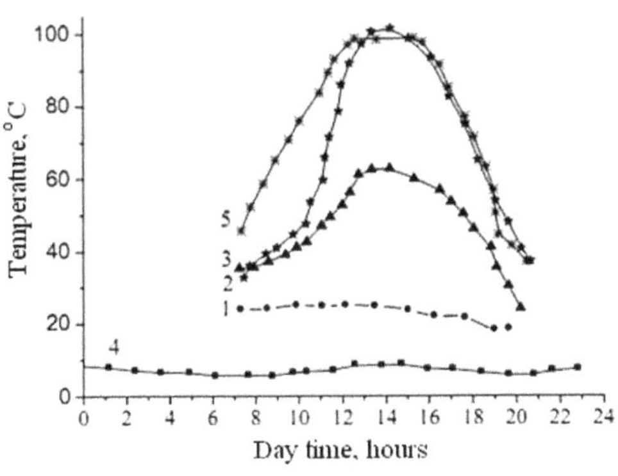

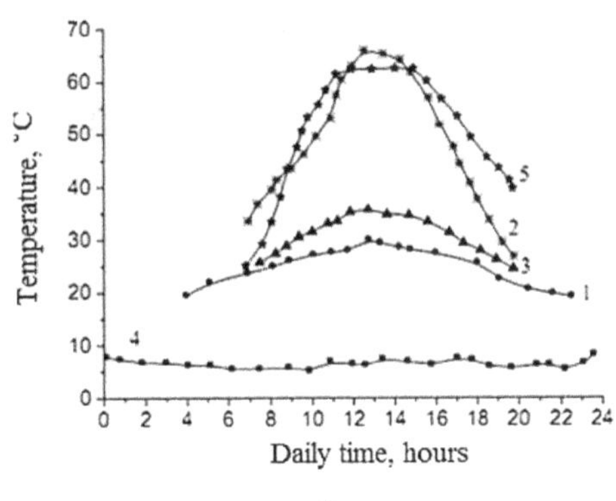

The temperature in the chilling chamber during the whole cycle is maintained at the level of 4 – 9 °C due to evaporation of water in the evaporator. The daily cycle of the adsorptive refrigerator can be described by several stages, which form a closed loop. At 8:00 am, adsorbent 3 adsorb totally or partially water from evaporator 4. The completeness of adsorption is affected by the mass of adsorbate and ambient temperature.

In order to ensure the independence of the adsorptive refrigerator from external meteorological conditions (cloudy weather or rain) during the day, the required mass of the composite sorbent ought to exceed the minimum required amount in 1.5 – 1.7 times. Evidently, increasing of an adsorbent mass negatively affects the rate of regeneration at a fixed area of the solar collector. So, at 8.00 a.m. the adsorbent is in a fully hydrated state. In the period from 8.00 to 11.00 a.m., a rapid increase in the temperature of the environment occurs due to the growth of the intensity of solar radiation, which heating the adsorber. The process of removal of adsorbed water begins at 10.00 a.m. when the adsorbent warmed up to the regeneration temperature. The most intensive desorption process occurs from 10.00 to 13.00. The cycle begins again at 20.00, when the complete dehydration of the adsorbent is achieved.

COEFFICIENT OF PERFORMANCE OF SOLAR ADSORPTIVE REFRIGERATOR

The coefficient of performance (COP) of the chilling system is calculated as the ratio of the heat withdrawn from the cold source to the energy required for this operation. So, it is a criterion of the efficiency of the cooling system or chilling device. COP estimates the measure of the energy conversion in the output cooling capacity of energy supplied to the system. This coefficient is very useful for comparing similar systems operating on the same base cycle and technology.

The coefficient of performance of the cycle (COP_c) of an adsorptive chilling system is calculated as follows (Belyanovskaya et al., 2019):

$$COP_c = \frac{Q_e}{Q_{reg}} \quad (9)$$

where Q_e corresponds to the cooling effect produced at the evaporator, Q_{reg} is only the energy that is effectively transferred to the adsorber to promote its regeneration.

The value of Q_e is determined as the difference between the refrigerant (i.e. chilling agent) latent heat of vaporization and the sensible heat required to cool down the refrigerant from the condensing temperature to the evaporation temperature:

$$Q_e = M \cdot [h_e - C_p \cdot (T_c - T_e)] \tag{10}$$

where M is the mass of refrigerant desorbed in one operation cycle, h_e is refrigerant latent heat of vaporization, C_p is the specific heat of the liquid chilling agent, i.e. refrigerant, T_c and T_e correspond to the temperatures of condensation and evaporation, respectively.

The value of Q_{reg} is determined as the sum of the heat Q_{d1} applied to the adsorbent when both temperature and pressure increased along the line 1 – 2 (Figure 4) and the heat Q_{d2} caused the adsorbent to release the refrigerant vapor (Figure 4, line 2 – 3):

$$Q_{reg} = Q_{d1} + Q_{d2} \tag{11}$$

Net coefficient of performance (COP_n) of a adsorptive cooling heliosystem is calculated as a ratio of the heat removed from the cold source (Q_e) and all the solar energy received by the solar collector surface (Q_s) (Sukhyy et al., 2018):

$$COP_n = \frac{Q_e}{Q_s} \tag{12}$$

This value is called as a solar or effective coefficient of performance.

The solar energy provided to the system can be determined as a product of the solar collector surface area F_k by the integration of the solar radiation flux on the collector surface (F_k) during one day of operation (Sukhyy et al., 2018)

$$Q_s = F_k \cdot \int_0^{24 \cdot 3600} G dt \tag{13}$$

On the other hand, net COP is calculated assuming that the cooling effect is used to cool down the water (Fernandes et al., 2014) or other substances contained in the chilling chamber:

$$COP_{s(n)} = \frac{m \cdot C_p \cdot \Delta T_1}{F_k \cdot \int_0^{24 \cdot 3600} G dt} \tag{14}$$

where m refers to the mass of water or other substances placed in the chilling chamber, c_p corresponds to their specific heat, ΔT_l is their temperature difference.

As quantity of heat received by composite taken into account, Eq. (8.14) is transformed to:

$$COP_{s(n)\eta} = \frac{m \cdot C_p \cdot \Delta T_1}{\eta \cdot F_k \cdot \int_0^{24 \cdot 3600} G dt} \tag{15}$$

where η is the efficiency factor of the solar collector.

Moreover, net solar COP can be estimated by experimental average monthly temperature dependences within the cold box and for external air during of 24 hours, i.e. in the course of the daily cycle (Sukhyy et al., 2018):

$$COP_n^{\exp} = \frac{k \cdot \left(\int_0^{24 \cdot 2600} T_{env} d\tau - \int_0^{24 \cdot 2600} T_{ch} d\tau \right)}{F_k \cdot \int_0^{24 \cdot 3600} G dt} \tag{16}$$

where *k* corresponds to specific heat of the material of the cold box and air in it, kJ/K:

$$\mathrm{k} = \mathrm{C_M} \cdot \mathrm{M_M} + \mathrm{C_A} \cdot \mathrm{M_A} + \mathrm{C_S} \cdot \mathrm{M_S} \tag{17}$$

where C_M, C_A and C_S corresponds to the specific heats of chilling chamber material, air in it and substances placed into chilling chamber, M_M, M_A, M_S refers to masses of chilling chamber material, air in it and substances placed into the cold box, $\left(\int_0^{24\cdot 2600} T_{env} d\tau\right)$ and $\left(\int_0^{24\cdot 2600} T_{ch} d\tau\right)$ are areas under temperature curves of external air and inside the cold box.

When solar collector efficiency taken into account, net solar COP is calculated as follows (Sukhyy et al., 2018)

$$COP_{\eta} = \frac{k \cdot \left(\int_0^{24\cdot 2600} T_{env} d\tau - \int_0^{24\cdot 2600} T_{ch} d\tau\right)}{\eta \cdot F_k \cdot \int_0^{24\cdot 3600} G dt} \quad (18)$$

Data on the intensity of the total daily average solar radiation for the period from May to September and COP of solar adsorptive refrigerators based on the composites 'silica gel – sodium acetate' and 'silica gel – sodium sulphate' calculated by the Eq. (8.12) – (8.18) as the solar collector installed at an angle of 30° up to the horizon in the southern orientation in Dnipro, Ukraine (Dnepropetrovsk) are given in the Table 2 and 3.

Table 4. Coefficients of performance of solar adsorptive chiller based on composite adsorbent 'silica gel – sodium acetate'

Month	Total Monthly Average Solar Radiation Value (kWh/m²)	COP_c	$COP_{s(n)}$	$COP_{s(n)\eta}$	COP_n^{exp}	COP_{η}
May	5,99	1.021	0.175	0.291	0.100	0.167
June	5,86	1.021	0.179	0.298	0.102	0.171
July	6,06	1.021	0.173	0.288	0.099	0.165
August	5,66	1.021	0.185	0.308	0.106	0.176
September	4,41	1.021	0.238	0.396	0.136	0.227

(Sukhyy et al., 2018)

According to the results presented in Tables 4 and 5, the coefficients of performance of the solar adsorptive refrigerator differ considerably.

Obviously, COP_c, which characterizes the efficiency of its own cycle is not affected by the total average monthly magnitude of solar radiation. On

Table 5. Coefficients of performance of solar adsorptive chiller based on composite adsorbent 'silica gel – sodium sulphate'

Month	Total Monthly Average Solar Radiation Value (kWh/m²)	COP_c	$COP_{s(n)}$	$COP_{s(n)\eta}$	COP_n^{exp}	COP_η
May	5,99	1.084	0.121	0.201	0.089	0.149
June	5,86	1.084	0.133	0.207	0.091	0.152
July	6,06	1.084	0.119	0.199	0.088	0.147
August	5,66	1.084	0.128	0.213	0.094	0.157
September	4,41	1.084	0.164	0.273	0.121	0.202

(Sukhyy et al., 2018)

the contrary, the values of net coefficients of performance, both experimental (COP and COP_η) and calculated ($COP_{s(n)}$ and $COP_{s(n)\eta}$) are observed to increase monotonously. The difference between the calculated and experimental values is a consequence of the fact that the fractional composition of the adsorbent, the change in the porosity of the layer during the operation, and the irregularity of adsorption-desorption behavior caused by these factors were not taken into account in theoretical calculations.

The coefficients of performance of the adsorptive refrigeration cycle as the composite adsorbent 'silica gel – sodium acetate' (1.021) are quite close to similar devices based on the composites 'silica gel – sodium sulphate' (1.084) synthesized according to (Sukhyy et al., 2018).

Table 6. Coefficient of performance of the adsorptive solar refrigerator based on the composite 'silica gel - sodium sulphate'

Adsorbent Composition, wt. %		Maximal Adsorption (Sukhyy et al., 2018), kg/kg	Mass of Adsorbent, kg	Coefficient of Energy Performance at ΔT_1		
Na_2SO_4	Silica Gel			$\Delta T_1 = 65°C$	$\Delta T_1 = 75°C$	$\Delta T_1 = 85°C$
80	20	1.349	0.58	1.11	1.09	1.07
60	40	1.060	0.74	1.10	1.08	1.06
40	60	0.771	1.01	1.09	1.07	1.05
20	80	0.482	1.62	1.06	1.04	1.02

(Belyanovskaya, Pustovoy, Sukhyy, Kolomiyets, & Sukhyy, 2018)

This results from the lower regeneration temperature of the composite 'silica gel – sodium acetate' (60°C) compared with 'silica gel – sodium sulphate' (90°C) and a lower mass of composite 'silica gel – sodium sulphate' required to maintain the same cooling power. At the same time, a significant difference in the temperature of the regeneration leads to the significant difference of the operating coefficients of performances. So, for adsorptive refrigerators based on the composite 'silica gel – CH_3COONa' is about 2 times higher as compared with the composite 'silica gel – Na_2SO_4'.

Thus, the efficiency of the operation of a solar adsorption refrigerator when the composite adsorbent 'silica gel – sodium acetate' applied is stated to be higher comparatively to a similar refrigerator based on a composite adsorbent 'silica gel – sodium sulphate'.

Characteristics of composite adsorbents strongly affect the coefficient of performance. For example, according to results given in the Table 6, maximal values of adsorptive refrigeration cycle COP_c are stated for composite containing, in mass. %: silica gel – 20 and sodium sulphate – 80. Evidently, increasing the content of sodium sulphate in the composite results in reducing the mass of the composite, and hence the amount of heat required to regeneration of the adsorbent. The amount of heat needed to regenerate the composite Q_3 depends on the temperature difference of the regeneration temperature and adsorbent after adsorption ΔT_1. So, its growth leads to a monotonous decrease of the coefficient of energy performance. The maximal values of COP_c are stated at ΔT_1 at most of 65 °C.

So, the coefficient of performance of adsorptive refrigerator and sodium sulphate content in the composite is stated to increase simultaneously. The growth of the COP_c occurs to result from decreasing the difference between adsorbent temperature and regeneration temperature. The maximal values of the coefficient of performance of studied solar adsorptive refrigerator about of 1.11 are determined to correspond to the difference between the adsorbent temperature and the regeneration temperature of 65 °C for composites containing about 20 wt. % silica gel and 80 wt.% % sodium sulphate.

The temperature in the chilling chamber is maintained for 24 hours below 10°C, which is favorable to the storage of certain types of agricultural products. The lower regeneration temperature of the composite 'silica gel – sodium acetate' (60°C) can significantly reduce the low-potential thermal emissions which accompany the operation of any industrial plant. For example, the temperature of waste gases during operation of boiler units is usually maintained at not lower than 120°C (Teplovoj raschet promyshlennyh parogeneratorov (1980), Kiev: Vysshaya shkola). As distinct from a vapor compression

Table 7. Net coefficients of performance of solar adsorptive refrigerator based on composite 'silica gel – sodium sulphate' (difference of the regeneration temperature and adsorbent after adsorption ΔT_1 = 55°C)

Month	Daily Solar Radiant Flux, MJ/m²	Heat Supplied by Solar Collector, MJ	Net Coefficient of Performance
May	21.56	214.56	0.25
June	21.09	209.90	0.26
July	21.81	217.06	0.25
August	20.37	202.74	0.27
September	15.87	157.96	0.34

(Belyanovskaya et al., 2019)

refrigerators, such a device does not require electricity consumption, the electric power consumed being at least 6 kW (Holodil'nye mashiny. Spravochnik (1982) / pod red. Bykova A. V., M.: Legkaya i pishchevaya promyshlennost'). Therefore, it contributes to a significant reduction in the cost of electricity in the operation of refrigeration equipment. Moreover, the regeneration temperature of the proposed composite sorbent meets the safety requirements.

According to the literary data on the coefficient of performance for adsorptive chilling devices based on various adsorbents are stated to be 0.136 –0.159 for activated charcoal (Khattab, 2014), 0.28 – 0.35 for zeolite (Kanoğlu, Çarpınlıoğlu, & Yıldırım, 2004), 0.3 – 0.32 for massive salts (Bao, Wang, & Wang, 2011), for composite materials, 0.33 – 0.59 [Li, 2010]. So, the obtained results are competitive as compared with the results obtained for similar devices.

Table 8. Results of calculation of water mass and temperature after adsorption

Δt°C	M_{H2O}, kg	Temperature of Water After Adsorption, °C
35	681.4	50
45	513.2	60
55	406	70
65	332.1	80
75	277.8	90

(Belyanovskaya et al., 2019)

Moreover, a higher adsorption capacity of the composite sorbent 'silica gel – sodium acetate' and sodium acetate which is at least 2 times greater than 0.2 g/g of silica gel (Askalany et al., 2013), allows significant area reducing of the solar collector.

Results of calculation of net coefficient of performance are presented in the Table 7.

The maximal values of COP_n are observed in September which corresponded to minimal solar radiant flux. According to data M.S. Fernandes et al. (2014) net coefficients of performance of chilling units based on pair 'silica gel – water' does not exceed 0.2. For pair 'zeolite – water' these values are at most 0.35. So, the efficiency of solar adsorptive refrigerator based on 'silica gel – sodium sulphate' conforms to the system with the same refrigerant.

Operating the adsorptive refrigerator is accompanied by releasing a significant amount of heat because of exothermic character of water vapor adsorption. It is expediently to consider possibility of its utilization.

Table 9. Test results of polymeric solar collectors

Grade	Design	Tests			
		Field		Laboratory	
		$F'(\tau\alpha)$	$F'U_L$	$F'(\tau\alpha)$	$F'U_L$
PSK-AB2-3	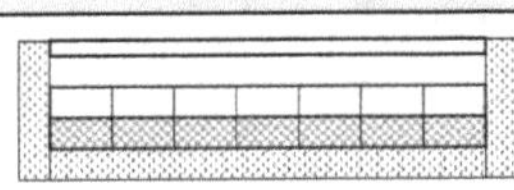	0,57	1,05	0,59	0,97
PSK-AB1-2	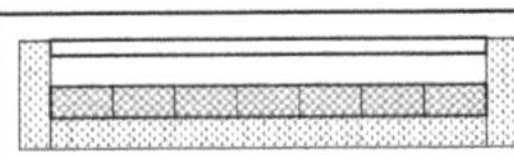	0,74	5,33	0,76	4,44
PSK-AB2-1	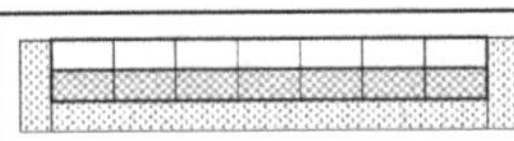	0,84	8,78	0,86	8,65
PSK-BC1-2	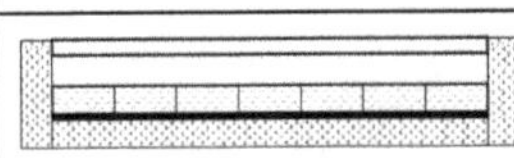	0,73	4,93	0,75	4,11
PSK-BC2-2	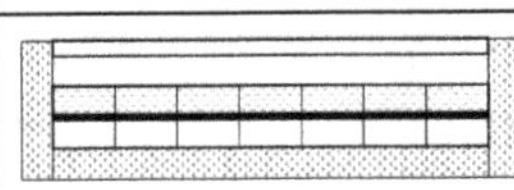	0,73	4,84	0,75	4,13
PSK-BC2-3	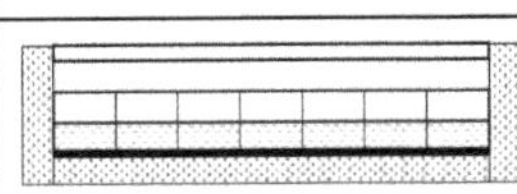	0,56	0,74	0,59	0,75
PSK-CT10-ChV	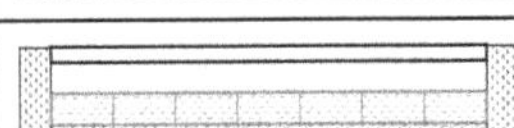	0,63	5,37	0,64	4,99

(Sukhyy, Kozlov, Belyanovskaya, Prokopenko, Sukha, & Doroshenko, 2018).

Obviously, a procedure can be warming a heat transfer media with the adsorption heat and during regeneration applied it for heating the adsorbent. Water mass and temperature after adsorption can be estimated by equations of heat balance. Results of calculations are given in the Table 6.

Obviously, water mass is decreased when temperature difference increased. The temperature difference 65 – 75°C corresponds to temperature of water after adsorption of 80 –90°C, water mass being 277,8 – 332,1 kg (277,8 – 332,1 l). So, obtained heat transfer can be used to warm the adsorbent in morning period.

DESIGN OPTIMIZATION OF SOLAR ADSORPTIVE CHILLER

Adsorptive refrigeration helioplants and devices are one of the most promising alternative systems for the adsorptive conversion of thermal energy (Sukhyy, Kozlov, Belyanovskaya, & Sukhyy, 2017). Their widespread implementation will significantly reduce the seasonal overload of power systems in the summer. In refrigeration solar installations, the supply of thermal energy is carried out by means of a solar collector. Obviously the design of the solar collector is one of the key factors that determine the performance of a solar adsorption refrigeration unit. The simplest in design and operation are flat metal solar collectors (Sukhyy et al. 2017). But their high cost stimulates the development of advanced solar collectors based on new materials, in particular polymers, which have proven effective in heating systems.

The experimental tests of the developed polymeric solar collectors involved a full-scale field and laboratory tests. Laboratory tests were carried out by means of a thermal-hydraulic stand (Sukhyy et al. 2017). The field tests were conducted on June 19, 2017 in the city of Dnipro (Dnepropetrovsk) at noon. The ambient temperature was t_0 = 25°C, and the temperature at the inlet of the solar collector was t_v = 27°C. The average value of total solar radiation on the horizontal surface was H = 761,514 W/m^2. The artificial radiation flux density was measured by a Kipp & Zonen CM21 pyranometer.

Instantaneous efficiency factor is determined by the ratio of the specific thermal output to the density of the solar radiation flux in the collector plane, and the optical efficiency factor $F'(\tau\alpha)$ i.e. the specific heat output to the flux density of the solar radiation at the same temperatures of the heat carrier and the environment, while the reduced factor of heat losses is calculated as product of the coefficient of efficiency factor of the solar collector and the full coefficient of heat losses.

Tests of collectors made of cellular polycarbonate plastics (table) protected by patents of Ukraine (Pat. 53855 A Ukraina, 2010, Pat. 53853 A, 2010, Pat. 53852 A,2010, Pat. 41237 A, 2009) have been carried out. As transparent insulation, single-layer transparent cellular polycarbonate plastic 4 mm thick and two-tier transparent cellular polycarbonate plastic 14 mm thick were used.

The absorber was made on the basis of a two-tier transparent cellular polycarbonate plastic with a thickness of 14 mm and a single-tier cellular polycarbonate plastic with a thickness of 10 mm of transparent and red hue, color tone 572′ and 625 nm with a purity of 3 and 38%, respectively. As the absorber of solar radiation used carbon tape, pellets of activated carbon and actually single-tier cellular polycarbonate plastic in red (color tone 625 nm, color purity 38%), and in the role of the housing - sheets of polyvinyl chloride (PVC). Polymeric solar collectors were insulated with sheet polyurethane foam (thickness 50 and 30 mm, density 30 kg/m^3, thermal conductivity coefficient is 0,04 W/(m · K)).

Tests results are given in Table 9.

The experimental results are presented in the form of dependences of the collector efficiency factor vs. the reduced temperature, which is equal to the ratio of the difference between the temperature of the heat transfer media T_a and the environment T_f to the intensity of the solar radiation, which at constant heat losses U_L = const and the coefficients of efficiency are linear.

Figure 9. The dependence of the efficiency of the developed polymeric solar collectors on the reduced temperature: a - field tests; b - laboratory tests (Sukhyy, Kozlov, et al., 2018).

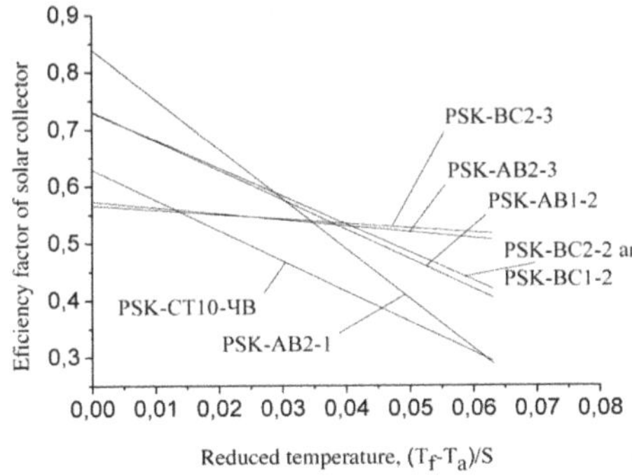

a

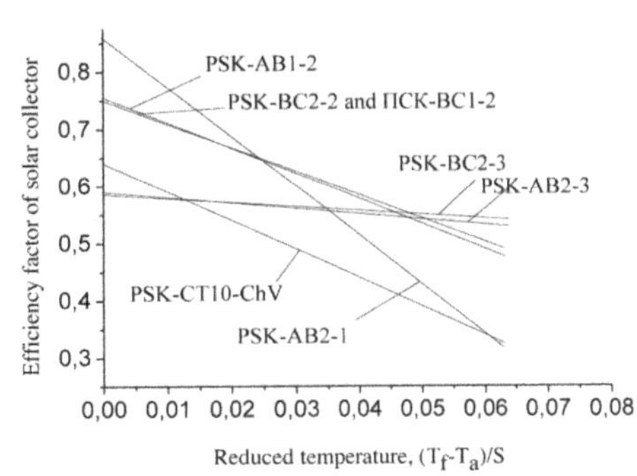

b

For regeneration of composites with a regeneration temperature of 50 – 60 °C, the average temperature of the absorber should be 20 – 35 K higher than the ambient temperature, which with an average solar radiation intensity of 850 W/m^2 corresponds to the characteristic temperature range of 0.024–0.041 (m^2·K)/W for these purposes. For desorption of water from adsorbents with a regeneration temperature greater than 80 ° C, the temperature of heat carrier should be higher than the ambient temperature by 40 – 50 K. Therefore, the characteristic temperature range can be considered of 0.047 – 0.059 (m^2·K) /W and above.

The thermal characteristics of the developed solar collectors presented in Figure 9 are obtained by approximation of the experimental data by the method of least squares.

The error of determining the instantaneous efficiency factor of the solar collector did not exceed 5 -–7%. The error of approximation of the experimental data by the method of the least squares for the optical efficiency of the solar collector was 0.16–0.25%, and for the resulted heat losses 6.96 – 12.5%. The correlation coefficient was R = 0.98 – 0.99.

The thermal parameters of solar collectors, which are determined according to the results of laboratory and field tests, are given in the Table 9. Differences between laboratory and field tests are the result of primarily the changing weather conditions. The relative error of laboratory test results relative to full-scale tests does not exceed 5-7%.

As in the summer months for solar collector the temperature difference between the absorber and the environment is relatively small, and the average intensity of solar radiation, on the contrary, acquires maximal values, the working area of the solar collector, according to Figure 9, corresponds to the decrease of the reduced temperature from 0,041 (m^2·K) / W. And for the winter months of work, it corresponds with increasing the reduced temperature, starting from 0,041 (m^2·K)/W.

Therefore, in the summer period according to the results of the test (Figure 9) will work most efficiently solar collectors PSK-AV2-1, PSK-ST10-CF because they have the maximal optical efficiency factor in the given operating range of reduced temperatures of 0,63 – 0,84 with reduced heat losses at the level of 5.37 – 8.78 W/(m^2·K).

In winter, the thermal characteristics are the best for PSK-BC2-3, PSK-AV2-3, which have the highest efficiency of 0.56 – 0.54 and the lowest rates of reduced heat loss of 0.74 – 1.05 W/(m^2·K).

Design of polymeric solar collectors PSK-AB1-2, PSK-BC2-2 and PSK-BC1-2 with optical efficiency of 0.73 – 0.74 and reduced heat losses of 4.84 – 5.33 W/ (m^2· K) can be effectively used for year-round solar systems.

The highest optical efficiency factor of 0.84 is stated for PSC-AB2-1. It results from the presence of only one layer of transparent insulation and a high absorption coefficient of activated carbon of 0.98. The low optical efficiency factor of PSK-CT10-ChV construction compared to PSK-AB2-1 is explained by a 23% less absorption coefficient and an increase of one layer of transparent insulation. This results in a 39% reduction in reduced heat loss of 5.37 W/(m^2·K). In the range of reduced temperatures of 0.024 – 0.041 (m^2·K) / W, these collectors have a rather high efficiency of 0.41 – 0.63. These collectors are promising for obtaining low temperatures and therefore they can be used when sorbents with regeneration temperatures in the range of 50 – 60 °C used.

The obtained thermophysical properties (Figure 9) allow concluding that structures such as PSK-BC2-3 and PSK-AB2-3 are promising for use as seasonal solar systems that can effectively work in winter, for heating systems and other technological needs, where the heat transfer media temperature exceeds the ambient temperature by more than 40-50°C. The low optical efficiency factors of PSK-BC2-3 of 0.56 and PSK-AV2-3 of 0.57 are explained by the presence of a triple layer of transparent insulation. Since the degree of blackness of the carbon band was assumed to be 0.77, the optical efficiency factor of PSK-BC2-3 is 2% less than the efficiency factor of PSK-AB2-3. On the other hand, triple transparent insulation reduces the reduced heat losses to 0.74 W/(m^2 · K) for PSC-BC2-3 and 1.05 W/(m^2 · K) for PSC-AB2-3.

Solar collectors PSK-AB1-2, PSK-BC2-2 and PSK-BC1-2 can be efficiently used in adsorption refrigeration solar systems based on adsorbents with a regeneration temperature of more than 80ºC, i.e. within the reduced temperature range from 0.041 to 0.059 (m^2·K) / W. Their thermophysical parameters are almost at the same level. The optical efficiency factor for PSK-AB1-2, PSK-BC2-2 and PSK-BC1-2 are stated to be at the level of 0.74. The experimental data prove the effectiveness of the developed design of PSK-BC2-2, which allows reducing the cost and overall dimensions of the solar collector without significantly changing its thermal characteristics. This can be done through the use of two-tier transparent cellular polycarbonate plastic, the lower tier of which is filled with air, which allows to reduce the thickness of thermal insulation material from 50 mm to 30 mm without substantially changing the thermal characteristics (which are practically the same values as the design of the PSK- BC1-2).

The developed designs of polymeric solar collectors in terms of thermal engineering parameters are not inferior to solar collectors of different manufacturers, in particular, the metal solar collector 'Sokol' produced by the Science and Production Association 'Mashinostroyeniye', Russian Federation (Suleimanov, 2007); vacuum Thermomax Ltd., model MAZDON TMA 600 S (Shestopalov, 2005). Moreover, collector designs PSK-AB1-2, PSK-AV2-1, PSK-BC2-2 PSK-BC1-2, PSK-BC2-3 and PSK-AB2- 3 exceed them.

So the optical efficiency factors for the investigated designs of polymeric solar collectors occur to be in the range from 0.59 to 0.86, and the reduced coefficients of thermal loss are changed within the interval from 0.75 to 8.65 W/(m^2·K). The optical efficiency factors of the existing polymeric solar collectors are stated to be of 0.62 – 0.78, for the best metal and vacuum ones they are 0.72–0.74 and 0.53 – 0.58, respectively.

In the range of reduced temperatures of 0.024–0.041 (m^2•K)/W, polymeric solar collectors based on cellular polycarbonate plastics (PSK-AB2-1, PSK-ST10-CH) exhibit an efficiency factor of 0.44 – 0.66, which surpass the efficiency of the most efficient industrial polymeric solar collector (0.3 – 0.54) [289] and corresponds to the efficiency of the best metal solar collectors (0.51 – 0.6) [288]. Moreover, it exceeds the efficiency factor of standard vacuum solar collectors (0.51 – 0.54) in these operating conditions (Shestopalov, 2005).

Other designs of the developed polymeric solar collectors (PSC-AB2-3, PSC-BC2-3), according to the results of laboratory tests, have rather low values of the reduced coefficient of heat loss, it being of 0.75 – 0.97 W/(m^2•K). This is accompanied by an increase in the efficiency in the characteristic range of reduced temperatures of 0.047 – 0.059 (m^2 •K)/W to 0.53–0.54, which indicates the feasibility of using such structures in seasonal solar systems that can effectively operate in conditions where the heat transfer media temperature exceeds ambient temperature of more than 40 – 50°C.

The efficiency of the developed collectors PSK-AB2-1 and PSK-CT10-ChV, in the range of reduced temperatures of 0.024 – 0.041 (m^2•K) / W, is in excess of 14 – 21% as compared with the already existing polymeric solar collectors. Some of them occur practically on the same level (PSK-ST10-ChV). The efficiency of PSK-AB2-1 is practically coincided with the efficiency of the metal solar collector 'Sokol' produced by the Science and Production Association 'Mashinostroyeniye", Russian Federation (0.56 – 0.63). The efficiency of standard vacuum solar collector (GREENone Tec, model VK 29) at a given temperature of 0.024 (m^2•K)/W is inferior to the efficiency of the design PSK-CT10-ChV by 1%, and PSK-AB2-1 by 14%.

In this case, the efficiency factor of the vacuum SK MAZDON TMA 600 S is less by 1% as compared with the efficiency of PSK-AB2-1 and more than the efficiency of PSK-CT10-ChV by 2%. At a reduced temperature of 0.041 (m^2•K)/W, the efficiency factor of the developed polymer collectors occurs to be almost at the same level.

When comparing year-round solar systems (PSK-AB1-2, PSK-BC2-2 and PSK-BC1-2) with known polymeric solar collectors, it can be concluded that at the given temperature of 0.024 (m^2 • K) /W, efficiency factor of these design surpass already existing polymeric solar collectors of different manufacturers by 11 – 16%, and at a reduced temperature of 0.059 (m^2•K)/W it being higher by 15–41%. As compared to the designs of the best metal solar collectors, in the range of reduced temperatures 0.024 – 0.059 (m^2 •K)/W these structures have an instantaneous efficiency higher by 5– 9% or almost at the same level as the efficiency of the metal solar collector 'Sokol' produced by the Scientific and Production Association 'Mashinostroyeniye', Russian Federation (Suleimanov, 2007). Having been compared the efficiency of the designed structures with vacuum solar colectors (GREENone Tec, model VK 29 and MAZDON TMA 600 S), the advantage of the polymeric collector efficiency can be noticed at the reduced temperature of 0.024 (m^2 • K)/W. At the reduced temperature of 0.059 (m^2 • K)/W efficiency of the developed polymeric solar collectors is practically at the same level.

Efficiency factor of polymeric solar collectors PSC-BC2-3 and PSC-AV2-3 surpass the existing polymeric solar colectors in the characteristic region of the reduced temperatures of 0.047 – 0.059 (m^2•K)/W by 13-43%. Thermal technical characteristics of PSK-BC2-3 and PSK-AB2-3 in this range of reduced temperatures exceed standard metal solar collector (Suleimanov, 2007) by 7 – 11%. At the reduced temperature of 0.047 (m^2•K) / W their efficiency factors coincide practically with the metal solar collector 'Sokol' produced by the Scientific and Production Association 'Mashinostroyeniye', Russian Federation. At the reduced temperature of no less than 0.059 (m^2•K)/W, the thermal technical characteristics of the developed solar collectors exceed the efficiency of the metallic collector 'Sokol' by 4 – 6%. In this range of reduced temperatures, vacuum solar collectors are inferior to the developed polymeric collectors by 3-6%.

The significant advantage of vacuum solar collectors begins to reveal only at the given temperatures above 0.06 (m^2•K)/W at which the total heat losses to the environment significantly increase.

So, at the regeneration temperature of adsorbents up to 50 – 60°C it is advisable to use collectors such as PSK-AB2-1, PSK-CT10-ChV, and at

Figure 10. An improved design of the solar adsorptive chiller (a): 1 – adsorber; 2 – coil (hydraulic accumulator); 3 – adsorbent; 4 – evaporator; 5 – condenser; 6 – chilling chamber; 7 - water cold accumulator; 8 – pipe; 9 – tap; 10 – flexible pipeline with vacuum isolation; 11 – metal plates with holes and applied calibration; 12 – transparent cellular polycarbonate plastic; 13 – photosensitive elements; 14 – electronics drive; 15 – bearings; 16 – support. Detailed picture of a flexible pipeline with vacuum isolation (b)
(Sukhyy, Belyanovskaya et al., 2018).

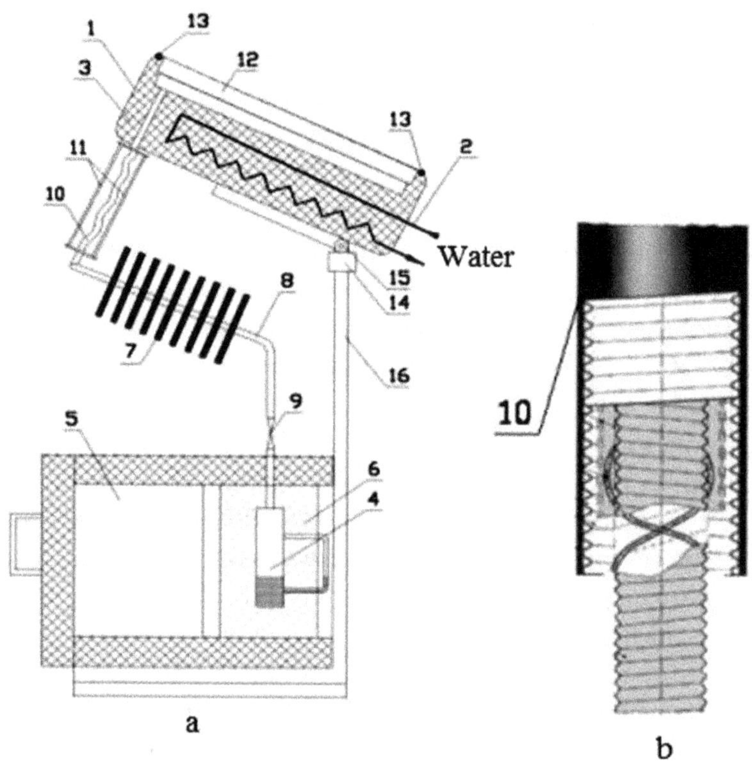

temperatures above 80°C PSK-AV2-3, PSK-BC2-3, PSK-AB1 -2, PSK-BC2-2 and PSK-BC1-2 become more promising.

According to Tables 4,5 and 7, the efficiency factor of a solar adsorptive refrigerator is strongly affected by the total average daily value of solar radiation. Obviously, it is determined by the solar collector tilt to the horizon in the southern orientation (Kozlov, Sukhyy, & Sukhyy, 2010). The angle of 30° will not be optimal for the whole operating period. Therefore, the existing design of the solar adsorptive refrigerator can be upgraded by equipping it with a photosensitive element and an electric drive, which will allow changing the angle of slope of the adsorber to the horizon depending on the intensity of the solar radiation. The design of an optimized solar refrigerator is given in Figure 10 (Sukhyy, Belyanovskaya et al., 2018).

The adsorptive chiller works in two stages. The sun position the sky is detected by the photosensitive element 13, which sends the signal to the electric drive 14, and the collector tilt to the horizon being changed. Such an improvement of the design allows maintaining the constant value of the coefficient of performance throughout the whole operational period. It is also possible to reconstruct ordinary car GAZelle into a transport adsorptive refrigerator, in which regeneration of the adsorbent proceed due to the solar collector installed on the chilling chamber and the heat of the exhaust gases emitted when driving the car (Patent 107051 Ukraina, 2016).

REFERENCES

Alghoul, M. A., Sulaiman, M. Y., Azmi, B. Z., & Wahab, M. A. (2007). Advances on multi/purpose solar adsorption systems for domestic refrigeration and water heating. *Applied Thermal Engineering*, *27*(5/6), 813–822. doi:10.1016/j.applthermaleng.2006.09.008

Askalany, A. A., Salem, A. A. M., Ismael, I. M., Ali, A. H. H., Morsy, M. G., & Saha, B. B. (2013). An overview on adsorption pairs for cooling. *Renewable & Sustainable Energy Reviews*, *19*, 565–572. doi:10.1016/j.rser.2012.11.037

Bao, H. S., Wang, R. Z., & Wang, L. W. (2011). A resorption refrigerator driven by low grade thermal energy. *Energy Conversion and Management*, *52*(6), 2339–2344. doi:10.1016/j.enconman.2010.12.045

Belyanovskaya, E. A., Pustovoy, G. N., Sergiyenko, Ya. O., Sukhyy, K. M., Yeromin, O. O., Prokopenko, E. M., ... Kizek, J. (2019). Performance of the Adsorptive Solar Refrigerators Based on Composite Adsobents 'Silica Gel –Sodium Sulphate'. *Advances in Thermal Processes and Energy Transformation*, *2*, 19–23.

Belyanovskaya, E.A., Pustovoy, G.N., Sukhyy, K.M., Kolomiyets, E.V., Sukhyy, M.P. (2018). Adsorbtsionnye kholodil'nye ustanovki na osnove kompozitnykh adsorbentov «silikagel' – sul'fat natriya», *Visnik Natsional'nogo tekhnichnogo universitetu «KHPI» Seriya: Khimiya, khimichna tekhnologiya ta ekologiya*, *39*(1315), 38 – 42.

Brown, J. S., & Domanski, P. A. (2014). Review of alternative cooling technologies. *Applied Thermal Engineering*, *64*(1-2), 252–262. doi:10.1016/j.applthermaleng.2013.12.014

Choudhury, B., Saha, B. B., Chatterjee, P. K., & Sarkar, J. P. (2013). An overview of developments in adsorption refrigeration systems towards a sustainable way of cooling. *Applied Energy*, *104*, 554–567. doi:10.1016/j.apenergy.2012.11.042

Fan, Y., Luo, L., & Souyri, B. (2007). Review of solar sorption refrigeration technologies: Development and applications. *Renewable & Sustainable Energy Reviews*, *11*(8), 1758–1775. doi:10.1016/j.rser.2006.01.007

Fernandes, M. S., Brites, G. J. V. N., Costa, J. J., Gaspar, A. R., & Costa, V. A. F. (2014). Review and future trends of solar adsorption refrigeration systems. *Renewable & Sustainable Energy Reviews*, *39*, 102–123. doi:10.1016/j.rser.2014.07.081

Freni, A., Maggio, G., Sapienza, A., Frazzica, A., Restuccia, G., & Vasta, S. (2016). Comparative analysis of promising adsorbent/adsorbate pairs for adsorptive heat pumping, air conditioning and refrigeration. *Applied Thermal Engineering*, *104*, 85–95. doi:10.1016/j.applthermaleng.2016.05.036

Goyal, P., Baredar, P., Mittal, A., & Siddiqui, A. R. (2016). Adsorption refrigeration technology – An overview of theory and its solar energy applications. *Renewable & Sustainable Energy Reviews*, *53*, 1389–1410. doi:10.1016/j.rser.2015.09.027

Holodil'nye mashiny. Spravochnik. (1982). pod red. Bykova A. V., M.: Legkaya i pishchevaya promyshlennost', 225.

Kanoğlu, M., Çarpınlıoğlu, M. Ö., & Yıldırım, M. (2004). Energy and exergy analyses of an experimental open-cycle desiccant cooling system. *Applied Thermal Engineering*, *24*(5-6), 5–6, 919–932. doi:10.1016/j.applthermaleng.2003.10.003

Khattab, N. M. (2004). A novel solar powered adsorption refrigeration module. *Applied Thermal Engineering*, *24*(17-18), 2747–2760. doi:10.1016/j.applthermaleng.2004.04.001

Kolomiyets, O.V., Belyanovskaya, E.A., Sukhyy, K.M., Prokopenko, O.M., Kozlov, Ya.M., Sukhyy, M.P. (2015). Osnovni robochi kharakteristiki sonyachnogo adsorbtsiynogo kholodilnika na osnovi kompositnogo sorbentu silikagel/Na2SO4. *Odeska natsionalna akademiya kharchovyh tekhnologiy. Naukovi pratsi, 2*(47), 176 – 181.

Kolomiyets, O.V., Kozlov, Ya.M, Belyanovskaya, E.A., Sukhyy, M.P., & Sukhyy, K.M. (2014). Doslidzhennya roboti sonyachnogo adsorbtsijnogo kholodil'nika na osnovi kompozitnogo sorbentu silikagel'/Na_2SO_4. *Integrovani tekhnologiï ta energozberezhennya, 4*, 86 – 88.

Kozlov, Ya. M., Sukhyy, M. P., & Sukhyy, K. M. (2010). Optymizatsiia roztashuvannia soniachnykh kolektoriv v heliosystemakh. *Skhidno-Yevropeiskyi zhurnal peredovykh tekhnolohii, 44*, 58 – 64.

Li, S. L., Xia, Z. Z., Wu, J. Y., Li, J., Wang, R. Z., & Wang, L. W. (2010). Experimental study of a novel CaCl2/expanded graphite/NH3 adsorption refrigerator. *International Journal of Refrigeration, 33*(1), 61–69. doi:10.1016/j.ijrefrig.2009.08.001

Morozyuk, T. V. (2006). Teoriya kholodil'nykh mashin i teplovykh nasosov. Odessa: Studiya «Negotsiant».

Muttakin, M., Ito, K., & Saha, B. B. (2020). Solar Thermal-Powered Adsorption Chiller. In Solar Energy (pp. 117–148). Springer. doi:10.1007/978-981-15-0675-8_8

Ostrov, M. (1999). *Rukovodstvo po raschetu teplovogo balansa kholodilnykh kamer i vyboru osnovnykh proektnykh parametrov kholodylnykh ustanovok.*

Rouf, R. A., Alam, K. C. A., Saha, B. B., & Kabir, K. M. A. (2018). Utilizing accessible heat enhancing cooling effect with three bed solar adsorption chiller. *Heat Transfer Engineering, 40*(12), 1049–1059. doi:10.1080/01457632.2018.1451244

Rouf, R. A., Jahan, N., Alam, K. C. A., Sultan, A. A., Saha, B. B., & Saha, S. C. (2020). Improved cooling capacity of a solar heat driven adsorption chiller. *Case Studies in Thermal Engineering, 17*, 100568. doi:10.1016/j.csite.2019.100568

Sangwan, S., & Chakraborty, P. R. (2020). Thermodynamic Analysis of Activated Carbon–Ethanol and Zeolite–Water Based Adsorption Cooling Systems. In Solar Energy (pp. 179–204). Springer. doi:10.1007/978-981-15-0675-8_10

Shestopalov, K.A. (2005). *Teplotekhnicheskie harakteristiki polimernyh solnechnyh kollektorov dlya sistem teplosnabzheniya [Tekst]: dis...kand.* tekhn. Nauk.

Sukhyy, K. M., Belyanovskaya, E. A., & Kolomiyets, E. V. (2018). *Design and performance of adsorptive transformers of heat energy*. Riga, Latvia: LAP Lambert Academic Publishing.

Sukhyy, K. M., Belyanovskaya, E. A., Sukha, I. V., Kolomiyets, E. V., Gavrilko, A. V., Sukhyy, M. P. (2015). Solar Adsorptive Chiller based on Composite Sorbent 'Silica Gel – Sodium Acetate', synthesized by Sol – Gel Method. *Voprosy khimii i khimicheskoy technologii, 6*(104), 80 – 86.

Sukhyy, K. M., Kozlov, Ya. M., Belyanovskaya, E. A., Prokopenko, E. M., Sukha, I. V., & Doroshenko, O. M. (2018). Ekspluatatsiini kharakterystyky polimernykh soniachnykh kolektoriv dlia adsorbtsiinykh kholodylnykh helioustanovok. *Kholodylna tekhnika i tekhnolohiia, 54*, 9 – 15. DOI doi:10.15673/ret.v54i1.984

Sukhyy, K. M., Kozlov, Ya. M., Belyanovskaya E. A., Sukhyy M. P. (2017). Systemy akumuliuvannia i transformatsii soniachnoi enerhii: *DVNZ UDKhTU, 168*.

Suleimanov, M. Zh. (2007). *Eksperymentalnoe yssledovanye teplotekhnycheskykh kharakterystyk solnechnykh kollektorov y vodonahrevatelnykh ustanovok [Tekst]: avtoref. dys...kand. tekhn. nauk: 05.14.01*. Obyedynennyi ynstytut vysokykh temperatur RAN.

Teplovoj raschet promyshlennyh parogeneratorov (1980). /pod red. CHastuhina V.I., Kiev: Vysshaya shkola, 181. (in Russian) Pat. 41237 A (2009). Soniachnyi kolektor [Tekst], Sukhyy M. P., Kozlov Ya. N., Sukhyy K. M. (Ukraina); zaiavnyk ta patentovlasnyk derzh. vyshch. navch. zaklad „Ukr. derzh. khim.-tekhn. un-t.", N° u 2008 14843; zaiavl. 23.12.08; opubl. 12.05.09, Biul. N° 9, 4 (in Ukrainian) Pat. 53852 A (2010). Soniachnyi kolektor [Tekst] / Sukhyy M. P., Kozlov Ya. N., Sukhyy K. M. (Ukraina); zaiavnyk ta patentovlasnyk derzh. vyshch. navch. zaklad „Ukr. derzh. khim.-tekhn. un-t.", N° u 2010 02818; zaiavl. 12.03.10; opubl. 25.10.10, Biul. N° 20, 4 (in Ukrainian) Pat. 53853 A (2010) Soniachnyi kolektor [Tekst] / Sukhyy, M. P., Kozlov, Ya. N., Sukhyy, K. M. (Ukraina); zaiavnyk ta patentovlasnyk derzh. vyshch. navch. zaklad „Ukr. derzh. khim.-tekhn. un-t.", N° u 2010 02819; zaiavl. 12.03.10; opubl. 25.10.10, Biul. N° 20, 6 (in Ukrainian) Pat. 53855 A Ukraina (2010). Soniachnyi kolektor [Tekst] / Sukhyy, M. P., Kozlov, Ya. N., Sukhyy, K. M. (Ukraina); zaiavnyk ta patentovlasnyk derzh. vyshch. navch. zaklad „Ukr. derzh. khim.-tekhn. un-t.", N° u 2010 02823; zaiavl. 12.03.10; opubl. 25.10.10, Biul. N° 20, 4 (in Ukrainian) Patent 107051 Ukraina (2016). Transportnyi refryzherator / Sukhyy, K. M., Sukhyy, M. P., Kolomiyets E.V. [ta in.]; zaiavnyk i patentovlasnyk Derzhavnyi vyshchyi navchalnyi zaklad —Ukrainskyi derzhavnyi khimiko/tekhnolohichnyi universytet, N° u 2015 08069; zaiavl. 13.08.2015; opubl. 25.05.2016, Biul. N°.10, 4 (in Ukrainian)

Conclusion

Adsorptive conversion is one of the most promising methods for storage and transformation of heat energy not only due to environmental safety of applied substances, but also because of high heat storage density which leads to minimal mass and volume for accumulation of 1 GJ of heat. Nevertheless, commercialization of adsorptive convertors of heat energy is strongly inhibited by properties of conventional adsorbents, such as silica gel or zeolites, e.g. regeneration temperature and maximal adsorbate uptake. Conventional engineering solutions of this problem are focused on multibed adsorptive heat converters, such as adsorptive heat storage devices or chillers, which involve at least two adsorptive-desorptive reactors. Obviously, such design can increase operating cycle time. However, performance of such systems is hugely complicated.

As another way to solve this problem, composite adsorbents 'salt inside porous matrix' can be considered. However, such materials are generally obtained by rather complicated and expensive technique of dry or wet impregnation.

Another obstacle for commercialization of technology of adsorptive heat energy transformation is a gap between laboratory prototype of adsorptive heat energy conversion device and its full-scale version which resulted from not only insufficient properties of adsorbent, but absence of algorithm that allows estimating the efficiency of adsorptive heat transformation unit according with full-cycle of performance. Despite the fact that adsorptive heat energy converters work in at least in two-stage mode, most of mathematical models considers only one operating stage. Generally, it is adsorption, the efficiency of the adsorptive device being estimated according only one stage, i.e. adsorption. Moreover, mathematical models leave out of accounts heat energy inputs for adsorbate evaporation and desorption which are of rather high values.

New composite adsorbents of water 'silica gel – sodium sulphate' and 'silica gel – sodium acetate' obtained by sol gel method are shown to be characterized with high adsorptive capacity 0.5 – 1.4 kg/kg and heat energy storage density more than 1.5 MJ/kg. Regeneration temperatures of the composite 'silica gel – Na_2SO_4' and 'silica gel – CH_3COONa' are 90°C and 60°C, respectively. Efficiency of adsorptive heat energy converters based on these materials was estimated.

Designs of adsorptive heat storage devices are developed for operating in heat supply systems. Algorithms for calculation their structural and operational parameters are proposed. They involve calculation of mass transfer coefficient, adsorption, useful adsorption heat, heat input for heating the adsorbent, device casing, water in the humidifier, evaporation of water, heating the adsorbed water, desorption, and efficiency factors. Suggested algorithms exhibit satisfactory agreement of calculation results and experimental data.

Device for regeneration of low-potential heat and moisture in ventilation and air-conditioning systems is suggested to provide fresh air intake into premises. The indoor relative humidity and temperature are shown to be maintained at 50 – 60% and 22°C.

The mathematical model and procedures for calculation of principal design characteristics and performance of adsorptive regenerator under the conditions of ventilation systems of residential premises are developed. The procedure to determine the mass and volume of the adsorbent for regeneration unit in ventilation systems has been developed. Maximal adsorption is confirmed to be the main factor affecting the volume of adsorbent.

The advantages of 'salt in the porous silica gel matrix' composites are proved as compared with traditional silica gels. The suggested procedure involves calculation of the volume of air passed through the layer of heat storage material, the absolute humidity of the inflow and outflow air, the adsorption (i.e. water uptake) and the heat of adsorption for inflow, the final temperature of the cold air, the air temperature after mixing the cold air from the street and the warm air in the room near the warm end of the regenerator during inflow, the volume of air passed through the layer of heat storage material, adsorption and heat of adsorption for outflow, the final temperature of the air at the cold end of the regenerator, the air temperature after mixing of the cold air from the street and the warm air from the room at the cold end of regenerator during outflow, determining the temperature efficiency factor, summarized adsorption and time to achieve maximal adsorption. The satisfactory fit of air temperatures near the warm and cold ends of the regenerator, as is also the temperature efficiency factors calculated according to the proposed algorithm and experimental way is shown.

The performance of adsorptive heat-moisture regenerators based on composites ‘silica gel – sodium sulphate’ and ‘silica gel – sodium acetate’ in the conditions of the typical ventilation system of residential premises is simulated. The temperature efficiency factor variation with airflow rate and flows switching time is stated. The maximal temperature efficiency factors are corresponded with the airflow velocity at most of 0.22 – 0.32 m/s and the flows switching time about of 5 – 10 minutes.

The temperature efficiency factors are stated to be affected by the temperatures of the external and internal air. The increasing of the efficiency factor is shown when the temperature of the outside air increased from -25 °C to -3 ° C.

Design of adsorptive heat pump based composite water adsorbent ‘silica gel – sodium sulphate’ or ‘silica gel – sodium acetate’ is developed. Novel data for temperature changes in various pump units during its operational run were obtained. Coefficients of performance of adsorptive heat pumps based on composite ‘silica gel – Na_2SO_4’ and ‘silica gel – CH_3COONa’ are 2.084 and 2.021.

Adsorptive refrigerator with additional hydraulic circuit to store heat energy released during water adsorption is developed. Suggested unit allows storing adsorption heat and using it for regeneration of an adsorbent which results in reducing significantly weather effects on duration of regeneration. Based on theoretical and experimental characteristics of solar adsorptive refrigerator, the calculation procedure of their principal structural and operational characteristics is suggested.

Related Readings

To continue IGI Global's long-standing tradition of advancing innovation through emerging research, please find below a compiled list of recommended IGI Global book chapters and journal articles in the areas of nanotechnology, coatings, and materials science. These related readings will provide additional information and guidance to further enrich your knowledge and assist you with your own research.

Agboola, O., Popoola, P. A., Sadiku, R., Sanni, S. E., Babatunde, D. E., Abatan, O. G., Fayomi, S. O., & Fasiku, O. V. (2019). Polymer Nanocomposites for Advanced Automobile Applications. In N. Ramdani (Ed.), *Polymer Nanocomposites for Advanced Engineering and Military Applications* (pp. 96–130). IGI Global. doi:10.4018/978-1-5225-7838-3.ch004

Ashrafiyan, O., Saremi, M., Pakseresht, A., & Ghasali, E. (2018). Oxidation-Protective Coatings for Carbon-Carbon Composites. In A. Pakseresht (Ed.), *Production, Properties, and Applications of High Temperature Coatings* (pp. 429–446). IGI Global. doi:10.4018/978-1-5225-4194-3.ch016

Aslam, S., & Ahmad, M. (2019). Nanoparticles for Degradation of Organic Pollutants. In R. Nazir (Ed.), *Nanotechnology Applications in Environmental Engineering* (pp. 241–267). IGI Global. doi:10.4018/978-1-5225-5745-6.ch011

Azadi, M. (2018). Deposition of TiN, TiC, and DLC Coatings by PACVD. In A. Pakseresht (Ed.), *Production, Properties, and Applications of High Temperature Coatings* (pp. 381–402). IGI Global. doi:10.4018/978-1-5225-4194-3.ch014

Babatunde, D. E., Denwigwe, I. H., Babatunde, O. M., Agboola, O., & Akinsipe, G. D. (2019). Relevance of Chemically Functionalized Nano-Fillers and Modified Nanocomposite in Energy Systems. In N. Ramdani (Ed.), *Polymer Nanocomposites for Advanced Engineering and Military Applications* (pp. 10–65). IGI Global. doi:10.4018/978-1-5225-7838-3.ch002

Bhattacharyya, S., & Majumdar, P. S. (2018). An Overview on Peak Shape Method for Thermoluminescence Glow Curve Analysis: Application on Tremolite and Actinolite Glow Peaks. In R. Tiwari, V. Dubey, & S. Dhoble (Eds.), *Emerging Synthesis Techniques for Luminescent Materials* (pp. 26–52). IGI Global. doi:10.4018/978-1-5225-5170-6.ch002

Bolleddu, V., Racherla, V., & Bandyopadhyay, P. P. (2018). Microstructural and Tribological Characteristics of Air Plasma Sprayed Alumina-Titania Coatings. In A. Pakseresht (Ed.), *Production, Properties, and Applications of High Temperature Coatings* (pp. 268–298). IGI Global. doi:10.4018/978-1-5225-4194-3.ch011

Borah, R., & Kumar, A. (2019). Enhanced Cellular Activity on Conducting Polymer: Chitosan Nanocomposite Scaffolds for Tissue Engineering Application. In N. Ramdani (Ed.), *Polymer Nanocomposites for Advanced Engineering and Military Applications* (pp. 150–189). IGI Global. doi:10.4018/978-1-5225-7838-3.ch006

Chawla, R., Singhal, P., & Garg, A. K. (2018). Investigation of Structural, Morphological and Electro-Optical Behavior of GO/rGO. *International Journal of Surface Engineering and Interdisciplinary Materials Science*, *6*(1), 32–43. doi:10.4018/IJSEIMS.2018010103

Chopra, S., Sreya, S., Babhulkar, R. V., Halde, S. P., Deshmukh, K. A., & Peshwe, D. R. (2019). Cryogenic Treatment of Polymer/MWCNT Nano-Composites for Mechanical and Tribological Applications. In N. Ramdani (Ed.), *Nanotechnology in Aerospace and Structural Mechanics* (pp. 103–161). IGI Global. doi:10.4018/978-1-5225-7921-2.ch004

Darwish, M. (2019). Lightweight Nanocomposites Polymers for Shielding Application. In N. Ramdani (Ed.), *Nanotechnology in Aerospace and Structural Mechanics* (pp. 206–233). IGI Global. doi:10.4018/978-1-5225-7921-2.ch006

Di Girolamo, G. (2018). Current Challenges and Future Perspectives in the Field of Thermal Barrier Coatings. In A. Pakseresht (Ed.), *Production, Properties, and Applications of High Temperature Coatings* (pp. 25–59). IGI Global. doi:10.4018/978-1-5225-4194-3.ch002

Dutta, M. M. (2019). Characterization of Nanomaterials. In R. Nazir (Ed.), *Nanotechnology Applications in Environmental Engineering* (pp. 44–80). IGI Global. doi:10.4018/978-1-5225-5745-6.ch003

Elehinafe, F. B., & Ayeni, A. O. (2019). Processing of Polymer-Based Nanocomposites in Advanced Engineering and Military Application. In N. Ramdani (Ed.), *Polymer Nanocomposites for Advanced Engineering and Military Applications* (pp. 1–9). IGI Global. doi:10.4018/978-1-5225-7838-3.ch001

Figueira, R. B. (2018). Hybrid Sol-Gel Coatings: Erosion-Corrosion Protection. In A. Pakseresht (Ed.), *Production, Properties, and Applications of High Temperature Coatings* (pp. 334–380). IGI Global. doi:10.4018/978-1-5225-4194-3.ch013

Gebai, S. S., Hallal, A. M., & Hammoud, M. S. (2018). *Mechanical Properties of Natural Fiber Reinforced Polymers: Emerging Research and Opportunities.* IGI Global. doi:10.4018/978-1-5225-4837-9

Hassanzadeh, M., Saremi, M., Valefi, Z., & Pakseresht, A. H. (2018). Investigation on Improved-Durability Thermal Barrier Coatings: An Overview of Nanostructured, Multilayered, and Self-Healing TBCs. In A. Pakseresht (Ed.), *Production, Properties, and Applications of High Temperature Coatings* (pp. 60–78). IGI Global. doi:10.4018/978-1-5225-4194-3.ch003

Henao, J., & Sotelo, O. (2018). Surface Engineering at High Temperature: Thermal Cycling and Corrosion Resistance. In A. Pakseresht (Ed.), *Production, Properties, and Applications of High Temperature Coatings* (pp. 131–159). IGI Global. doi:10.4018/978-1-5225-4194-3.ch006

Horst, D. J., & Junior, P. P. (2019). 3D-Printed Conductive Filaments Based on Carbon Nanostructures Embedded in a Polymer Matrix: A Review. *International Journal of Applied Nanotechnology Research*, *4*(1), 26–40. doi:10.4018/IJANR.2019010103

Irzhak, T. F., & Irzhak, V. I. (2019). Synthesis of Epoxy Nanocomposites. In N. Ramdani (Ed.), *Nanotechnology in Aerospace and Structural Mechanics* (pp. 34–79). IGI Global. doi:10.4018/978-1-5225-7921-2.ch002

Ismar, E., & Sarac, A. (2019). Carbon Nanomaterials: Carbon Nanotubes, Graphene, and Carbon Nanofibers. In N. Ramdani (Ed.), *Nanotechnology in Aerospace and Structural Mechanics* (pp. 1–33). IGI Global. doi:10.4018/978-1-5225-7921-2.ch001

Jucha, S. O., & Moskal, G. J. (2018). Thermal Barrier Coatings on the Base of Samarium Zirconates. In A. Pakseresht (Ed.), *Production, Properties, and Applications of High Temperature Coatings* (pp. 79–106). IGI Global. doi:10.4018/978-1-5225-4194-3.ch004

Katrenipadu, S. P., & Gurugubelli, S. N. (2018). Regression Modeling and Experimental Investigations on Ageing Behavior of Nano-Fly Ash Reinforced Al-10wt%Mg Alloy Matrix Composites. *International Journal of Surface Engineering and Interdisciplinary Materials Science*, *6*(2), 36–49. doi:10.4018/IJSEIMS.2018070103

Kausar, A. (2019). Aeronautical Impact of Epoxy/Carbon Nanotube Nanocomposite. In N. Ramdani (Ed.), *Nanotechnology in Aerospace and Structural Mechanics* (pp. 80–102). IGI Global. doi:10.4018/978-1-5225-7921-2.ch003

Kausar, A. (2019). Scope of Polymer/Graphene Nanocomposite in Defense Relevance: Defense Application of Polymer/Graphene. In N. Ramdani (Ed.), *Polymer Nanocomposites for Advanced Engineering and Military Applications* (pp. 296–315). IGI Global. doi:10.4018/978-1-5225-7838-3.ch010

Keshavamurthy, R., Naveena, B. E., & Sekhar, N. (2018). Thermal Spray Coatings for Erosion-Corrosion Protection. In A. Pakseresht (Ed.), *Production, Properties, and Applications of High Temperature Coatings* (pp. 246–267). IGI Global. doi:10.4018/978-1-5225-4194-3.ch010

Khelfi, A. (2019). Toxicocinetic and Mechanisms of Action of Nanoparticles. In R. Nazir (Ed.), *Nanotechnology Applications in Environmental Engineering* (pp. 344–368). IGI Global. doi:10.4018/978-1-5225-5745-6.ch014

Kosti, S. (2019). Nanomaterials and Nanocomposites Thermal and Mechanical Properties Modelling. In N. Ramdani (Ed.), *Nanotechnology in Aerospace and Structural Mechanics* (pp. 234–256). IGI Global. doi:10.4018/978-1-5225-7921-2.ch007

Kumar, I. (2019). Simulation and Modeling of Nanotechnology Aircraft Using MATLAB. In N. Ramdani (Ed.), *Nanotechnology in Aerospace and Structural Mechanics* (pp. 257–290). IGI Global. doi:10.4018/978-1-5225-7921-2.ch008

Kumar, S., Krishnan, S., Samal, S. K., Mohanty, S., & Nayak, S. K. (2019). Polymer Nanocomposites Coating for Anticorrosion Application. In N. Ramdani (Ed.), *Polymer Nanocomposites for Advanced Engineering and Military Applications* (pp. 254–294). IGI Global. doi:10.4018/978-1-5225-7838-3.ch009

Kuz'min, V. E., Ognichenko, L. N., Sizochenko, N., Chapkin, V. A., Stelmakh, S. I., Shyrykalova, A. O., & Leszczynski, J. (2019). Combining Features of Metal Oxide Nanoparticles: Nano-QSAR for Cytotoxicity. *International Journal of Quantitative Structure-Property Relationships, 4*(1), 28–40. doi:10.4018/IJQSPR.2019010103

Lahariya, V., & Ramrakhiani, M. (2018). Electroluminescence Principle and Novel Electroluminescent Materials. In R. Tiwari, V. Dubey, & S. Dhoble (Eds.), *Emerging Synthesis Techniques for Luminescent Materials* (pp. 415–438). IGI Global. doi:10.4018/978-1-5225-5170-6.ch011

Lateef, A., & Nazir, R. (2019). Application of Magnetic Nanomaterials for Water Treatment. In R. Nazir (Ed.), *Nanotechnology Applications in Environmental Engineering* (pp. 201–219). IGI Global. doi:10.4018/978-1-5225-5745-6.ch009

Majumdar, D. (2019). Polyaniline as Proficient Electrode Material for Supercapacitor Applications: PANI Nanocomposites for Supercapacitor Applications. In N. Ramdani (Ed.), *Polymer Nanocomposites for Advanced Engineering and Military Applications* (pp. 190–219). IGI Global. doi:10.4018/978-1-5225-7838-3.ch007

Majumdar, D. (2019). Polyaniline Nanocomposites: Innovative Materials for Supercapacitor Applications – PANI Nanocomposites for Supercapacitor Applications. In N. Ramdani (Ed.), *Polymer Nanocomposites for Advanced Engineering and Military Applications* (pp. 220–253). IGI Global. doi:10.4018/978-1-5225-7838-3.ch008

Mubarak, S. M. (2019). Nanotechnology for Air Remediation. In R. Nazir (Ed.), *Nanotechnology Applications in Environmental Engineering* (pp. 121–142). IGI Global. doi:10.4018/978-1-5225-5745-6.ch006

Naik, R., Revathi, V., Prashantha, S. C., & Nagabhushana, H. (2018). Synthesis of Magnesium Based Nanophosphors and Nanocomposites by Different Techniques: Silicates and Ferrites. In R. Tiwari, V. Dubey, & S. Dhoble (Eds.), *Emerging Synthesis Techniques for Luminescent Materials* (pp. 251–276). IGI Global. doi:10.4018/978-1-5225-5170-6.ch007

Nasreen, S., & Rafique, U. (2019). Nanotechnology for Water Environmental Application: Functionalized Silica Hybrids as Nano-Sorbents. In R. Nazir (Ed.), *Nanotechnology Applications in Environmental Engineering* (pp. 171–200). IGI Global. doi:10.4018/978-1-5225-5745-6.ch008

Nazir, R. (2019). Introduction to Environmental Nanotechnology: E-Nano. In R. Nazir (Ed.), *Nanotechnology Applications in Environmental Engineering* (pp. 1–27). IGI Global. doi:10.4018/978-1-5225-5745-6.ch001

Noman, D., & Owens, G. (2019). Wastewater Treatment and Role of Green Synthesized Metal Oxide Nanocomposites. In R. Nazir (Ed.), *Nanotechnology Applications in Environmental Engineering* (pp. 268–307). IGI Global. doi:10.4018/978-1-5225-5745-6.ch012

Nosheen, S. (2019). Nanomembrane Applications in Environmental Engineering. In R. Nazir (Ed.), *Nanotechnology Applications in Environmental Engineering* (pp. 103–120). IGI Global. doi:10.4018/978-1-5225-5745-6.ch005

Odetola, P. I., Popoola, P. A., & Oladijo, P. (2018). Thin Coating Deposition by Magnetron Sputtering. In A. Pakseresht (Ed.), *Production, Properties, and Applications of High Temperature Coatings* (pp. 403–428). IGI Global. doi:10.4018/978-1-5225-4194-3.ch015

Pandey, A., Tiwari, S. P., Dutta, V., & Kumar, V. (2018). Photon Upconversion in Lanthanide-Activated Inorganic Luminescent Materials. In R. Tiwari, V. Dubey, & S. Dhoble (Eds.), *Emerging Synthesis Techniques for Luminescent Materials* (pp. 86–116). IGI Global. doi:10.4018/978-1-5225-5170-6.ch004

Pandey, J. (2019). Recent Progresses in Membranes for Proton Exchange Membrane Fuel Cell (PEMFC) for Clean and Environmentally Friendly Applications. In R. Nazir (Ed.), *Nanotechnology Applications in Environmental Engineering* (pp. 308–343). IGI Global. doi:10.4018/978-1-5225-5745-6.ch013

Pesetskii, S. S., Bogdanovich, S. P., & Aderikha, V. N. (2019). Polymer/Clay Nanocomposites Produced by Dispersing Layered Silicates in Thermoplastic Melts. In N. Ramdani (Ed.), *Polymer Nanocomposites for Advanced Engineering and Military Applications* (pp. 66–94). IGI Global. doi:10.4018/978-1-5225-7838-3.ch003

Purohit, P., & Vagge, S. T. (2018). Oxidation Behavior of LTA/YSZ Intermixed Layer Coating. In A. Pakseresht (Ed.), *Production, Properties, and Applications of High Temperature Coatings* (pp. 107–130). IGI Global. doi:10.4018/978-1-5225-4194-3.ch005

Sadegh, H., Ali, G. A., Nia, H. J., & Mahmoodi, Z. (2019). Nanomaterial Surface Modifications for Enhancement of the Pollutant Adsorption From Wastewater: Adsorption of Nanomaterials. In R. Nazir (Ed.), *Nanotechnology Applications in Environmental Engineering* (pp. 143–170). IGI Global. doi:10.4018/978-1-5225-5745-6.ch007

Sadiku, E. R., Agboola, O., Mochane, M. J., Fasiku, V. O., Owonubi, S. J., Ibrahim, I. D., Abbavaram, B. R., Kupolati, W. K., Jayaramudu, T., Uwa, C. A., Daramola, O. O., Iheaturu, N. C., Nkuna, C., Sanni, S. E., Biotidara, O. F., Eze, A. A., Varaprasad, K., Adeyeye, O. A., Selatile, K. M., & Ndamase, A. S. (2019). The Use of Polymer Nanocomposites in the Aerospace and the Military/Defence Industries. In N. Ramdani (Ed.), *Polymer Nanocomposites for Advanced Engineering and Military Applications* (pp. 316–349). IGI Global. doi:10.4018/978-1-5225-7838-3.ch011

Sadri, E., & Ashrafizadeh, F. (2018). High Temperature Nanocomposite Coatings by Plasma Spraying for Friction and Wear Applications. In A. Pakseresht (Ed.), *Production, Properties, and Applications of High Temperature Coatings* (pp. 216–245). IGI Global. doi:10.4018/978-1-5225-4194-3.ch009

Saeedi, F., Montazeri, A., Bahari, Y., Pishvaee, M., & Ranjbar, M. (2018). Synthesis and Characterization of Chitosan-Poly Vinyl Alcohol-Graphene Oxide Nanocomposites. *International Journal of Chemoinformatics and Chemical Engineering*, *7*(1), 1–12. doi:10.4018/IJCCE.2018010101

Shahien, M. (2018). Reactive Plasma Spray: A Method for Nitride Coatings Deposition in Thermal Spray. In A. Pakseresht (Ed.), *Production, Properties, and Applications of High Temperature Coatings* (pp. 299–332). IGI Global. doi:10.4018/978-1-5225-4194-3.ch012

Shahzadi, P. (2019). Properties of Nanomaterials and Environment. In R. Nazir (Ed.), *Nanotechnology Applications in Environmental Engineering* (pp. 28–43). IGI Global. doi:10.4018/978-1-5225-5745-6.ch002

Shamim, A., Ahmad, S., Khitab, A., Anwar, W., Khushnood, R. A., & Usman, M. (2018). Applications of Nano Technology in Civil Engineering: A Review. *International Journal of Strategic Engineering*, *1*(1), 48–64. doi:10.4018/IJoSE.2018010104

Shikuku, V. O., & Sylvain, T. (2019). Application of Geopolymer Composites in Wastewater Treatment: Trends, Opportunities, and Challenges. In N. Ramdani (Ed.), *Polymer Nanocomposites for Advanced Engineering and Military Applications* (pp. 131–149). IGI Global. doi:10.4018/978-1-5225-7838-3.ch005

Singh, J., Bhadane, M. S., Dubey, V., Dhole, S. D., Manam, J., & Dahiwale, S. S. (2018). Phosphors for Various Dosimetry Applications Derived by Different Synthesis Routes. In R. Tiwari, V. Dubey, & S. Dhoble (Eds.), *Emerging Synthesis Techniques for Luminescent Materials* (pp. 53–84). IGI Global. doi:10.4018/978-1-5225-5170-6.ch003

Som, S., Dutta, S., Kumar, V., & Swart, H. C. (2018). Swift Heavy Ion Synthesis and Modifications of Nanophosphors for Dosimetric Application: Effect of Swift Heavy Ion Irradiation. In R. Tiwari, V. Dubey, & S. Dhoble (Eds.), *Emerging Synthesis Techniques for Luminescent Materials* (pp. 1–25). IGI Global. doi:10.4018/978-1-5225-5170-6.ch001

Suresh, S. S., George, K., Mohanty, S., & Nayak, S. K. (2019). Architect of Polymer Nanocomposites for Aerospace Applications. In N. Ramdani (Ed.), *Nanotechnology in Aerospace and Structural Mechanics* (pp. 163–205). IGI Global. doi:10.4018/978-1-5225-7921-2.ch005

Tahir, L. (2019). Green Approaches to Environmental Sustainability. In R. Nazir (Ed.), *Nanotechnology Applications in Environmental Engineering* (pp. 81–101). IGI Global. doi:10.4018/978-1-5225-5745-6.ch004

Tripathi, S., Mishra, M. K., Jain, V. K., Tiwari, R., & Dubey, N. (2018). Thermo, Photo and Mechanoluminescence Studies of Eu3+ Doped Y4Al2O9 Phosphors: TL, PL, and ML Analysis of YAM: Eu3. In R. Tiwari, V. Dubey, & S. Dhoble (Eds.), *Emerging Synthesis Techniques for Luminescent Materials* (pp. 389–414). IGI Global. doi:10.4018/978-1-5225-5170-6.ch010

Ugemuge, N. S., Gayner, C., Natrajan, V., & Dhoble, S. J. (2018). Perspectives of Hydrothermal Synthesis of Fluorides for Luminescence Applications: Fluorides Phosphors for Luminescence. In R. Tiwari, V. Dubey, & S. Dhoble (Eds.), *Emerging Synthesis Techniques for Luminescent Materials* (pp. 277–303). IGI Global. doi:10.4018/978-1-5225-5170-6.ch008

Vaidyanathan, S., & Tagare, J. (2018). Phenanthroimidazole-Based Organic Fluorophores for Organic Light-Emitting Diodes: An Overview. In R. Tiwari, V. Dubey, & S. Dhoble (Eds.), *Emerging Synthesis Techniques for Luminescent Materials* (pp. 305–388). IGI Global. doi:10.4018/978-1-5225-5170-6.ch009

Van Miert, S., Creylman, J., & Verheyen, G. R. (2019). Mining a Nanoparticle Dataset, Compiled Within the MODENA-COST Action. *International Journal of Quantitative Structure-Property Relationships*, *4*(1), 1–17. doi:10.4018/IJQSPR.2019010101

Verma, R., Suri, N. M., & Kant, S. (2018). Slurry Sprayed Mullite Coatings and Their Corrosion Performances. In A. Pakseresht (Ed.), *Production, Properties, and Applications of High Temperature Coatings* (pp. 187–214). IGI Global. doi:10.4018/978-1-5225-4194-3.ch008

Xie, W. (2018). Thermal Spray Methods and Splat Formation. In A. Pakseresht (Ed.), *Production, Properties, and Applications of High Temperature Coatings* (pp. 1–24). IGI Global. doi:10.4018/978-1-5225-4194-3.ch001

Yaseen, T., & Yaseen, A. (2019). Wastewater Treatment in Removal of Heavy Metals: Nanotechnology Applications in Environmental Engineering. In R. Nazir (Ed.), *Nanotechnology Applications in Environmental Engineering* (pp. 220–240). IGI Global. doi:10.4018/978-1-5225-5745-6.ch010

Zhizhin, G. V. (2019). Higher Dimensions of Clusters of Intermetallic Compounds: Dimensions of Metallic Nanoclusters. *International Journal of Applied Nanotechnology Research*, *4*(1), 8–25. doi:10.4018/IJANR.2019010102

Zhu, S., & Wang, F. (2018). Nanocrystalline, Enamel and Composite Coatings for Superalloys. In A. Pakseresht (Ed.), *Production, Properties, and Applications of High Temperature Coatings* (pp. 160–186). IGI Global. doi:10.4018/978-1-5225-4194-3.ch007

About the Authors

Kostyantyn M. Sukhyy was born on 1976. Gained a PhD in 2002 and SciD in 2014 from Ukrainian State University of Chemical Engineering. Acquired the academic titles of associate professor in 2004 and professor in 2015. Has been the Rector of Ukrainian State University of Chemical Engineering since 2019. Scientific interests: development of methods for the synthesis of new polymer quaternary ammonium salts, polymer nanocomposites and polymeric electrolytes in solid state.

Elena A. Belyanovskaya was born on 1972. Gained a PhD in 2003 from Ukrainian State University of Chemical Engineering. Acquired the academic title of associate professor in 2015. Has been the associate professor of the Department of Power Engineering of Ukrainian State University of Chemical Engineering since 2014. Scientific interests: processes and devices for adsorptive heat energy storage and conversion, utilization of low-grade energy resources.

Mikhailo P. Sukhyy was born on 1943. Gained a PhD 1978 from Ukrainian State University of Chemical Engineering. Acquired the academic title of professor in 2007. Has been the Head of the Department of Power Engineering of Ukrainian State University of Chemical Engineering since 2002. Scientific interests: materials for adsorptive storage and conversion of thermal energy, utilization of low-grade energy resources.

Index

I

K

L

M

O

P

R

S

T

V

W

X

www.ingramcontent.com/pod-product-compliance
Lightning Source LLC
LaVergne TN
LVHW081336300926
849161LV00008B/1985